Covariant Loop Quantum Gravity

Quantum gravity is among the most fascinating problems in physics. It modifies our understanding of time, space, and matter. The recent development of the loop approach has allowed us to explore domains ranging from black hole thermodynamics to the early universe.

This book provides readers with a simple introduction to loop quantum gravity, centred on its covariant approach. It focuses on the physical and conceptual aspects of the problem and includes the background material needed to enter this lively domain of research, making it ideal for researchers and graduate students.

Topics covered include quanta of space, classical and quantum physics without time, tetrad formalism, Holst action, lattice gauge theory, Regge calculus, ADM and Ashtekar variables, Ponzano–Regge and Turaev–Viro amplitudes, kinematics and dynamics of 4d Lorentzian quantum gravity, spectrum of area and volume, coherent states, classical limit, matter couplings, graviton propagator, spinfoam cosmology and black hole thermodynamics.

Carlo Rovelli is Professor of Physics at Aix-Marseille Université, where he directs the gravity research group. He is one of the founders of loop quantum gravity.

Francesca Vidotto is NWO Veni Fellow at the Radboud Universiteit Nijmegen and initiated the spinfoam approach to cosmology.

"The approach to quantum gravity known as loop quantum gravity has progressed enormously in the last decade and this book by Carlo Rovelli and Francesca Vidotto admirably fills the need for an up-to-date textbook in this area. It will serve well to bring beginning students and established researchers alike up-to-date on developments in this fast moving area. It is the only book presenting key results of the theory, including those related to black holes, quantum cosmology and the derivation of general relativity from the fundamental theory of quantum spacetime. The authors achieve a good balance of big ideas and principles with the technical details."

Lee Smolin, Perimeter Institute for Theoretical Physics

"This is an excellent introduction to spinfoams, an area of loop quantum gravity that draws ideas also from Regge calculus, topological field theory and group field theory. It fills an important gap in the literature offering both a pedagogical overview and a platform for further developments in a forefront area of research that is advancing rapidly."

Abhay Ashtekar, The Pennsylvania State University

Covariant Loop Quantum Gravity

An Elementary Introduction to Quantum Gravity and Spinfoam Theory

CARLO ROVELLI
Université d' Aix-Marseille

FRANCESCA VIDOTTO
Radboud Universiteit Nijmegen

CAMBRIDGE
UNIVERSITY PRESS

CAMBRIDGE
UNIVERSITY PRESS

University Printing House, Cambridge CB2 8BS, United Kingdom

One Liberty Plaza, 20th Floor, New York, NY 10006, USA

477 Williamstown Road, Port Melbourne, VIC 3207, Australia

314-321, 3rd Floor, Plot 3, Splendor Forum, Jasola District Centre, New Delhi - 110025, India

103 Penang Road, #05-06/07, Visioncrest Commercial, Singapore 238467

Cambridge University Press is part of the University of Cambridge.

It furthers the University's mission by disseminating knowledge in the pursuit of education, learning and research at the highest international levels of excellence.

www.cambridge.org
Information on this title: www.cambridge.org/9781107069626

First published 2015
5th printing 2019
First paperback edition 2020

A catalogue record for this publication is available from the British Library

ISBN 978-1-107-06962-6 Hardback
ISBN 978-1-108-81025-8 Paperback

To our teachers and to all those who teach children to question our knowledge, learn through collaboration, and feel the joy of discovery

Contents

Part III   THE REAL WORLD

Preface

This book is an introduction to loop quantum gravity (LQG) focusing on its covariant formulation. The book has grown from a series of lectures given by Carlo Rovelli and Eugenio Bianchi at Perimeter Institute during April 2012 and a course given by Rovelli in Marseille in the winter of 2013. The book is introductory, and assumes only some basic knowledge of general relativity, quantum mechanics, and quantum field theory. It is simpler and far more readable than the loop quantum gravity text *Quantum Gravity* (Rovelli 2004), and the advanced and condensed "Zakopane lectures" (Rovelli 2011), but it covers, and in fact focuses on, the momentous advances in the covariant theory developed in the last few years, which have lead to finite transition amplitudes and were only foreshadowed in Rovelli (2004).

There is a rich literature on LQG, to which we refer for all the topics not covered in this book. On quantum gravity in general, Claus Kiefer has a recent general introduction Kiefer (2007). At the time of writing, Ashtekar and Petkov are editing a *Springer Handbook of Spacetime*, with numerous useful contributions, including that of John Engle's article on spinfoams.

A fine book with much useful background material is that of John Baez and Javier Munian (1994). See also Baez (1994b), with many ideas and a nice introduction to the subject. An undergraduate-level introduction to LQG is provided by Rodolfo Gambini and Jorge Pullin (Gambini and Pullin 2010). A punctilious and comprehensive text on the canonical formulation of the theory, rich in mathematical details, is given by Thomas Thiemann (2007). The very early form of the theory and the first ideas giving rise to it can be found in the 1991 book by Abhay Ashtekar.

A good recent reference is the collection of the proceedings of the 3rd Zakopane school on loop quantum gravity, organized by Jerzy Lewandowski (Barrett *et al.* 2011a). It contains the introduction to LQG by Abhay Ashtekar (Ashtekar 2011), Rovelli's "Zakopane lectures" (Rovelli 2011), and the introduction by Kristina Giesel and Hanno Sahlmann to the canonical theory, and John Barrett *et al.*'s review on the semiclassical approximation to the spinfoam dynamics (Barrett *et al.* 2011b). We also recommend Alejandro Perez's spinfoam review (Perez 2012), which is complementary to this book in several ways. Finally, we recommend Hal Haggard's thesis online (Haggard 2011), for a careful and useful introduction to and reference for the mathematics of spin networks.

We are very grateful to Klaas Landsman, Gabriele Stagno, Marco Finocchiaro, Hal Haggard, Tim Kittel, Thomas Krajewski, Cedrick Miranda Mello, Aldo Riello, Tapio

Salminem and, come sempre, Leonard Cottrell, for careful reading of the notes, corrections, and clarifications. Several tutorials have been prepared by David Kubiznak and Jonathan Ziprick for the students of the International Perimeter Scholars: Andrzej, Grisha, Lance, Lucas, Mark, Pavel, Brenda, Jacob, Linging, Robert, Rosa; thanks also to them!

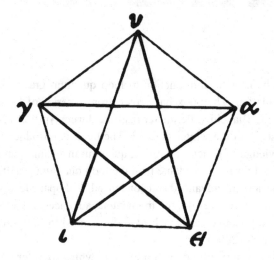

PART I

FOUNDATIONS

1 Spacetime as a quantum object

This book introduces the reader to a theory of quantum gravity. The theory is covariant loop quantum gravity (covariant LQG). It is a theory that has grown historically via a long, indirect path, briefly summarized at the end of this chapter. The book does not follow the historical path. Rather, it is pedagogical, taking the reader through the steps needed to learn the theory.

The theory is still tentative for two reasons. First, some questions about its consistency remain open; these will be discussed later in the book. Second, a scientific theory must pass the test of experience before becoming a reliable description of a domain of the world; no direct empirical corroboration of the theory is available yet. The book is written in the hope that some of you, our readers, will be able to fill these gaps.

This first chapter clarifies what is the problem addressed by the theory and gives a simple and sketchy derivation of the core physical content of the theory, including its general consequences.

1.1 The problem

After the detection at CERN of a particle that appears to match the expected properties of the Higgs [ATLAS Collaboration (2012); CMS Collaboration (2012)], the demarcation line separating what we know about the elementary physical world from what we do not know is now traced in a particularly clear-cut way. What we know is encapsulated into three major theories:

- *Quantum mechanics*, which is the general theoretical framework for describing dynamics
- *The* $SU(3) \times SU(2) \times U(1)$ *standard model of particle physics*, which describes all matter we have so far observed directly, with its non-gravitational interactions
- *General relativity* (GR), which describes gravity, space, and time.

In spite of the decades-long continuous expectation of violations of these theories, in spite of the initial implausibility of many of their predictions (long-distance entanglement, fundamental scalar particles, expansion of the universe, black holes, ...), and in spite of the bad press suffered by the standard model, often put down as an incoherent patchwork, so far Nature has steadily continued to say "Yes" to *all* predictions of these theories and "No" to *all* predictions of alternative theories (proton decay, signatures of extra dimensions, supersymmetric particles, new short-range forces, black holes at LHC, ...). Anything beyond

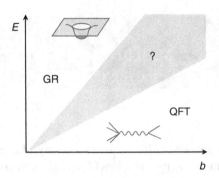

Figure 1.1 Regimes for the gravitational scattering of neutral particles in Planck units $c = \hbar = G = 1$. E is the energy in the center of the mass reference system and b the impact parameter (how close to one another the two particles come). At low energy, effective quantum field theory (QFT) is sufficient to predict the scattering amplitude. At high energy, classical general relativity (GR) is generally sufficient. In (at least parts of) the intermediate region (gray wedge) we do not have any predictive theory.

these theories is speculative. It is good to try and to dream: all good theories were attempts and dreams, before becoming credible. But lots of attempts and dreams go nowhere. The success of the above package of theories has gone far beyond anybody's expectation, and should be taken at its face value.

These theories are not the final story about the elementary world, of course. Among the open problems, three stand out:

- Dark matter
- Unification
- Quantum gravity.

These are problems of very different kind.[1] The first of these[2] is due to converging elements of empirical evidence indicating that about 85% of the galactic and cosmological matter is likely *not* to be of the kind described by the standard model. Many tentative alternative explanations are on the table, so far none convincing (Bertone 2010). The second is the old hope of reducing the number of free parameter and independent elements in our elementary description of Nature. The third, quantum gravity, is the problem we discuss here. It is not necessarily related to the first two.

The problem of quantum gravity is simply the fact that the current theories are not capable of describing the quantum behavior of the gravitational field. Because of this, we lack a predictive theory capable of describing phenomena where both gravity and quantum theory play a role. Examples are the center of a black hole, very early cosmology, the structure of Nature at very short scale, or simply the scattering amplitude of two neutral particles at small impact parameter and high energy. See Figure 1.1.

---

[1] To these one can add the problem of the interpretation of quantum mechanics, which is probably of still another kind.

[2] Not to be confused with the improperly called "dark energy mystery," much less of a mystery than usually advertised (Bianchi and Rovelli 2010a,b).

Observational technology has recently began to reach and probe some aspects of this regime, for instance its Lorentz invariance, and has already empirically ruled out some tentative theoretical ideas (Liberati and Maccione 2009). This is a major advance from a few years ago, when the quantum-gravitational domain appeared completely unreachable by our observation. But for the moment direct empirical information on this regime is minimal. This would be a problem if we had *many* alternative complete theories of quantum gravity to select from. But we are not in this situation: we have very few, if any. We are not at all in a situation of excessive theoretical freedom: the shortcoming in the set of fundamental laws is strident and calls for a solution, but consistency with what we know dramatically limits our freedom – which is good, since freedom is just another word for "nothing left to lose".

The problem is even more serious: our successful theories are based on contradictory hypotheses. A good student following a general-relativity class in the morning and a quantum-field-theory class in the afternoon must think her teachers are chumps, or haven't been talking to one another for decades. They teach two totally different worlds. In the morning, spacetime is curved and everything is smooth and deterministic. In the afternoon, the world is formed by discrete quanta jumping over a flat spacetime, governed by global symmetries (Poincaré) that the morning teacher has carefully explained *not* to be features of our world.

Contradiction between empirically successful theories is not a curse: it is a terrific opportunity. Several of the major jumps ahead in physics have been the result of efforts to resolve precisely such contradictions. Newton discovered universal gravitation by combining Galileo's parabolas with Kepler's ellipses. Einstein discovered special relativity to solve the "irreconcilable" contradiction between mechanics and electrodynamics. Ten years later, he discovered that spacetime is curved in an effort to reconcile newtonian gravitation with special relativity. Notice that these and other major steps in science have been achieved virtually without any *new* empirical data. Copernicus, for instance, constructed the heliocentric model and was able to compute the distances of the planets from the Sun using only the data in the book of Ptolemy.[3]

This is precisely the situation with quantum gravity. The scarcity of direct empirical information about the Planck scale is not dramatic: Copernicus, Einstein, and, to a lesser extent, Newton, have understood something new about the world without new data – just comparing apparently contradictory successful theories. We are in the same privileged situation. We lack their stature, but we are not excused from trying hard.[4]

---

[3] This is not in contradiction with the fact that scientific knowledge is grounded on an empirical basis. First, a theory becomes reliable only after *new* empirical support. But also the discovery itself of a new theory is based on an empirical basis even when there are no *new* data: the empirical basis is the empirical content of the previous *theories*. The advance is obtained from the effort of finding the overall conceptual structure wherein these can be framed. The scientific enterprise is still finding theories explaining observations, also when *new* observations are not available. Copernicus and Einstein were scientists even when they did not make use of new data. (Even Newton, though obsessed by getting good and recent data, found universal gravitation essentially by merging Galileo's and Kepler's laws.) Their example shows that the common claim that there is no advance in physics without new data is patently false.

[4] And we stand on their shoulders.

1.2 The end of space and time

The reason for the difficulty, but also the source of the beauty and the fascination of the problem, is that GR is not just a theory of gravity. It is a modification of our understanding of the nature of space and time. Einstein's discovery is that spacetime and the gravitational field are the same physical entity.[5] Spacetime is a manifestation of a physical field. All fields we know exhibit quantum properties at some scale, therefore we believe space and time to have quantum properties as well.

We must thus modify our understanding of the nature of space and time, in order to take these quantum properties into account. The description of spacetime as a (pseudo-) Riemannian manifold cannot survive quantum gravity. We have to learn a new language for describing the world: a language which is neither that of standard field theory on flat spacetime, nor that of Riemannian geometry. We have to understand what quantum space and quantum time are. This is the difficult side of quantum gravity, but also the source of its beauty.

The way this was first understood is enlightening. It all started with a mistake by Lev Landau. Shortly after Heisenberg introduced his commutation relations

$$[q,p] = i\hbar \tag{1.1}$$

and the ensuing uncertainty relations, the problem on the table was extending quantum theory to the electromagnetic field. In a 1931 paper with Peierls (Landau and Peierls 1931), Landau suggested that once applied to the electromagnetic field, the uncertainty relation would imply that no component of the field at a given spacetime point could be measured with arbitrary precision. The intuition was that an arbitrarily sharp spatiotemporal localization would be in contradiction with the Heisenberg uncertainty relations.

Niels Bohr guessed immediately, and correctly, that Landau was wrong. To prove him wrong, he embarked on a research program with Léon Rosenfeld, which led to a classic paper (Bohr and Rosenfeld 1933) proving that in the quantum theory of the electromagnetic field the Heisenberg uncertainty relations *do not* prevent a single component of the field at a spacetime point from being measured with arbitrary precision.

But Landau being Landau, even his mistakes have bite. Landau, indeed, had a younger friend, Matvei Petrovich Bronstein (Gorelik and Frenkel 1994), a brilliant young Russian theoretical physicist. Bronstein repeated the Bohr–Rosenfeld analysis using the gravitational field rather than the electromagnetic field. And here, surprise, Landau's

---

[5] In the mathematics of Riemannian geometry one might distinguish the metric field from the manifold and identify spacetime with the second. But in the physics of general relativity this terminology is misleading, because of the peculiar gauge invariance of the theory. If by "spacetime" we denote the manifold, then, using Einstein's words, "The requirement of general covariance takes away from space and time the last remnant of physical objectivity" (Einstein 1916). A detailed discussion of this point is given in Sections 2.2 and 2.3 of Rovelli (2004).

Figure 1.2   The last picture of Matvei Bronstein, the scientist who understood that quantum gravity affects the nature of spacetime. Matvei was arrested on the night of August 6, 1937. He was thirty. He was executed in a Leningrad prison in February 1938.

intuition turned out to be correct (Bronstein 1936a,b). If we do not disregard general relativity, quantum theory *does* prevent the measurability of the field in an arbitrarily small region.

In August 1937, Matvei Bronstein was arrested in the context of Stalin's Great Purge; he was convicted in a brief trial and executed. His fault was to believe in communism without Stalinism (Figure 1.2).

Let us give a modern and simplified version of Bronstein's argument, because it is not just the beginning, it is also the core of quantum gravity.

Say you want to measure some field value at a location x. For this, you have to mark this location. Say you want to determine it with precision L. Say you do this by having a particle at x. Since any particle is a quantum particle, there will be uncertainties Δx and Δp associated with position and momentum of the particle. To have localization determined with precision L, you want $\Delta x < L$, and since Heisenberg uncertainty gives $\Delta x > \hbar/\Delta p$, it follows that $\Delta p > \hbar/L$. The mean value of p^2 is larger than $(\Delta p)^2$, therefore $p^2 > (\hbar/L)^2$. This is a well-known consequence of Heisenberg uncertainty: sharp location requires large momentum; which is the reason why at CERN high-momentum particles are used to investigate small scales. In turn, large momentum implies large energy E. In the relativistic limit, where rest mass is negligible, $E \sim cp$. Sharp localization requires large energy.

Now let us add GR. In GR, any form of energy E acts as a gravitational mass $M \sim E/c^2$ and distorts spacetime around itself. The distortion increases when energy is concentrated, to the point that a black hole forms when a mass M is concentrated in a sphere of radius $R \sim GM/c^2$, where G is the Newton constant. If we take L arbitrarily small, to get a sharper localization, the concentrated energy will grow to the point where R becomes larger than L. But in this case the region of size L that we wanted to mark will be hidden beyond a black hole horizon, and we lose localization. Therefore we can decrease L only up to a minimum value, which clearly is reached when the horizon radius reaches L, that is when $R = L$.

Combining the relations above, we obtain that the minimal size where we can localize a quantum particle without having it hidden by its own horizon is

$$L = \frac{MG}{c^2} = \frac{EG}{c^4} = \frac{pG}{c^3} = \frac{\hbar G}{Lc^3}. \tag{1.2}$$

Solving this for L, we find that it is not possible to localize anything with a precision better than the length

$$L_{\text{Planck}} = \sqrt{\frac{\hbar G}{c^3}} \sim 10^{-33} \text{ cm},\tag{1.3}$$

which is called the Planck length. Well above this length scale, we can treat spacetime as a smooth space. Below, it makes no sense to talk about distance. What happens at this scale is that the quantum fluctuations of the gravitational field, namely the metric, become wide, and spacetime can no longer be viewed as a smooth manifold: anything smaller than L_{Planck} is "hidden inside its own mini-black hole."

This simple derivation is obtained by extrapolating semiclassical physics. But the conclusion is correct, and characterizes the physics of quantum spacetime.

In Bronstein's words: "Without a deep revision of classical notions it seems hardly possible to extend the quantum theory of gravity also to [the short-distance] domain" (Bronstein 1936b). Bronstein's result forces us to take seriously the connection between gravity and geometry. It shows that the Bohr–Rosenfeld argument, according to which quantum fields can be defined in arbitrary small regions of space, fails in the presence of gravity. Therefore we cannot treat the quantum gravitational field simply as a quantum field in space. The smooth metric geometry of physical space, which is the ground needed to define a standard quantum field, is itself affected by quantum theory. What we need is a genuine quantum theory of geometry.

This implies that the conventional intuition provided by quantum field theory fails for quantum gravity. The worldview where quantum fields are defined over spacetime is the common world-picture in quantum field theory, but it needs to be abandoned for quantum gravity. We need a genuinely new way of doing physics, where space and time come *after*, and not *before*, the quantum states. Space and time are semiclassical approximations to quantum configurations. The quantum states are not quantum states *on* spacetime. They are quantum states *of* spacetime. This is what loop quantum gravity provides (Figure 1.3).

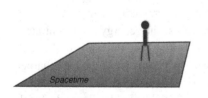

Conventional quantum field theorist   Post-Maldacena string theorist   Genuine quantum-gravity physicist

Figure 1.3   Pre-general-relativistic physics is conceived on spacetime. The recent developments of string theory, with bulk physics described in terms of a boundary theory, are a step toward the same direction. Genuine full quantum gravity requires no spacetime at all.

1.3 Geometry quantized

The best guide we have toward quantum gravity is provided by our current quantum theory and our current gravity theory. We cannot be sure whether the basic physics on which each of these theories is grounded still applies at the Planck scale, but the history of physics teaches that vast extrapolation of good theories often works very well. The Maxwell equations, discovered with experiments in a small lab, turn out to be extremely good from nuclear to galactic scale, some 35 orders of magnitudes away, more than our distance from the Planck scale. General relativity, found at the solar system scale, appears to work remarkably well at cosmological scales, some 20 orders of magnitudes larger, and so on. In science, the best hypothesis, until something new appears empirically, is that what we know extends.

The problem, therefore, is not to *guess* what happens at the Planck scale. The problem is: is there a consistent theory that merges general relativity and quantum theory? This is the form of thinking that has been extraordinarily productive in the past. The physics of guessing, the physics of "why not try this?" is a waste of time. No great idea came from the blue sky in the past: good ideas come either from experiments or from taking seriously the empirically successful theories. Let us therefore take seriously geometry and the quantum and see, in the simplest possible terms, what a "quantum geometry" implies.

General relativity teaches us that geometry is a manifestation of the gravitational field. Geometry deals with quantities such as area, volume, length, angles, ... These are quantities determined by the gravitational field. Quantum theory teaches us that fields have quantum properties. The problem of quantum gravity is therefore to understand what are the quantum proprieties of geometrical quantities such as area, volume, et cetera.

The quantum nature of a physical quantity is manifested in three forms:

1. In the possible discretization (or "quantization") of the quantity itself
2. In the short-scale "fuzziness" implied by the uncertainty relations
3. In the probabilistic nature of its evolution (given by the transition amplitudes).

We focus here on the first two of these (probabilistic evolution in a gravitational context is discussed in the next chapter), and consider a simple example of how they can come about, namely, how space can become discrete and fuzzy. This example is elementary and is going to leave some points out, but it is illustrative and it leads to the most characteristic aspect of loop quantum gravity: the existence of "quanta of space."

Let us start by reviewing basic quantum theory in three very elementary examples; then we describe an elementary geometrical object; and finally we see how the combination of these two languages leads directly to the quanta of space.

Harmonic oscillator

Consider a mass m attached to a spring with elastic constant k. We describe its motion in terms of the position q, the velocity v and the momentum $p = mv$. The energy $E = \frac{1}{2}mv^2 + \frac{1}{2}kq^2$ is a positive real number and is conserved. The "quantization postulate" from which the quantum theory follows is the existence of a Hilbert space $\mathcal{H}$ where (p,q) are non-commuting (essentially) self-adjoint operators satisfying (Born and Jordan 1925)

$$[q,p] = i\hbar. \tag{1.4}$$

This is the "new law of nature" (Heisenberg 1925) from which discretization can be computed. These commutation relations imply that the energy operator $E(p,q) = \frac{p^2}{2m} + \frac{k}{2}q^2$ has discrete spectrum with eigenvalues ($E\psi^{(n)} = E_n\psi^{(n)}$)

$$E_n = \hbar\omega\left(n + \frac{1}{2}\right), \tag{1.5}$$

where $\omega = \sqrt{k/m}$. That is, energy is "quantized": it comes in discrete quanta. Since a free field is a collection of oscillators, one per mode, a quantum field is a collection of discrete quanta (Einstein 1905a). The quanta of the electromagnetic field are the photons. The quanta of Dirac fields are the particles that make up ordinary matter. We are interested in the elementary quanta of gravity.

The magic circle: discreteness is kinematics

Consider a particle moving on a circle, subject to a potential $V(\alpha)$. Let its position be an angular variable $\alpha \in S^1 \sim [0,2\pi]$ and its hamiltonian $H = \frac{p^2}{2c} + V(\alpha)$, where $p = c\,d\alpha/dt$ is the momentum and c is a constant (with dimensions ML^2). The quantum behavior of the particle is described by the Hilbert space $L_2[S^1]$ of the square integrable functions $\psi(\alpha)$ on the circle and the momentum operator is $p = -i\hbar d/d\alpha$. This operator has a discrete spectrum, with eigenvalues

$$p_n = n\hbar, \tag{1.6}$$

independently from the potential. We call "kinematic" the properties of a system that depend only on its basic variables, such as its coordinates and momenta, and "dynamic" the properties that depend on the hamiltonian, or, in general, on the evolution. Then it is clear that, in general, discreteness is a *kinematic* property.[6]

The discreteness of p is a direct consequence of the fact that α is in a compact domain. (The same happens for a particle in a box.) Notice that $[\alpha,p] \neq i\hbar$ because the derivative of the function α on the circle diverges at $\alpha = 0 \sim 2\pi$: indeed, α is a discontinuous function on S^1. Quantization must take into account the global topology of phase space. One of the

---

[6] Not so for the discreteness of the *energy*, as in the previous example, which of course depends on the form of the hamiltonian.

many ways to do so is to avoid using a discontinuous function like α and use instead a continuous function like $s = \sin(\alpha)$ or/and $c = \cos(\alpha)$. The three observables s, c, p have closed Poisson brackets $\{s, c\} = 0, \{p, s\} = c, \{p, c\} = -s$ correctly represented by the commutators of the operator $-i\hbar d/d\alpha$, and the multiplication operators $s = \sin(\alpha)$ and $c = \cos(\alpha)$. The last two operators can be combined into the complex operator $h = e^{i\alpha}$. In this sense, the correct elementary operator of this system is not α, but rather $h = e^{i\alpha}$. (We shall see that for the same reason the correct operator in quantum gravity is not the gravitational connection but rather its exponentiation along "loops." This is the first hint of the "loops" of LQG.)

Angular momentum

Let $\vec{L} = (L^1, L^2, L^3)$ be the angular momentum of a system that can rotate, with components $\{L^i\}$, with $i = 1, 2, 3$. The total angular momentum is $L = |\vec{L}| = \sqrt{L^i L^i}$ (summation on repeated indices always understood unless stated). Classical mechanics teaches us that $\vec{L}$ is the generator (in the sense of Poisson brackets) of infinitesimal rotations. Postulating that the corresponding quantum operator is also the generator of rotations in the Hilbert space, we have the quantization law (Born *et al.* 1926)

$$[L^i, L^j] = i\hbar\, \varepsilon^{ij}{}_k L^k, \tag{1.7}$$

where $\varepsilon^{ij}{}_k$ is the totally antisymmetric (Levi-Civita) symbol. SU(2) representation theory (reviewed in the Complements to this chapter) then immediately gives the eigenvalues of L, if the operators $\vec{L}$ satisfy the above commutation relations. These are

$$L_j = \hbar\sqrt{j(j+1)}, \quad j = 0, \frac{1}{2}, 1, \frac{3}{2}, 2, \ldots \tag{1.8}$$

That is, total angular momentum is quantized. Notice that the quantization of angular momentum is a purely kinematical prediction of quantum theory: it remains the same irrespective of the form of the hamiltonian, and in particular irrespective of whether or not angular momentum is conserved. Notice also that, as for the magic circle, discreteness is a consequence of compact directions in phase space: here the space of the orientations of the body.

This is all we need from quantum theory. Let us move on to geometry.

Geometry

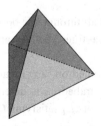

Pick a simple geometrical object, an elementary portion of space. Say we pick a small tetrahedron τ, not necessarily regular.

The geometry of a tetrahedron is characterized by the length of its sides, the area of its faces, its volume, the dihedral angles at its edges, the angles at the vertices of its faces, and so on. These are all local functions of the gravitational field, because geometry is the same thing as the gravitational field. These geometrical quantities are related to one

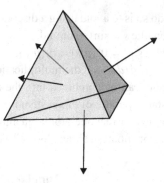

Figure 1.4   The four vectors $\vec{L}_a$, normals to the faces.

another. A set of independent quantities is provided, for instance, by the six lengths of the sides, but these are not appropriate for studying quantization, because they are constrained by inequalities. The length of the three sides of a triangle, for instance, cannot be chosen arbitrarily: it must satisfy the triangle inequalities. Non-trivial inequalities between dynamical variables, like all global features of phase space, are generally difficult to implement in quantum theory.

Instead, we choose the four vectors $\vec{L}_a$, $a = 1, \ldots, 4$ defined for each triangle a as $\frac{1}{2}$ of the (outward oriented) vector-product of two edges bounding the triangle. See Figure 1.4. These four vectors have several nice properties. Elementary geometry shows that they can be equivalently defined in one of the two following ways:

- The vectors $\vec{L}_a$ are (outgoing) normals to the faces of the tetrahedron and their norm is equal to the area of the face.
- The matrix of the components L_a^i for $a = 1, 2, 3$ (notice that only 3 edges are involved) is $L^T = -\frac{1}{2}(\det M)M^{-1}$, where M is the matrix formed by the components of three edges of the tetrahedron that emanate from a common vertex.

Exercise 1.1 *Show that the definitions above are all equivalent.*

The vectors $\vec{L}_a$ have the following properties:

- They satisfy the "closure" relation

$$\vec{C} := \sum_{a=1}^{4} \vec{L}_a = 0. \tag{1.9}$$

(Keep this in mind, because this equation will reappear all through the book.)
- The quantities $\vec{L}_a$ determine all other geometrical quantities such as areas, volume, angles between edges and dihedral angles between faces.
- All these quantities, that is, the geometry of the tetrahedron, are invariant under a common SO(3) rotation of the four $\vec{L}_a$. Therefore a tetrahedron is determined by an equivalence class under rotations of a quadruplet of vectors $\vec{L}_a$ satisfying (1.9).
- The resulting number of degrees of freedom is correct.
- The area A_a of the face a is $|\vec{L}_a|$.

- The volume V is determined by the (properly oriented) triple product of any three faces:

$$V^2 = \frac{2}{9}(\vec{L}_1 \times \vec{L}_2) \cdot \vec{L}_3 = \frac{2}{9}\epsilon_{ijk}L_1^i L_2^j L_3^k = \frac{2}{9} \det L. \qquad (1.10)$$

(Defined in this manner, a negative value of V^2 is simply the indication of a change of the order of the triple product, that is, an inversion in the orientation.)

Exercise 1.2 *Prove these relations. Hint: choose a tetrahedron determined by a triple of orthonormal edges, and then argue that the results are general because they are invariant under linear transformations. Derive the $\frac{2}{9}$ factor.*[7]

Quantization of the geometry

We have all the ingredients for jumping to quantum gravity. The geometry of a real physical tetrahedron is determined by the gravitational field, which is a quantum field. Therefore the normals $\vec{L}_a$ are to be described by quantum operators, if we take the quantum nature of gravity into account. These will obey commutation relations. The commutation relation can be obtained from the hamiltonian analysis of GR, by promoting Poisson brackets to operators, in the same manner in which (1.1) and (1.7) can; but ultimately they are quantization postulates, like (1.1) and (1.7). Let us therefore just postulate them here. The simplest possibility (Barbieri 1998) is to mimic (1.7), namely, to write

$$[L_a^i, L_b^j] = i\delta_{ab}l_o^2 \, \varepsilon^{ij}{}_k \, L_a^k, \qquad (1.12)$$

where l_o^2 is a constant proportional to $\hbar$ and with the dimension of an area. These commutation relations are again realizations of the algebra of SU(2), like in the case of the rotator, reflecting again the rotational symmetry in the description of the tetrahedron. This is good: for instance, we see that $\vec{C}$ defined in (1.9) is precisely the generator of common rotations and therefore the closure condition (1.9) is an immediate condition of rotational invariance, which is what we want: the geometry is determined by the $\vec{L}_a$ up to rotations, which here are gauge. Let us thus fix (1.12) as the quantization postulate.

The constant l_o must be related to the Planck scale L_{Planck}, which is the only dimensional constant in quantum gravity. Leaving the exact relation open for the moment, we pose

$$l_o^2 = 8\pi\gamma \, L_{\text{Planck}}^2 = \gamma \, \frac{\hbar(8\pi G)}{c^3}, \qquad (1.13)$$

---

[7] In Chapter 3 we describe the gravitational field in terms of triads and tetrads. Let us anticipate here the relation between the vectors $\vec{L}_a$ and the triad. If the tetrahedron is small compared to the scale of the local curvature, so that the metric can be assumed to be locally flat, then

$$L_a^i = \frac{1}{2}\epsilon^i{}_{jk} \int_a e^j \wedge e^k. \qquad (1.11)$$

Equivalently, $\vec{L}_a$ can be identified with the flux of the densitized inverse triad field $E_i^a = \frac{1}{2}\epsilon^{abc}\epsilon_{ijk}e_b^j e_c^k$, which is Ashtekar's electric field, across the face. Since the triad is the gravitational field, this gives the explicit relation between $\vec{L}_a$ and the gravitational field. Here the triad is defined in the three-dimensional hyperplane determined by the tetrahedron.

AREA DEMOCRITI.

Figure 1.5   An image from *De triplici minimo et mensura* (1591) by Giordano Bruno, showing discrete elements and referring to Democritus's atomism. Bruno's images played an important role as sources of the modern intuition about the atomic structure of matter.

where 8π is just there for historical reasons (the coupling constant on the right of Einstein's equations is not G but $8\pi G$) and γ is a dimensionless parameter (presumably of the order of unity) that fixes the precise scale of the theory. Let us study the consequences of this quantization law.

1.3.1 Quanta of area and volume

One consequence of the commutation relations (1.12) is immediate (Rovelli and Smolin 1995a; Ashtekar and Lewandowski 1997): the quantity $A_a = |\vec{L}_a|$ behaves as total angular momentum. As this quantity is the area, it follows immediately that the area of the triangles bounding any tetrahedron is quantized with eigenvalues

$$A = l_o^2\,\sqrt{j(j+1)}, \qquad j = 0, \frac{1}{2}, 1, \frac{3}{2}, 2, \ldots \tag{1.14}$$

This is the gist of loop quantum gravity. Space is not divisible into arbitrarily small bits.[8] As we shall see, the result extends to any surface, not just the area of the triangles bounding a tetrahedron.

Say that the quantum geometry is in a state with area eigenvalues $j_1, \ldots, j_4$. The four vector operators $\vec{L}_a$ act on the tensor product $\mathcal{H}$ of four representations of SU(2), with respective spins $j_1, \ldots, j_4$. That is, the Hilbert space of the quantum states of the geometry of the tetrahedron at fixed values of the area of its faces is

$$\mathcal{H} = \mathcal{H}_{j_1} \otimes \mathcal{H}_{j_2} \otimes \mathcal{H}_{j_3} \otimes \mathcal{H}_{j_4}. \tag{1.15}$$

We have to take also into account the closure equation (1.9), which is a condition the states must satisfy, if they are to describe a tetrahedron. $\vec{C}$ is nothing else than the generator of the

---

[8] The intuition that this property holds for matter can be traced to the ancient atomism (Figure 1.5).

global (diagonal) action of SU(2) on the four representation spaces. The states that solve Eq. (1.9) (strongly, namely, $\vec{C}\psi = 0$) are the states that are invariant under this action, namely, the states in the subspace

$$\mathcal{K} = \text{Inv}_{\text{SU}(2)}[\mathcal{H}_{j_1} \otimes \mathcal{H}_{j_2} \otimes \mathcal{H}_{j_3} \otimes \mathcal{H}_{j_4}] \,. \tag{1.16}$$

Thus we find that, as anticipated, the physical states are invariant under (common) rotations, because the geometry is defined only by the equivalence classes of $\vec{L}_a$ under rotations. (This connection between the closure equation (1.9) and invariance under rotations is nice and encouraging; it confirms that our quantization postulate is reasonable. When we come back to this in the context of full general relativity, we will see that this gauge is nothing else than the local-rotation gauge invariance of the tetrad formulation of general relativity.)

Consider now the volume operator V defined by (1.10). This is well defined in $\mathcal{K}$ because it commutes with $\vec{C}$, that is, it is rotationally invariant. Therefore we have a well-posed eigenvalue problem for the self-adjoint volume operator on the Hilbert space $\mathcal{K}$. As this space is finite dimensional, it follows that its eigenvalues are discrete (Rovelli and Smolin 1995a; Ashtekar and Lewandowski 1998). Therefore we have the result that the volume has discrete eigenvalues as well. In other words, there are "quanta of volume" or "quanta of space": the volume of our tetrahedron can grow only in discrete steps, precisely as the amplitude of a mode of the electromagnetic field. In the Complement 1.7.3 to this chapter we compute some eigenvalues of the volume explicitly.

It is important not to confuse this discretization of geometry, namely, the fact that area and volume are quantized, with the discretization of space implied by focusing on a single tetrahedron. The first is the analog of the fact that the energy of a mode of the electromagnetic field comes in discrete quanta. It is a quantum phenomenon. The second is the analog of the fact that it is convenient to decompose a field into discrete modes and study one mode at a time: it is a convenient isolation of degrees of freedom, completely independent of quantum theory. Geometry is not discrete because we focused on a tetrahedron: geometry is discrete because area and volume of *any* tetrahedron (in fact, any polyhedron, as we shall see) take only quantized values. The quantum discretization of geometry is determined by the spectral properties of area and volume.

The astute reader may wonder whether the fact that we have started with a fixed chunk of space plays a role in the argument. Had we chosen a smaller tetrahedron to start with, would we have obtained smaller geometric quanta? The answer is no, and the reason is at the core of the physics of general relativity: *there is no notion of size (length, area, volume) independent from the one provided by the gravitational field itself.* The coordinates used in general relativity carry no metrical meaning. In fact, they carry no physical meaning at all. If we repeat the above calculation starting from a "smaller" tetrahedron in coordinate space, we are not dealing with a physically smaller tetrahedron, only with a different choice of coordinates. This is apparent in the fact that the coordinates play no role in the derivation. Whatever coordinate tetrahedron we may wish to draw, however small, its *physical* size will be determined by the gravitational field on it, and this is quantized, so that its physical size will be quantized with the *same* eigenvalues. Digesting this point is the first step to understanding quantum gravity. There is no way to cut a minimal tetrahedron in half, just as there is no way to split the minimal angular momentum in quantum mechanics. Space itself has a "granular structure" formed by individual quanta.

The shape of the quanta of space and the fuzziness of the geometry

As we shall prove later, the four areas A_a of the four faces and the volume V form a maximally commuting set of operators in the sense of Dirac. Therefore they can be diagonalized together and quantum states of the geometry of the tetrahedron are uniquely characterized by their eigenvalues $|j_a, v\rangle$.

Is the shape of such a *quantum* state truly a tetrahedron? The answer is no, for the following reason. The geometry of a classical tetrahedron is determined by six numbers, for instance, the six lengths of its edges. (Equivalently, the 4×3 quantities L_a^i constrained by the 3 closure equations, up to 3 rotations.) But the corresponding quantum numbers that determine the quantum states of the tetrahedron are not six; they are only five: four areas and one volume.

The situation is exactly analogous to angular momentum, where the classical system is determined by three numbers, the three components of the angular momentum, but only two quantities (say L^2, L_z) form a complete set. Because of this fact, as is well known, a quantum rotator never has a definite angular momentum $\vec{L}$, and we cannot really think of an electron as a small rotating stone: if L_x is sharp, necessarily L_y is fuzzy, is quantum-spread.

For the very same reason, therefore, geometry can never be sharp in the quantum theory, in the same sense in which the three components of angular momentum can never be all sharp. In any real quantum state there will be residual quantum fuzziness of the geometry: it is not possible to have all dihedral angles, all areas, and all lengths sharply determined. Geometry is fuzzy at the Planck scale.

We have found two characteristic features of quantum geometry:

- Areas and volumes have discrete eigenvalues.
- Geometry is spread quantum mechanically at the Planck scale.

Quantum geometry differs from Riemannian geometry on both these grounds.

1.4 Physical consequences of the existence of the Planck scale

1.4.1 Discreteness: scaling is finite

The existence of the Planck length sets quantum GR aside from standard quantum field theory for two reasons. First, we cannot expect quantum gravity to be described by a local quantum field theory, in the strict sense of this term (Haag 1996). Local quantum theory requires quantum fields to be described by observables at arbitrarily small regions in a continuous manifold. This is not going to happen in quantum gravity, because physical regions cannot be arbitrarily small.

Second, the quantum field theories of the standard model are defined in terms of an *infinite* renormalization group. The existence of the Planck length indicates that this is not going to be the case for quantum gravity. Let us see this in more detail.

When computing transition amplitudes for a field theory using perturbation methods, infinite quantities appear due to the effect of modes of the field at arbitrary small wavelengths. Infinities can be removed by introducing a cut-off and adjusting the definition of the theory to make it cut-off-dependent in such a manner that physical observables are cut-off-independent and match experimental observations. The cut-off can be regarded as a technical trick; alternatively, the quantum field theory can be viewed as effective, useful at energy scales much lower than the (unknown) natural scale of the physics. Accordingly, care must be taken for the final amplitudes not to depend on the cut-off. Condensed matter offers a prototypical example of independence from short scale: second-order phase transitions. At the critical point of a second-order phase transition, the behavior of the system becomes scale-independent, and large-scale physics is largely independent of the microscopic dynamics. Conventional quantum field theories are modeled on condensed matter phase transitions: they are defined using a cut-off, but this is then taken to infinity, and the theory is defined in such a manner that the final result remains finite and independent of the details of the cut-off chosen.

This framework has proven effective for describing particle physics, but it is not likely to work for quantum gravity. The alternative is also indicated by condensed matter physics: consider a generic matter system *not* at a critical point; say, a bar of iron at room temperature. Its behavior at macroscopic scales is described by a low-energy theory, characterized by certain physical constants. This behavior includes wave propagation and finite correlation functions. The high-frequency modes of the bar have an effect on the value of the macroscopic physics, and can be explored using a renormalization group equation describing the dependence of physical parameters on the scale. But the system is characterized by a physical and *finite* cut-off scale – the atomic scale – and there are no modes of the bar beyond this scale. The bar can be described as a system with a large but *finite* number of degrees of freedom. The short-distance cut-off in the modes is not a mathematical trick for removing infinities, nor a way for hiding unknown physics: it is a genuine physical feature of the system. Quantum gravity is similar: the Planck-scale cut-off is a genuine physical feature of the system formed by quantum spacetime.[9]

The existence of a minimal length scale gives quantum gravity universal character, analogous to special relativity and quantum mechanics: Special relativity can be seen as the discovery of the existence of a maximal local physical velocity, the speed of light c. Quantum mechanics can be interpreted as the discovery of a minimal action, $\hbar$, in all physical interactions, or, equivalently, the fact that a compact region of phase space contains only a finite number of distinguishable (orthogonal) quantum states, and therefore there is a

---

[9] As Douglas Stone puts it to illustrate the early work by Einstein on quantum theory: "Discreteness is not a mathematical trick; it is the way of the atomic world. Get used to it" (Stone 2013).

minimal amount of information[10] in the state of a system. Quantum gravity yields the discovery that there is a minimal length l_o at the Planck scale. This leads to a fundamental finiteness and discreteness of the world.

Natural physical units are obtained by measuring speed as the ratio to the maximal speed c, action in multiples of the minimal action $\hbar$, and lengths in multiples of the minimal length l_o. In these "natural units" $c = \hbar = l_o = 1$. To avoid confusion with γ, we shall not use these units in the first part of the book, and rather use the more conventional Planck units $c = \hbar = 8\pi G = 1$.

1.4.2 Fuzziness: disappearance of classical space and time

The absence of the conventional notions of space and time at small scales forces us to rethink the basis of physics. For instance, hamiltonian mechanics is about evolution in time; so is the conventional formulation of quantum mechanics and so is QFT, where time evolution is captured by the unitary representations of the Poincaré group. All this must change in quantum gravity. Understanding quantum spacetime requires therefore a substantial conceptual revolution. The physics of quantum gravity is not the physics of the gravitational field *in* spacetime. It is the physics of the quantum fields that *build up* spacetime. The basic ontology of physics, which has evolved during the last century, simplifies.

According to Descartes, who in this was essentially still following Aristotle, matter, moving in time, was the only component of the physical universe and extension was just a property of matter ("res extensa"). Newton introduced a description of the world in terms of particles located in space and moving in time. Faraday and Maxwell showed that this ontology needed to be supplemented by a new entity: the field. Special relativity showed that space and time must be thought of as aspects of a single entity, spacetime. General relativity showed that spacetime is itself a field: the gravitational field. Finally, quantum theory showed that particles are quanta of quantum fields. This is summarized in Table 1.1. Bringing all these results together implies that, as far as we know today, all that exist in nature are general-covariant quantum fields.

1.5 Graphs, loops, and quantum Faraday lines

Above we have described the quantum geometry of a single grain of physical space. A region of possibly curved physical space can be described by a set of interconnected grains of space. These can be represented by a graph, where each node is a grain of space and the links relate adjacent grains (Figure 1.6).

---

[10] "Information" is used here in the sense of Shannon: number of distinguishable alternatives (Shannon 1948). It has no relation to semantics, meaning, significance, consciousness, records, storage, or mental, cognitive, idealistic, or subjectivistic ideas.

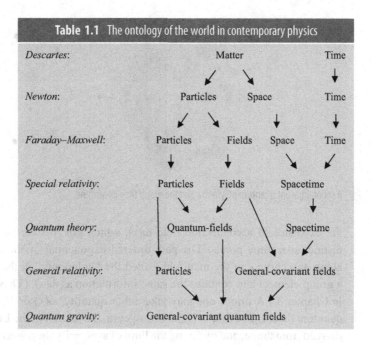

Table 1.1 The ontology of the world in contemporary physics

	Matter			Time
Descartes:				
Newton:	Particles	Space		Time
Faraday–Maxwell:	Particles	Fields	Space	Time
Special relativity:	Particles	Fields	Spacetime	
Quantum theory:	Quantum-fields	Spacetime		
General relativity:	Particles	General-covariant fields		
Quantum gravity:	General-covariant quantum fields			

We shall see below that this picture emerges naturally from the quantization of the gravitational field. The quantum states of the theory will have a natural graph structure of this
kind.

The "loops" in "loop quantum gravity" (LQG) refers to the loops formed by closed
sequences of links in such a graph.[11] The individual lines in the graph can be viewed as
discrete Faraday lines of the gravitational field. The Faraday lines, or "lines of force,"
introduced by Faraday, form the initial intuition at the root of the modern notion of field.
LQG grew on the intuition that the quantum discreteness makes these lines discrete in the
quantum theory. This idea was anticipated by Dirac:

> We can assume that when we go to the quantum theory, the lines of force become all dis-
> crete and separate from one another. [...] We have so a model where the basic physical
> entity is a line of force... Paul Dirac (Dirac 1956)

Since the gravitational field is spacetime, its discrete quantum lines of force are not in
space, but rather themselves form the texture of space. This is the physical intuition of
LQG.

Here is a more complete and condensed account of what happens in the theory (if you
find the rest of this subsection incomprehensible, do not worry, just skip it; it will become
clear after studying the book). The existence of fermions shows that the metric field is not
sufficient to describe the gravitational field. Tetrads, recalled in Chapter 3, are needed. This
introduces a local Lorentz gauge invariance, related to the freedom of choosing an independent Lorentz frame at each point of spacetime. This local gauge invariance implies

---

[11] Historically, the first states constructed had no nodes (Rovelli and Smolin 1988, 1990).

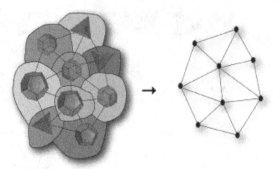

Figure 1.6   A set of adjacent quantum polyhedra and the graph they determine.

the existence of a connection field $\omega(x)$, which governs the parallel transport between distinct spacetime points. The path-ordered exponential of the connection $U_e = Pe^{\int_e \omega}$ along any curve e in the manifold, called the "holonomy" in the jargon of the theory, is a group element and contains the same information as $\omega(x)$. (These quantities are defined in Chapter 3.) A priori, one may take either quantity, $\omega(x)$ or U_e, as the variable for the quantum theory. The first is in the Lie algebra, the second in the Lie group; the first can be derived from the second by taking the limit of arbitrarily short e, as $U_e = \mathbb{1} + \omega(\dot{e}) + O(|e|^2)$. However, the Planck-scale discreteness that we expect in quantum gravity breaks the relation between the two. If space is discrete, there is no meaning in *infinitesimal* shifts in space, and therefore U_e remains well defined, while its derivative $\omega(x)$ is not. Therefore we are led to forget $\omega(x)$ and seek a quantization using the group variables U_e instead, as mentioned above for the quantization of a variable on a circle. These are called "loop" variables in the jargon of the theory. The corresponding quantum operators are the Wilson loop operators in quantum chromodynamics (QCD), which are also exponentials of the connection.[12] The fact that the rotation group is compact is the origin of the discreteness, precisely as in the case of the particle on a circle discussed above.

To see this more clearly, the analogy with lattice QCD (which we review in Chapter 4) is enlightening. In lattice QCD, one takes a lattice in spacetime with lattice spacing (the length of the links) a and describes the field in terms of group elements U_e associated to the links of the lattice. The physical theory is recovered in the limit where a goes to zero, and in this limit the group elements associated to individual e's all become close to the identity. The limit defines the Yang–Mills connection. In gravity, we can equally start from a discretization with group elements U_e associated to the links. But the length of these links is not an external parameter to be taken to zero: it is determined by the field itself, because geometry is determined by gravity. Since quantization renders geometry discrete, the theory does not have a limit where the e's become infinitesimal. Therefore there is no connection $\omega(x)$ defined in the quantum theory. A connection is defined only in the classical limit, where we look at the theory at scales much larger than the Planck scale, and therefore we can formally take the length of the e's to zero.

---

[12] The name "loop" is proper only when e is a closed curve, or a loop, in which case the trace of U_e is gauge-invariant.

1.6 The landscape

At present we have two theories that incorporate the ultraviolet finiteness following from the existence of the Planck scale into their foundation, which are well developed, and give well-defined definitions for the transition amplitudes in quantum gravity: these are loop quantum gravity and string theory.

String theory has evolved from attempts to quantize gravity by splitting the gravitational field into a background fixed metric field, used to fix the causal structure of spacetime, and a quantum fluctuating "gravitational field" $h_{\mu\nu}(x)$. The non-renormalizability of the resulting theory has pushed the theorists into a quest for a larger renormalizable or finite theory, following the path indicated by the weak interaction. The quest has wandered through modifications of GR with curvature-square terms in the action, Kaluza–Klein-like theories, supergravity, ... Merging with the search for a unified theory of all interactions, it has eventually led to string theory, a presumably finite quantum theory of all interactions including gravity, defined in 10 dimensions, including supersymmetry, so far difficult to reconcile with the observed world.

The canonical version of LQG was born from the discovery of "loop" solutions of the Wheeler–deWitt equation, namely the formal quantization of canonical GR, rewritten in the Ashtekar variables. The quantization of geometry was derived within this theory and led to the spin network description of quantum geometry (Chapters 4 and 5). The canonical theory branched into a "sum over geometries form," à la Feynman, inspired by the functional integral euclidean formulation developed by Hawking and his group in the 1970s. This "spinfoam" theory (Chapter 7) merged with the canonical LQG kinematics and evolved into the current covariant theory described in this book. The historical development of these theories is sketched in Table 1.2. For a historical reconstruction and references, see the appendix in Rovelli (2004).

The rest of the book describes the covariant formulation of loop quantum gravity.

1.7 Complements

We recall some basic SU(2) representation theory. This plays an important role in quantum gravity. Then we compute the eigenvalues of the volume for a minimal quantum of space.

1.7.1 SU(2) representations and spinors

Definition

SU(2) is the group of unitary 2×2 complex matrices U. They satisfy $U^{-1} = U^\dagger$ and $\det U = 1$. These conditions fix the form of the matrix U as:

$$U = \begin{pmatrix} a & -\overline{b} \\ b & \overline{a} \end{pmatrix}, \tag{1.17}$$

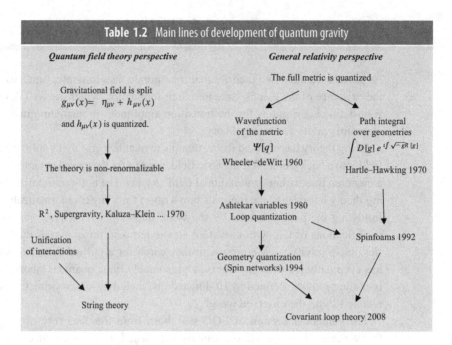

Table 1.2   Main lines of development of quantum gravity

where (because of the unit determinant)

$$|a|^2 + |b|^2 = 1. \tag{1.18}$$

We write the matrix elements as $U^A{}_B$, with indices $A, B, C, \ldots$ taking the values $0, 1$. SU(2) has the same algebra as the rotation group SO(3) (the group SU(2) is universal cover of SO(3)). One can view SU(2) as a minimal building block from many points of view, including, as we shall see, quantum spacetime.

Measure

Equation (1.18) defines a sphere of unit radius in $\mathbb{C}^2 \sim \mathbb{R}^4$. Thus, the topology of the group is that of the 3-sphere S^3. The euclidean metric of $\mathbb{R}^4$ restricted to this sphere defines an invariant measure on the group. Normalized by

$$\int_{\mathrm{SU}(2)} dU = 1 \,, \tag{1.19}$$

this is the Haar measure, invariant under both right and left multiplication: $dU = d(UV) = d(VU)$, $\forall V \in \mathrm{SU}(2)$. The Hilbert space $L_2[\mathrm{SU}(2)]$ formed by the functions $\psi(U)$ on the group square-integrable in this measure plays an important role in what follows.

Spinors

The space of the fundamental representation of SU(2) is the space of *spinors*, i.e., complex vectors $\mathbf{z}$ with two components,

$$\mathbf{z} = \begin{pmatrix} z^0 \\ z^1 \end{pmatrix} \in \mathbb{C}^2 \,. \tag{1.20}$$

We shall commonly use the "abstract index notation" that is implicitly used by physicists and is explicit for instance in Wald's book (Wald 1984); that is, the notation with an index does not indicate a component, but rather the full vector, so z^A is synonymous of $\mathbf{z}$ for spinors and v^i is synonymous of $\vec{v}$ for vectors.

In Chapter 8 we study the geometrical properties of spinors in better detail, and we reinterpret spinors as *spacetime* objects. Here we only introduce some basic facts about them and their role in SU(2) representation theory.

Representations and spin

The vector space of the completely symmetric n-index spinors

$$z^{A_1...A_n} = z^{(A_1...A_n)} \tag{1.21}$$

transforms under the action of SU(2), $z^{A'_1...A'_n} = U^{A'_1}{}_{A_1} ... U^{A'_n}{}_{A_n} z^{A_1...A_n}$, and therefore defines a representation of the group. This space is denoted $\mathcal{H}_j$, where $j = n/2$ (so that $j = 0, \frac{1}{2}, 1, \frac{3}{2}, ...$) and the representation it defines is called the *spin-j* representation. This representation is irreducible.

Exercise 1.3 *Show that $\mathcal{H}_j$ has dimension $2j + 1$.*

Let us review some properties of these representations.

1. Consider the two antisymmetric tensors

$$\epsilon^{AB} = \begin{pmatrix} 0 & 1 \\ -1 & 0 \end{pmatrix}, \quad \epsilon_{AB} = \begin{pmatrix} 0 & 1 \\ -1 & 0 \end{pmatrix}. \tag{1.22}$$

These can be used for raising or lowering indices of spinors, in a way analogous to g^{ab} and g_{ab} for tensors but being careful about the order: using the down-left-up-right rule, or $_A/^A$ rule:

$$z_A = \epsilon_{AB} z^B, \quad z^A = z_B \epsilon^{BA}, \tag{1.23}$$

For example, we have a contraction $z^A{}_A = \epsilon_{AB} z^{AB} = -\epsilon_{AB} z^{BA} = -z_A{}^A$. Show that

(1) $\epsilon_{AC} \epsilon^{CB} = -\delta^B_A$, $\epsilon_{BA} \epsilon^{AB} = -2$, $\epsilon_{AB} \epsilon^{AB} = 2$.

(2) ϵ^{AB} is invariant under the action of SU(2), i.e., $U^A{}_C U^B{}_D \epsilon^{CD} = \epsilon^{AB}$.

(3) $\det U = \epsilon^{BD} U^0{}_B U^1{}_D = \frac{1}{2} \epsilon_{AC} \epsilon^{BD} U^A{}_B U^C{}_D = 1$.

(4) $U^{-1} = -\epsilon U \epsilon$; that is, $(U^{-1})^A{}_B = -\epsilon_{BD} U^D{}_C \epsilon^{CA}$.

2. There are two SU(2) invariant quadratic forms defined on $\mathbb{C}^2$, which should not be confused. The first is the (sesquilinear) scalar product

$$\langle \mathbf{z} | \mathbf{y} \rangle = \sum_A \overline{z^A} y^A = \overline{z^0} y^0 + \overline{z^1} y^1, \tag{1.24}$$

where the bar indicates the complex conjugate. This scalar product is what promotes $\mathbb{C}^2$ to a Hilbert space, and therefore is what makes the SU(2) representations unitary.

The second is the (bilinear) antisymmetric quadratic form[13]

$$(\mathbf{z}, \mathbf{y}) = \epsilon_{AB} z^A y^B = z^0 y^1 - z^1 y^0. \tag{1.25}$$

---

[13] This is also indicated as $[\mathbf{z} | \mathbf{y} \rangle$, a notation that emphasizes its antisymmetry

The two can be related by defining the antilinear map $J: \mathbb{C} \to \mathbb{C}$,

$$(J\mathbf{z})^A = \begin{pmatrix} \overline{z^1} \\ -\overline{z^0} \end{pmatrix} \qquad\qquad (1.26)$$

so that

$$\langle \mathbf{z} | \mathbf{y} \rangle = (J\mathbf{z}, \mathbf{y}). \qquad\qquad (1.27)$$

All these structures are SU(2) invariant, but, as we shall see later on, (,) is also SL(2,C) invariant, whereas J and $\langle\, |\, \rangle$ are not. Therefore spinors are "spacetime" objects, since they carry also a representation of the Lorentz group, but as a representation of the Lorentz group $\mathbb{C}^2$ is *not* a unitary representation. In a precise sense, the scalar product and J depend on a choice of Lorentz frame in spacetime.

3. Most of SU(2) representation theory follows directly from the invariance of ϵ_{AB}. Consider first a tensor product of two fundamental ($j = 1/2$) representations $(\mathbf{z} \otimes \mathbf{y})^{AB} = z^A y^B$. Show that any two-index spinor z^{AB} can be decomposed into its symmetric and antisymmetric parts

$$z^{AB} = z_0 \epsilon^{AB} + z_1^{AB}, \quad z_0 = \frac{1}{2} z^A_{\ A}, \quad z_1^{AB} = z^{(AB)}, \qquad\qquad (1.28)$$

which are invariant under the action of SU(2). Because of the invariance of ϵ_{AB}, this decomposition is SU(2) invariant; scalars z_0 define the trivial representation $j = 0$, whereas z_1^{AB} defines the *adjoint* representation $j = 1$. Hence you have proved that the tensor product of two spin-1/2 representations is the sum of spin-0 and spin-1 representation: $\frac{1}{2} \otimes \frac{1}{2} = 0 \oplus 1$.

4. In general, if we tensor two representations of spin j_1 and j_2, we obtain a space of spinors with $(2j_1 + 2j_2)$ indices, symmetric in the first $2j_1$ and the last $2j_2$ indices. By symmetrizing all the indices, we obtain an invariant subspace transforming in the $(j_1 + j_2)$ representation. Alternatively, we can contract k indices of the first group with k indices of the second group using k times the tensor ϵ_{AB}, and then symmetrize the remaining $2(j_1 + j_2 - k)$ indices to obtain the spin-j_3 representation. Show that

$$j_1 + j_2 + j_3 \in \mathbb{N}, \quad |j_1 - j_2| \leq j_3 \leq (j_1 + j_2). \qquad\qquad (1.29)$$

These two conditions are called *Clebsch–Gordon conditions*. Does this ring a bell? They are equivalent to the fact that there exist three non-negative integers a, b, c such that

$$2j_1 = b + c, \quad 2j_2 = c + a, \quad 2j_3 = a + b. \qquad\qquad (1.30)$$

This has a nice graphical interpretation; see Figure 1.7.

Exercise 1.4 *Show that* (1.30) *implies* (1.29).
Draw similar pictures for (j_1, j_2, j_3) *given by i:* $(1/2, 1/2, 1)$, *ii:* $(5/2, 5/2, 2)$, *iii:* $(1, 3/2, 5/2)$, *iv:* $(5/2, 2, 5)$, *v:* $(5/2, 5/2, 7/2)$, *vi:* $(1, 1, 1)$. *Find the corresponding* a, b, c.

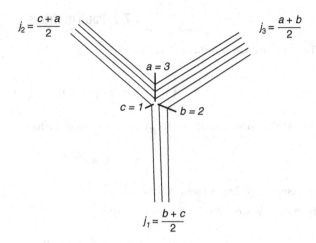

$j_2 = \dfrac{c+a}{2}$ $j_3 = \dfrac{a+b}{2}$

$a = 3$

$c = 1$ $b = 2$

$j_1 = \dfrac{b+c}{2}$

Figure 1.7   Elementary recoupling.

5. The spinor basis is not always the most convenient for the SU(2) representations. If we diagonalize L_z in $\mathcal{H}_j$ we obtain the well-known basis $|j,m\rangle$ where $m = -j,\ldots,j$, described in all quantum mechanics textbooks. In this basis, the representation matrices $D^j_{nm}(U)$ are called the Wigner matrices. *Mathematica* gives them explicitly; they are called `WignerD[{j,m,n},`ψ,θ,ϕ`]` and are given in terms of the Euler angles parametrization of SU(2) (given below in (1.42)).

6. The two bilinear forms of the fundamental representation extend to all irreducible representations. Given two vectors $\mathbf{z}$ and $\mathbf{y}$ in $\mathcal{H}_j$, we can either take their invariant contraction or their scalar product. In integer representations the two bilinear forms turn out to be the same. In half-integer representations, they are different. The contraction is defined by the projection $\mathcal{H}_j \otimes \mathcal{H}_j \to \mathcal{H}_0$ and in the $z^{A_1\ldots A_{2j}}$ representation, it is given by

$$(\mathbf{z},\mathbf{y}) = z^{A_1\ldots A_{2j}}\, y^{B_1\ldots B_{2j}}\, \epsilon_{A_1 B_1} \cdots \epsilon_{A_{2j}B_{2j}}, \tag{1.31}$$

while the scalar product is given by

$$\langle \mathbf{z}|\mathbf{y}\rangle = \overline{z^{A_1\ldots A_{2j}}}\, y^{B_1\ldots B_{2j}}\, \delta_{A_1 B_1} \cdots \delta_{A_{2j}B_{2j}}. \tag{1.32}$$

The basis that diagonalizes L_z is of course orthonormal, since L_z is self-adjoint. Therefore

$$\langle j,m|j,n\rangle = \delta_{mn}, \tag{1.33}$$

while a direct calculation (see for instance Landau and Lifshitz (1959)) gives

$$(j,m,j,n) = (-1)^{j-m}\delta_{m,-n}. \tag{1.34}$$

The factor $\delta_{m,-n}$ is easy to understand: the singlet must have vanishing total L_z.

1.7.2 Pauli matrices

The Pauli matrices σ_i with $i = 1, 2, 3$ are

$$\sigma_{i\,B}^A = \left\{ \begin{pmatrix} 0 & 1 \\ 1 & 0 \end{pmatrix}, \begin{pmatrix} 0 & -i \\ i & 0 \end{pmatrix}, \begin{pmatrix} 1 & 0 \\ 0 & -1 \end{pmatrix} \right\}. \tag{1.35}$$

Any SU(2) group element $U = (U^A_B)$ can be written in the form

$$U = e^{i\alpha\hat{n}\cdot\hat{\sigma}} \tag{1.36}$$

for some $\alpha \in [0, 2\pi[$ and some unit vector $\hat{n}$.

Exercise 1.5 *Show that the Pauli matrices obey*

$$\sigma_i\sigma_j = \delta_{ij} + i\epsilon_{ijk}\sigma_k \tag{1.37}$$

and

$$\epsilon\sigma_i\epsilon = \sigma_i^T = \sigma_i^*, \tag{1.38}$$

where ^{T} indicate the transpose and * the complex conjugate.

Exercise 1.6 *Show that*

$$e^{i\alpha\hat{n}\cdot\hat{\sigma}} = \cos\alpha + i\hat{n}\cdot\hat{\sigma}\sin\alpha. \tag{1.39}$$

The three matrices

$$\tau_i = -\frac{i}{2}\sigma_i \tag{1.40}$$

are the generators of SU(2) in its fundamental representation.

Exercise 1.7 *Verify that they satisfy the relation*

$$[\tau_i, \tau_j] = \epsilon_{ij}{}^k \tau_k, \tag{1.41}$$

which defines the algebra su(2).

The Euler angle parametrization of SU(2) is defined by

$$U(\psi, \theta, \phi) = e^{\psi\tau_3} e^{\theta\tau_2} e^{\phi\tau_3}, \tag{1.42}$$

where

$$\psi \in [0, 2\pi[, \quad \theta \in [0, \pi[, \quad \phi \in [0, 4\pi[. \tag{1.43}$$

In terms of these coordinates, the Haar measure reads

$$\int dU = \frac{1}{16\pi^2} \int_0^{2\pi} d\psi \int_0^{\pi} \sin\theta\, d\theta \int_0^{4\pi} d\phi. \tag{1.44}$$

Exercise 1.8 *Verify the relations*

$$Tr(\tau_i \tau_j) = -\frac{1}{2}\delta_{ij}, \quad Tr(\tau_i \tau_j \tau_k) = -\frac{1}{4}\epsilon_{ijk} \tag{1.45}$$

$$\delta^{ij}\tau_i{}^A{}_B \tau_j{}^C{}_D = -\frac{1}{4}(\delta^A_D \delta^C_B - \epsilon^{AC}\epsilon_{BD}), \tag{1.46}$$

$$\delta^A_B \delta^D_C = \delta^A_C \delta^D_B + \epsilon^{AD}\epsilon_{BC} \tag{1.47}$$

$$\delta^{ij} Tr(A\tau_i) Tr(B\tau_j) = -\frac{1}{4}\left[Tr(AB) - Tr(AB^{-1}) \right], \tag{1.48}$$

$$Tr(A)Tr(B) = Tr(AB) + Tr(AB^{-1}), \tag{1.49}$$

for A and B SL(2, $\mathbb{C}$) matrices.

The relation (1.47) is of particular importance. Notice that it can be written graphically in the form

$$\times \;=\; A^{-1}\Big) \Big(\;+\; A \;\smile \tag{1.50}$$

with $A = 1$. The reason for this writing will become clearer in Chapter 6.

Exercise 1.9 *If we raise the index of the Pauli matrices with ϵ we obtain the 2-index spinors $(\sigma_i)^{AB} = (\sigma_i)^A{}_C \epsilon^{CB}$. Show that these are invariant tensors in the representation $\frac{1}{2} \otimes \frac{1}{2} \otimes 1$.*

1.7.3 Eigenvalues of the volume

Problem

Equipped with this SU(2) math, compute the volume eigenvalues for a quantum of space whose sides have minimal (non-vanishing) area.

Solution

Recall (Eq. 1.10) that the volume operator V is determined by

$$V^2 = \frac{2}{9}\epsilon_{ijk} L_1^i L_2^j L_3^k. \tag{1.51}$$

where the operators $\vec{L}_a$ satisfy the commutation relations (1.12).

If the faces of the quantum of space have minimal area, the Casimir of the corresponding representations has minimal non-vanishing value. Therefore the four operators L_a act on the fundamental representations $j_1 = j_2 = j_3 = j_4 = \frac{1}{2}$. Therefore they are proportional to the (self-adjoint) generators of SU(2), which in the fundamental representation are Pauli matrices. That is,

$$L_f^i = \alpha \frac{\sigma^i}{2}. \tag{1.52}$$

The proportionality constant has the dimension of length squared, is of Planck scale, and is fixed by comparing the commutation relations of the Pauli matrices with (1.12). This gives $\alpha = l_o^2 = \frac{8\pi\gamma\hbar G}{c^3}$.

The Hilbert space on which these operators act is therefore $\mathcal{H} = \mathcal{H}_{\frac{1}{2}} \otimes \mathcal{H}_{\frac{1}{2}} \otimes \mathcal{H}_{\frac{1}{2}} \otimes \mathcal{H}_{\frac{1}{2}}$. This is the space of objects with four spinor indices $A, B = 0, 1$, each being in the $\frac{1}{2}$-representation of SU(2),

$$\mathcal{H}_{j_1} \otimes \mathcal{H}_{j_2} \otimes \mathcal{H}_{j_3} \otimes \mathcal{H}_{j_4} \ni z^{ABCD}. \tag{1.53}$$

The operator $\vec{L}_a$ acts on the a-th index. Therefore the volume operator acts as

$$(V^2 \mathbf{z})^{ABCD} = \frac{2}{9} \left(\frac{\alpha}{2} \right)^3 \epsilon^{ijk} \sigma_i{}^A{}_{A'} \sigma_j{}^B{}_{B'} \sigma_k{}^C{}_{C'} z^{A'B'C'D}. \tag{1.54}$$

Let us now implement the closure condition (1.9). Let

$$\mathcal{H}_{\frac{1}{2}\frac{1}{2}\frac{1}{2}\frac{1}{2}} = \mathcal{H}_{\frac{1}{2}} \otimes \mathcal{H}_{\frac{1}{2}} \otimes \mathcal{H}_{\frac{1}{2}} \otimes \mathcal{H}_{\frac{1}{2}}, \tag{1.55}$$

$$\mathcal{K}_{\frac{1}{2}\frac{1}{2}\frac{1}{2}\frac{1}{2}} = Inv_{\mathrm{SU}(2)}[\mathcal{H}_{\frac{1}{2}} \otimes \mathcal{H}_{\frac{1}{2}} \otimes \mathcal{H}_{\frac{1}{2}} \otimes \mathcal{H}_{\frac{1}{2}}]. \tag{1.56}$$

We have to look only for subspaces that are invariant under a common rotation for each space $\mathcal{H}_{j_i}$; namely, we should look for a quantity with four spinor indices that are invariant under rotations. What is the dimension of this space? Remember that for SU(2) representations $\frac{1}{2} \otimes \frac{1}{2} = 0 \oplus 1$, that implies:

$$\mathcal{H}_{\frac{1}{2}\frac{1}{2}\frac{1}{2}\frac{1}{2}} = (0 \oplus 1) \otimes (0 \oplus 1) = 0 \oplus 1 \oplus 1 \oplus (0 \oplus 1 \oplus 2). \tag{1.57}$$

Since the trivial representation appears twice, the dimension of $\mathcal{K}_{\frac{1}{2}\frac{1}{2}\frac{1}{2}\frac{1}{2}}$ is 2. Therefore there must be two independent invariant tensors with four indices. These are easy to guess, because the only invariant objects available are ε^{AB} and $\sigma_i^{AB} = (\sigma_i \epsilon)^{AB}$, obtained by raising the indices of the Pauli matrices $\sigma_i{}^A{}_B$,

$$\sigma_i{}^{AB} = \sigma_i{}^A{}_C \varepsilon^{CB}. \tag{1.58}$$

Therefore two states that span $\mathcal{K}_{\frac{1}{2}\frac{1}{2}\frac{1}{2}\frac{1}{2}}$ are

$$z_1^{ABCD} = \varepsilon^{AB} \varepsilon^{CD}, \tag{1.59}$$

$$z_2^{ABCD} = \sigma_i^{AB} \sigma_i^{CD}. \tag{1.60}$$

These form a (non-orthogonal) basis in $\mathcal{K}_{\frac{1}{2}\frac{1}{2}\frac{1}{2}\frac{1}{2}}$. These two states span the physical SU(2)-invariant part of the Hilbert space, which gives all the shapes of our quantum of space with a given area. To find the eigenvalues of the volume it suffices to diagonalize the 2×2 matrix V^2_{nm}

$$V^2 \mathbf{z}_n = V^2_{nm} \mathbf{z}_m. \tag{1.61}$$

Let us compute this matrix. A straightforward calculation with Pauli matrices (do it! Eqs. (1.37) and (1.38) are useful) gives

$$V^2 \mathbf{z}_1 = -\frac{i\alpha^3}{18} \mathbf{z}_2, \qquad V^2 \mathbf{z}_2 = \frac{i\alpha^3}{6} \mathbf{z}_1. \tag{1.62}$$

so that

$$V^2 = -\frac{i\alpha^3}{18} \begin{pmatrix} 0 & 1 \\ -3 & 0 \end{pmatrix} \tag{1.63}$$

and the diagonalization gives the eigenvalues

$$V^2 = \pm \frac{\alpha^6}{6\sqrt{3}}. \tag{1.64}$$

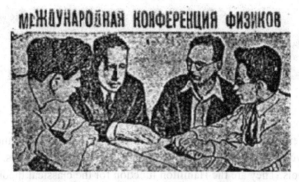

МЕЖДУНАРОДНАЯ КОНФЕРЕНЦИЯ ФИЗИКОВ

Figure 1.8   The four characters of the discussion from which quantum gravity has emerged. From the left: Landau, Bohr, Rosenfeld and Bronstein. The original photograph was taken in Kharkov and published in the newspaper *Khar'kovskii rabochii* (*The Kharkov Worker*) on May 20, 1934.

The sign depends on the fact that this is the *oriented* volume squared, which depends on the relative orientation of the triple of normals chosen. Inserting the value $\alpha = \frac{8\pi\gamma\hbar G}{c^3}$ determined above, into the last equation, we finally have the eigenvalue of the (non-oriented) volume

$$V = \frac{1}{\sqrt{6\sqrt{3}}} \left(\frac{8\pi\gamma\hbar G}{c^3} \right)^{\frac{3}{2}}. \tag{1.65}$$

About 10^{100} quanta of volume of this size fit into a cm^3. In Chapter 7 we give a general algorithm for computing volume eigenvalues.

2

Physics without time

General relativity has modified the way we think of space and time and the way we describe evolution theoretically. This change requires adapted tools. In this chapter, we study these tools. They are the Hamilton function for the classical theory (not to be confused with the hamiltonian) and the transition amplitude for the quantum theory. Unlike most of the other tools of mechanics (hamiltonian, quantum states at fixed time, Schrödinger equation, ...), these quantities remain meaningful in quantum gravity. Let us start by introducing them in a familiar context.

2.1 Hamilton function

> *"I feel that there will always be something missing from other methods which we can only get by working from a hamiltonian (or maybe from some generalization of the concept of hamiltonian)."*
>
> Paul Dirac (2001)

Let us have a system described by a configuration variable $q \in \mathcal{C}$ where $\mathcal{C}$ is the configuration space (of arbitrary dimension). We describe the evolution of this system in the time variable $t \in \mathbb{R}$. This means that evolution is going to be given by $q(t)$, namely functions $\mathbb{R} \to \mathcal{C}$ describing the possible motions. These quantities can be interpreted operationally as follows. We have two kinds of measuring apparatus at our disposal: a clock on which we read t, and other devices for the variables q.

The physical motions $q(t)$ that the system can follow are determined by a lagrangian $\mathcal{L}(q, \dot{q})$, where $\dot{q} = dq/dt$, as the ones that minimize the action

$$\mathcal{S}[q] = \int dt \, \mathcal{L}(q(t), \dot{q}(t)) . \tag{2.1}$$

The action is a functional on the functions $q(t)$. This is mechanics.

Now let us define a key object that plays a major role in what follows. The **Hamilton function** $\mathcal{S}(q, t, q', t')$ is a function of four variables, q, t, q', t', defined as the value of the action on a physical trajectory (namely a solution of the equations of motion) that starts at

q at time t and ends at q' at time t'.[1] Do not confuse the Hamilton function with the action: the first is a function of four variables, the second is a functional of the full trajectory, giving a number for any function $q(t)$. The two objects are related: the Hamilton function is the action of a particular trajectory determined by its boundaries.

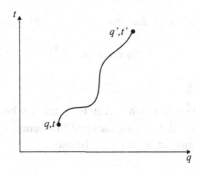

Let us denote $q_{qt,q't'}(\tilde{t})$ as a physical motion that starts at (q,t) and ends at (q',t'). That is, a function of time that solves the equations of motion and such that

$$q_{qt,q't'}(t) = q , \tag{2.2}$$

$$q_{qt,q't'}(t') = q' , \tag{2.3}$$

and assume for the moment that it is unique. The Hamilton function is defined by

$$S(q,t,q',t') = \int_t^{t'} d\tilde{t}\, \mathcal{L}\left(q_{qt,q't'}(\tilde{t}), \dot{q}_{qt,q't'}(\tilde{t})\right). \tag{2.4}$$

Example 2.1: Free particle

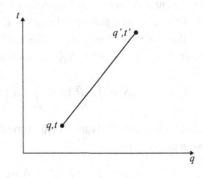

The action is

$$S[q] = \int dt\, \frac{1}{2}m\dot{q}^2. \tag{2.5}$$

The solutions of the equations of motion are the straight motions

$$v \equiv \dot{q} = \frac{q'-q}{t'-t}, \tag{2.6}$$

---

[1] If such trajectory is unique, the Hamilton function is a proper function. If there are many, it is multiple-valued; if there is none, it is not defined for the corresponding values. For a cleaner treatment, see the Complements to this chapter.

so that we immediately have the Hamilton function

$$S(q,t,q',t') = \int_t^{t'} d\tilde{t}\, \frac{1}{2}m\left(\frac{q'-q}{t'-t}\right)^2 = \frac{m(q'-q)^2}{2(t'-t)}. \tag{2.7}$$

Exercise 2.1 *Show that the Hamilton function of an harmonic oscillator is*

$$S(q,t,q',t') = m\omega\, \frac{(q^2+q'^2)\cos\omega(t'-t) - 2qq'}{2\sin\omega(t'-t)}. \tag{2.8}$$

The Hamilton function is easy to compute when we know the solution of the equations of motion. Indeed, knowing the Hamilton function amounts to knowing the solution of the equations of motion, as we show below.

Properties of the Hamilton function

1. Recall that the *momentum* is

$$p(q,\dot{q}) = \frac{\partial \mathcal{L}(q,\dot{q})}{\partial \dot{q}}. \tag{2.9}$$

It is not difficult to show that

$$\frac{\partial S(q,t,q',t')}{\partial q} = -p(q,t,q',t'), \qquad \frac{\partial S(q,t,q',t')}{\partial q'} = p'(q,t,q',t'), \tag{2.10}$$

where $p(q,t,q',t') = p(q,\dot{q}(q,t,q',t'))$ and $p'(q,t,q',t') = p(q',\dot{q}'(q,t,q',t'))$ are the initial and final momenta, expressed as functions of initial and final positions and times, via the actual solution of the equations of motion, determined by these initial and final data.

To show this, vary the final point q (keeping the time fixed) in the definition of the Hamilton function: $q(t) \to q(t) + \delta q(t)$ (see Figure 2.1). This gives

$$\delta S = \int_t^{t'} d\tilde{t}\, \delta \mathcal{L} = \int_t^{t'} d\tilde{t}\left(\frac{\partial \mathcal{L}}{\partial q}\delta q + \frac{\partial \mathcal{L}}{\partial \dot{q}}\delta \dot{q}\right). \tag{2.11}$$

The variation of the time derivative is the time derivative of the variation $\delta\dot{q} = \frac{d}{dt}\delta q$, so we can integrate by parts:

$$\delta S = \int_t^{t'} d\tilde{t}\left(\frac{\partial \mathcal{L}}{\partial q} + \frac{d}{dt}\frac{\partial \mathcal{L}}{\partial \dot{q}}\right)\delta q + \left.\frac{\partial \mathcal{L}}{\partial \dot{q}}\delta q\right|_t^{t'}. \tag{2.12}$$

The parentheses in the first term in (2.12) enclose the Euler–Lagrange equation and vanish because by definition q is a solution of the equations of motion. The boundary terms give

$$\delta S = \left.\frac{\partial \mathcal{L}}{\partial \dot{q}}\delta q\right|_t^{t'} = p'\,\delta q' - p\,\delta q, \tag{2.13}$$

which is precisely (2.10).

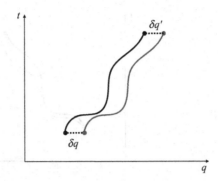

Figure 2.1   Variation of the motion with fixed boundary time.

2. This result shows that the solutions of the equations of motion can be obtained directly from the Hamilton function by taking derivatives and inverting. Indeed, inverting $p(q,t,q',t')$ with respect to q' we have $q'(t';q,p,t)$, namely, the final position as a function of the time t' and the initial data q,p,t.

   Hamilton wrote: "Mr Lagrange's function describes the dynamics, Mr Hamilton's function solves it." Very British.

   But we are not at the real beauty yet!

3. What if we take the derivative of the Hamilton function with respect to time rather than to q? It is easy to show that

$$\frac{\partial S(q,t,q',t')}{\partial t} = E(q,t,q',t'), \qquad \frac{\partial S(q,t,q',t')}{\partial t'} = -E'(q,t,q',t'). \tag{2.14}$$

(Note the sign switch with respect to (2.10). The similarity with special relativity is surprising and mysterious.)

   To derive (2.14), we vary S by varying the boundary time but not the boundary position (see Figure 2.2). Let us for simplicity vary only at the final point:

$$\delta S = \delta \int_t^{t'} d\tilde{t}\,\mathcal{L} = \mathcal{L}|_{t'}\,\delta t' - \left.\frac{\partial \mathcal{L}}{\partial \dot{q}}\right|_{t'} \dot{q}'\delta t' = -\left(p'\dot{q}' - \mathcal{L}(t')\right)\delta t' = -E'\delta t' \tag{2.15}$$

and we recognize the energy $\left(p'\dot{q}' - \mathcal{L}\right) = E$. Thus,

$$\vec{\nabla}_{(q',t')}\,S = (p',-E'). \tag{2.16}$$

and similarly

$$-\vec{\nabla}_{(q,t)}\,S = (p,-E). \tag{2.17}$$

Therefore we learn that the variables q and t get a sort of equal status in this language. And this is precisely what we need for quantum gravity, where, as we will see, the distinction between dependent (q) and independent (t) variables loses meaning. The Hamilton function is the natural object when we want to treat t and q on an equal footing, as we will soon be forced to.

4. Finally, the Hamilton function is a solution of the Hamilton–Jacobi equation (in both sets of variables, that is, in (q,t) as well as in (q',t')). This follows immediately from

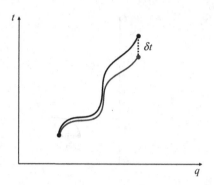

Figure 2.2   Variation of the motion with fixed boundary position.

the equations above. If energy is expressed by the hamiltonian as a function of position and momentum, $E = H(p,q)$, then, inserting (2.16) we immediately have

$$\frac{\partial S}{\partial t} + H\left(\frac{\partial S}{\partial q}, q\right) = 0, \qquad (2.18)$$

which is the Hamilton–Jacobi equation. The Hamilton–Jacobi equation has many solutions; the Hamilton function is a peculiar solution of this equation, where the Hamilton–Jacobi integration constants are taken to be initial position and time.

Example 2.2: Free particle   From the Hamilton function (2.7)

$$S(q,t,q',t') = \frac{m(q'-q)^2}{2(t'-t)} \qquad (2.19)$$

we have immediately

$$\frac{\partial S}{\partial q} = -m\frac{q'-q}{t'-t} = -mv = -p,$$

$$\frac{\partial S}{\partial t} = \frac{m}{2}v^2 = E.$$

Inverting the first gives the general solution of the equations of motion

$$q' = q + \frac{p}{m}(t'-t). \qquad (2.20)$$

The Hamilton–Jacobi equation is

$$\frac{\partial S}{\partial t} = \frac{1}{2m}\left(\frac{\partial S}{\partial q}\right)^2, \qquad (2.21)$$

and it is easy to see it is solved by (2.19).

2.1.1 Boundary terms

Care should be taken about boundary terms in the action, especially when the action contains second derivatives or is in first-order form, as is the case in gravity. An action with second derivatives for a free particle is

$$S[q] = \int dt \, \frac{1}{2} m \, q \ddot{q}, \tag{2.22}$$

while a first-order action is

$$S[q,p] = \int dt \left(p\dot{q} - \frac{1}{2m} p^2 \right). \tag{2.23}$$

If we disregard boundary terms, all these actions give the same equations of motion as for a free particle: $\ddot{q} = 0$. In fact, they differ from the conventional action (2.5) only by total derivatives. But their values on a physical motion differ. The Hamilton function is the value of the "right" action on the physical motions. The "right" action is determined by which quantities we want the Hamilton function to depend upon; that is, which quantities we are keeping fixed at the boundary in the variational principle. Suppose these are q (and t). Then the variation of the motion $\delta q(t)$ is chosen such that $\delta q(t) = \delta q(t') = 0$. It is easy then to see that δS turns out to be independent of $\delta \dot{q}(t)$ for the action (2.5). Then the variational principle is well defined. If we instead use (2.22), keeping q (and t) fixed at the boundary, we have to supplement it with a boundary term

$$S_{boundary} = m q \dot{q}|_{boundary} = pq|_{boundary} \tag{2.24}$$

for the variation to be independent of $\delta \dot{q}(0)$.

Exercise 2.2 *Try!*

This is the rationale for which a boundary term is also needed in gravity when dealing with gravitational actions. In the case of gravity the correct boundary term was written by York and then Gibbons and Hawking (York 1972; Gibbons and Hawking 1977). We will write it later.

Notice that the action (2.22) vanishes on the physical motions. Then the Hamilton function is given entirely by the (difference between the initial and final) boundary terms (2.24). The same happens in pure gravity. This does not mean that the Hamilton function becomes trivial, because the momenta p on the boundary must be expressed as a function of all the boundary $q's$, and for this the equations of motion must still be solved.

2.2 Transition amplitude

Let us now leap to the quantum theory. In the quantum theory, the object that corresponds to the Hamilton function is the transition amplitude. A quantum theory is defined by

- A Hilbert space $\mathcal{H}$
- Operators $\hat{q}, \hat{p}$ corresponding to classical variables[2]
- The time variable t
- A hamiltonian $\hat{H}$, or, equivalently,[3] the transition amplitude it defines.

Let $\hat{q}$ be a set of operators that commute, are complete in the sense of Dirac,[4] and whose corresponding classical variables coordinatize the configuration space. Consider the basis that diagonalizes these operators: $\hat{q}|q\rangle = q|q\rangle$. The **transition amplitude** is defined by

$$W(q,t,q',t') = \langle q'|e^{-\frac{i}{\hbar}H(t'-t)}|q\rangle. \tag{2.25}$$

That is, the transition amplitude is given by the matrix elements of the evolution operator

$$U(t) = e^{-\frac{i}{\hbar}Ht} \tag{2.26}$$

in the $|q\rangle$ basis.

Notice that the transition amplitude is a function of the same variables as the Hamilton function. The transition amplitude gives the dynamics of quantum theory. In classical physics, given something initial we compute something final. In quantum physics, given the initial and final something, we compute the probability of the pair.

Example 2.3: Free particle For a free particle:

$$W(q,t,q',t') = \langle q'|e^{-\frac{i}{\hbar}\frac{\hat{p}^2}{2m}(t'-t)}|q\rangle. \tag{2.27}$$

Inserting two resolutions of the identity $\mathbf{1} = \int dp \, |p\rangle\langle p|$,

$$W(q,t,q',t') = \int dp \int dp' \, \langle q'|p'\rangle \langle p'|e^{-\frac{i}{\hbar}\frac{\hat{p}^2}{2m}(t'-t)}|p\rangle \langle p|q\rangle$$

$$= \frac{1}{2\pi\hbar} \int dp \, e^{\frac{i}{\hbar}p(q-q')-\frac{i}{\hbar}\frac{p^2}{2m}(t'-t)}, \tag{2.28}$$

where we have used $\langle p|q\rangle = \frac{1}{\sqrt{2\pi\hbar}}e^{ipq/\hbar}$. Solving the gaussian integral we get

$$W(q,t,q',t') = A \, e^{\frac{i}{\hbar}\frac{m(q'-q)^2}{2(t'-t)}}, \tag{2.29}$$

where the amplitude is $A = \sqrt{\frac{m}{2\pi\hbar i(t'-t)}}$. Recall that $\frac{m(q'-q)^2}{2(t'-t)}$ was the Hamilton function of this system. Therefore

$$W(q,t,q',t') \propto e^{\frac{i}{\hbar}S(q,t,q',t')}. \tag{2.30}$$

---

[2] We put hats over operators only when we need to stress that they are such. Whenever it is clear, we drop the hat and write H, q, p for $\hat{H}, \hat{q}, \hat{p}$.

[3] By Stone's theorem.

[4] They form a maximally commuting set.

This relation between the Hamilton function and the transition amplitude is exact for the free particle, but in general it is still true to lowest order in $\hbar$. If we can consider $\hbar$ small, the Hamilton function works as a phase that makes the exponential oscillate rapidly, while the prefactor in front of the exponential varies slowly.

The rapidly varying part of the transition amplitude is given by the exponential of the Hamilton function. This is the general way in which quantum theory encodes classical dynamics.

2.2.1 Transition amplitude as an integral over paths

The above result is general, which can be shown as follows. Write the transition amplitude à la Feynman. That is, start from

$$W(q,t,q',t') = \langle q' | U(t'-t) | q \rangle , \tag{2.31}$$

and use the fact that the evolution operator defines a group

$$U(t-t')U(t'-t'') = U(t-t'') \tag{2.32}$$

to write it as a product of short-time evolution operators

$$W(q,t,q',t') = \langle q' | U(\epsilon) \dots U(\epsilon) | q \rangle , \tag{2.33}$$

where $\epsilon = (t'-t)/N$.

At each step insert a resolution of the identity $\mathbb{1} = \int dq_n |q_n\rangle\langle q_n|$. The transition amplitudes becomes a multiple integral:

$$W(q,t,q',t') = \int dq_n \prod_{n=1}^{N} \langle q_n | U(\epsilon) | q_{n-1} \rangle . \tag{2.34}$$

This expression is exact for any N, therefore it is also true if we take $N \to \infty$:

$$W(q,t,q',t') = \lim_{N\to\infty} \int dq_n \prod_{n=1}^{N} \langle q_n | U(\epsilon) | q_{n-1} \rangle . \tag{2.35}$$

Now, consider in particular the case with hamiltonian $H = \frac{p^2}{2m} + V(q)$. The evolution operator is

$$U(\epsilon) = e^{-\frac{i}{\hbar}(\frac{p^2}{2m}+V(q))\epsilon} , \tag{2.36}$$

namely, the exponent of a sum of two operators. This can be rewritten as the exponent of the first, times the exponent of the second, times corrections given by commutator of the two, which are of order $\sim \epsilon^2$ or higher and may be disregarded in the large N limit since in this limit ϵ is small. (This is of course far from being a rigorous step, and the resulting equalities must be checked case by case.) For small ϵ

$$U(\epsilon) \sim e^{-\frac{i}{\hbar}\frac{p^2}{2m}\epsilon} e^{-\frac{i}{\hbar}V(q)\epsilon} . \tag{2.37}$$

In the $|q\rangle$ basis the second exponential gives just a number. The first was computed above, in (2.28–2.29). The two together give

$$\langle q_{n+1} | U(\epsilon) | q_n \rangle \sim e^{\frac{i}{\hbar}\left(\frac{m(q_{n+1}-q_n)^2}{2(t_{n+1}-t_n)^2} - V(q_n) \right)\epsilon} \tag{2.38}$$

multiplied by an amplitude which we absorb in an overall multiplicative factor $\mathcal{N}$. But $t_{n+1} - t_n = \epsilon$, so bringing everything together we obtain:

$$W(q,t,q',t') = \lim_{N \to \infty} \mathcal{N} \int dq_n \, e^{\frac{i}{\hbar} \sum_{n=1}^{N}\left(\frac{m(q_{n+1}-q_n)^2}{2\epsilon^2} - V(q_n) \right)\epsilon}$$

$$\equiv \lim_{N \to \infty} \mathcal{N} \int dq_n \, e^{\frac{i}{\hbar} S_N(q_n)} . \tag{2.39}$$

The exponent in the last equation is a discretization of the classical action. The transition amplitude can therefore be written as a multiple integral of the discretization of the action in the limit for the discretization going to zero, namely, $\epsilon \to 0$. This limit is the definition of the functional integral

$$W(q,t,q',t') = \int D[q(t)] \, e^{\frac{i}{\hbar} S[q]} . \tag{2.40}$$

This nice construction was developed in Feynman's PhD thesis (Feynman 1948). Its interest is twofold. First, it provides a new intuition for quantum theory as a "sum over paths." Second, in the absence of a well-defined hamiltonian operator we can take an expression like (2.39) as a tentative ansatz for *defining* the quantum theory, and study if *this* theory is physically interesting. This is for instance the logic in lattice QCD, and we shall adopt a similar logic in quantum gravity.

If there is no potential term or if the potential has a simple quadratic form, we have a gaussian integral, which can be computed explicitly. If the rest can be treated as a perturbation, we can compute the transition amplitude expanding around its gaussian part. Functional integrals in fundamental physics are almost always either a way to keep track of the perturbation expansion (for instance in QED) or a limit of multiple integrals (as in lattice QCD).

In a regime where $\hbar$ can be considered small (where classical physics is a good approximation) we have an oscillating integral with a small parameter in front, and this is dominated by its saddle-point expansion. The oscillations cancel everywhere except where the variation of $S[q]$ is zero, i.e., on the classical solutions. Assuming this is unique:

$$W(q,t,q',t') = \int D[q(t)] e^{\frac{i}{\hbar} S[q]} \sim e^{\frac{i}{\hbar} S[q_{tq't'}]} \sim e^{\frac{i}{\hbar} S(q,t,q',t')}; \tag{2.41}$$

that is, the transition amplitudes are dominated by the exponential of the Hamilton function.[5] This result is therefore general, and provides a tool for studying the classical limit of a quantum theory.

---

[5] This clarifies also the meaning of the cases where there is no classical solution or there is more than one: these are simply determined by the configurations where the transition amplitude is suppressed in the classical limit, or where there is more than a single saddle point in the integral. In other words, $S(q,t,q',t')$ may be ill defined as a proper function, but $W(q,t,q',t')$ is not.

The formal classical limit of a quantum field theory may not turn out to be particularly physically relevant, as in the case of QCD, whose interesting low-energy phenomenology is not well described by a field theory. But general relativity, on the contrary, works well at large distance (like electromagnetism); therefore to have a classical regime described by general relativity is a *necessary* condition for a good theory of quantum gravity. We will use the technique described above, appropriately generalized to the general covariant context, to connect the quantum theory of gravity that we shall define to classical general relativity.

2.2.2 General properties of the transition amplitude

The transition amplitude $W(q, t, q', t')$ is also (as a function of q') the wavefunction at time t' for a state that at time t was a delta function concentrated at q. Therefore it satisfies the Schrödinger equation, indeed obviously, in both sets of variables:

$$i\hbar \frac{\partial}{\partial t} W + H\left(i\hbar \frac{\partial}{\partial q}, q\right) W = 0. \tag{2.42}$$

The Hamilton–Jacobi equation is the eikonal approximation (the equation that gives the optical approximation to a wave equation) to the Schrödinger equation.

In fact, in his first paper on wave mechanics (Schrödinger 1926), Schrödinger introduced his equation precisely by backtracking from this approximation: he started from the Hamilton–Jacobi formulation of classical mechanics, and interpreted it as the eikonal approximation to a wave theory. In other words, he found the Schrödinger equation by seeking a wave equation whose eikonal approximation would give the Hamilton–Jacobi equation. No surprise then that the phase of the transition amplitudes is given by the Hamilton function.

There is one important difference between $S(q, t, q', t')$ and $W(q, t, q', t')$ that must not be overlooked. The similarity of the notation can sometimes be misleading. The first depends on the configuration space variable q. The second is defined in (2.31) where q does not necessarily indicate a classical variable: it labels the eigenstates of the $\hat{q}$ operator. Classical variables and labels of eigenstates can be identified only if the operator has continuum spectrum. If the spectrum is discrete, then, by definition, the q in $W(q, t, q', t')$ are not classical variables. They are the discrete labels of the eigenstates of the $\hat{q}$ operator, that is, its quantum numbers. For instance, for a particle moving on a line we can define the transition amplitude on the momentum eigenbasis,

$$W(p, t, p', t') = \langle p' | U(t' - t) | p \rangle. \tag{2.43}$$

This is simply going to be given by the Fourier transform of $W(q, t, q', t')$ expressing the amplitude for changing momentum. But if the particle moves on a circle, as in the example in the first chapter, the eigenstates of the momentum are discrete $\hat{p}|n\rangle = p_n|n\rangle = n\hbar|n\rangle$ and therefore the amplitude is

$$W(n, t, n', t') = \langle n' | U(t' - t) | n \rangle, \tag{2.44}$$

that is, it is not a function of the (continuous) classical momentum variable p. It has to be so, because the amplitude determines probabilities for quantities having certain values, and some quantities can have only discrete values in quantum theory. This observation will be important in quantum gravity, where the transition amplitude will not be a function of classical configuration-space variables but rather a function of the corresponding quantum numbers, which, as we shall see, are given by spin networks.

The transition amplitude has a direct physical interpretation: it determines the amplitude of a process. A process is characterized by its boundary quantities, which are (q, t, q', t'). For the theory to make sense, there must be a way to measure, that is to assign numbers to, these quantities: they must have some operational meaning.[6]

The Hamilton function and the transition amplitude are tools that maintain their meaning in a general covariant context and in quantum gravity, where many of the other tools of classical and quantum theory are no longer available. We have introduced them in a familiar context; now it is time to step into physics without time.

---

[6] This does not mean that we adopt an instrumentalist interpretation of quantum theory. A process is what happens to a system S between interactions with other physical systems. The manner in which S affects the physical systems it interact with, is described by the quantities (q, t, q', t'). This is discussed in detail in Rovelli (1996b), to which we refer the interested reader for an interpretation of quantum mechanics that makes sense in the exacting context of quantum gravity.

2.3 General covariant form of mechanics

The quantities characterizing the initial and final state of affairs of a process are of two kinds: the time t, and the variables q. From an instrumentalist perspective, the tools we need to study the dynamics of the system are of two kinds: a *clock* that measures t, and other instruments that measure q. We now build a formalism in which these two kinds of quantity, t and q, are treated on the same footing.

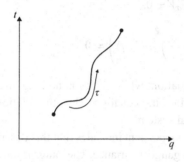

By analogy, recall that a function $q(t)$ can be expressed "parametrically" in terms of a couple of functions

$$q(t) \to \begin{cases} q(\tau), \\ t(\tau). \end{cases} \tag{2.45}$$

For instance, we can write the function $q = t^{2/3}$ (where t is treated as the independent variable) in terms of two functions $q = \tau^2, t = \tau^3$ (where t is treated as one of the dependent variables). The opposite is not always true: if $q = \sin\tau, t = \cos\tau$, then q is not a proper function of t. Therefore the parametric representation is more general than the $q(t)$ representation of possible relations between the clock variable t and the other variables q. We will see that this wider generality is demanded by physics.

Of course, a parametric representation is largely redundant. In the example above, $q = f(\tau)^2, t = f(\tau)^3$ defines the same motion $q(t)$ for any invertible function $f(\tau)$. Therefore a parametric representation of the motions carries a large redundancy, which is to say, a large gauge invariance. This is the root of the large gauge invariance of general relativity, namely diffeomorphism invariance.

Let us write the dynamics of a simple system in parametric form. The action

$$S[q] = \int_t^{t'} d\tilde{t}\, \mathcal{L}\left(q(\tilde{t}), \dot{q}(\tilde{t})\right) \tag{2.46}$$

can be rewritten as a functional of two functions, q and t, by changing variable $t \to t(\tau)$

$$S[q,t] = \int_\tau^{\tau'} d\tilde{\tau}\, \frac{dt(\tilde{\tau})}{d\tilde{\tau}}\, \mathcal{L}\left(q(\tilde{\tau}), \frac{dq(\tilde{\tau})/d\tilde{\tau}}{dt(\tilde{\tau})/d\tilde{\tau}}\right). \tag{2.47}$$

The motions $(q(\tau), t(\tau))$ that minimize this action determine motions $q(t)$ that minimize the original action.

Let us do this more concretely. Take for simplicity a newtonian system, with lagrangian

$$\mathcal{L}(q, \dot{q}) = \frac{1}{2}m\dot{q}^2 - V(q). \tag{2.48}$$

This gives the Newton equations of motion

$$\frac{\mathrm{d}}{\mathrm{d}t}m\dot{q} = -\nabla_q V \tag{2.49}$$

(that is, $F = ma$). The parametric form of this system is given by the two variables $q(\tau)$ and $t(\tau)$ evolving in τ with lagrangian

$$\mathcal{L}(q,t,\dot{q},\dot{t}) = \frac{1}{2}m\frac{\dot{q}^2}{\dot{t}} - \dot{t}V(q), \qquad (2.50)$$

where now the dot indicates the τ derivative. The equations of motion of this lagrangian are

$$q : \frac{d}{d\tau}m\frac{\dot{q}}{\dot{t}} + \dot{t}\,\nabla_q V = 0, \qquad (2.51)$$

$$t : \frac{d}{d\tau}\left(-\frac{1}{2}m\left(\frac{\dot{q}}{\dot{t}}\right)^2 - V(q)\right) = 0. \qquad (2.52)$$

Equation (2.51) is exactly the Newton equation, while (2.52) is nothing else but energy conservation, which is a consequence of the first equation. Thus the relation between q and t is precisely the same as in the original system.

The fact that the two equations are not independent indicates that this description of the system is partly redundant, namely, there is gauge invariance. The gauge is the arbitrariness in the choice of the parameter τ along the motions. The equation of motion and the action are invariant under the gauge transformations

$$q(\tau) \to q(\tau'(\tau)) \qquad \text{and} \qquad t(\tau) \to t(\tau'(\tau)) \qquad (2.53)$$

for any differentiable invertible function $\tau'(\tau)$. This means that τ is pure gauge: the physics is not in the specific function $q(\tau)$ and $t(\tau)$, but in the *relation* between q and t determined parametrically by these functions.

The hamiltonian structure of this system is important. The momenta are

$$p_t = \frac{\partial \mathcal{L}}{\partial \dot{t}} = -\frac{1}{2}m\left(\frac{\dot{q}}{\dot{t}}\right)^2 - V(q), \qquad (2.54)$$

$$p_q = \frac{\partial \mathcal{L}}{\partial \dot{q}} = m\frac{\dot{q}}{\dot{t}}, \qquad (2.55)$$

and if you try to invert these equations to express the velocities in terms of the momenta, you see that this is not possible: the map $(\dot{t},\dot{q}) \to (p_t,p_q)$ is not invertible. The reason is that the image of this map is not the full (p_t,p_q) space, but a subspace, determined by a constraint $C(t,q,p_t,p_q) = 0$. This is easily found from the definition of the momenta

$$C = p_t + H_o(p_q,q) = 0, \qquad (2.56)$$

where

$$H_o(p_q,q) = \frac{p_q^2}{2m} + V(q) \qquad (2.57)$$

is easily recognized as the hamiltonian of the unparametrized newtonian system we started from. The constraint states that the momentum conjugate to t is (minus) the energy.

What is the canonical hamiltonian H of the parametrized system defined by the lagrangian (2.50)? The Legendre transform from the lagrangian gives the hamiltonian

$$H = \dot{q} \frac{\partial \mathcal{L}}{\partial \dot{q}} + \dot{t} \frac{\partial \mathcal{L}}{\partial \dot{t}} - \mathcal{L} \qquad (2.58)$$

and using the constraint we see immediately that

$$H = 0, \qquad (2.59)$$

where the constraint is verified. More precisely, the hamiltonian is proportional to the constraint

$$H \sim C. \qquad (2.60)$$

This should not be a real surprise, since the hamiltonian generates evolution in the evolution parameter in the action, namely in τ, but a change in τ is pure gauge, and in the hamiltonian formalism the generator of a gauge transformation vanishes (weakly).

This does not mean in any sense that the dynamics is "frozen," or other similar absurdities that one sometimes reads. The dynamics of this system is the one described by the Newton equation above. The vanishing of the canonical hamiltonian H only means that the dynamics is expressed in this formalism by the relation between the dependent variables q and t, rather than by the individual evolution of these in the gauge parameter τ.

So, how does the hamiltonian formalism keep track of the information about the physical evolution if the hamiltonian vanishes? It does so by means of the constraint

$$C(q, t, p_q, p_t) = 0 \qquad (2.61)$$

as follows. For any function on phase space we can compute the equations of motion in τ by taking the Poisson brackets with the constraint

$$\frac{dA}{d\tau} = \{A, C\}, \qquad (2.62)$$

and we must supplement these with the constraint equation (2.61) itself and remember that the physics is not in the dependence of the variables on τ but in their relative dependence when τ is eliminated.[7] Thus $C(q, t, p_q, p_t)$ allows us to derive all observable correlations between variables. This is why this constraint is sometimes called the "hamiltonian constraint."

It is important to emphasize that in this formulation it is *not* necessary to identify one of the variables as the physical time in order to compute the observable correlations and derive predictions for the theory. The physical phase space is interpreted as the space of the possible (solutions of the equations of) motions, rather than the space of the initial data, and the time variable is treated on the same footing as all other variables.

---

[7] For the reader who likes more mathematical elegance, a formal symplectic treatment of this generalized form of the dynamics is described for instance in Sundermeyer (1982), and briefly developed in the Complements of this chapter.

Is this construction artificial? It is not. In fact, we are already used to it: recall for instance the case of a relativistic particle. We write the action in the form

$$S = m \int d\tau \sqrt{\dot{x}^\mu \dot{x}_\mu} . \tag{2.63}$$

The indices μ label four variables, but the system has only three degrees of freedom, and in fact this action is invariant under reparametrization of τ. The hamiltonian is zero and the constraint reads

$$C = p^2 - m^2 = 0. \tag{2.64}$$

In general relativity the action is

$$S[g] = \int d^4x \sqrt{-\det g} \, R[g] \tag{2.65}$$

and is invariant under any reparametrization of x. The canonical hamiltonian vanishes and the information about the dynamics is coded in the constraints. This means that the dynamics does not describe the evolution of the gravitational field $g_{\mu\nu}(x)$, and other matter fields, as functions of x (this is just gauge), but rather the relative evolution of the fields with respect to one another.

We call "generally covariant," or simply "covariant," this generalized formulation of mechanics. The first person who understood the need generalizing mechanics in this manner was Dirac (Dirac 1950).

2.3.1 Hamilton function of a general covariant system

Let us get back to the parametrized system lagrangian (2.50), keeping in mind that this describes the same physics of a conventional newtonian system with lagrangian $\mathcal{L}(q,\dot{q}) = \frac{1}{2}m\dot{q}^2 - V(q)$. What is the Hamilton function of the lagrangian (2.50)? By definition,

$$S(q,t,\tau,q',t',\tau') = \int_{\tau_i}^{\tau_f} d\tau \, \mathcal{L}(q_{qt\tau,q't'\tau'}, t_{qt\tau,q't'\tau'}, \dot{q}_{qt\tau,q't'\tau'}, \dot{t}_{qt\tau,q't'\tau'}), \tag{2.66}$$

but we can change variables in the integral and rewrite this as

$$S(q,t,\tau,q',t',\tau') = \int_t^{t'} d\tilde{t} \left(\frac{1}{2}m \left(\frac{dq}{dt} \right)^2 - V(q) \right) = S(q,t,q',t'). \tag{2.67}$$

That is: (i) $S(q,t,\tau,q',t',\tau')$ is independent of τ and τ'

$$\frac{\partial S}{\partial \tau} = 0 \tag{2.68}$$

and (ii) its value is precisely equal to the Hamilton function $S(q,t,q',t')$ of the original newtonian system! The Hamilton function of a general covariant system does not depend on the evolution parameter, but only on the boundary values of two variables q and t. The Hamilton function is well defined both for systems that evolve in time and for systems that are generally covariant: the Hamilton function is always the same object, it depends only on the physical quantities such as q and t.

Notice also that the Hamilton–Jacobi equation for the Hamilton function of such a system,

$$\frac{\partial S}{\partial \tau} + H\left(\frac{\partial S}{\partial t}, \frac{\partial S}{\partial q}, q, t\right) S(q, t, q', t') = 0, \tag{2.69}$$

loses the first term because of (2.68), and because of (2.60) the second term reads

$$C\left(\frac{\partial S}{\partial t}, \frac{\partial S}{\partial q}, q, t\right) S(q, t, q', t') = 0, \tag{2.70}$$

where C is the constraint. This is the general covariant form of the Hamilton–Jacobi equation. It treats q and t on equal footing and is determined by the constraint C.

If the constraint C has the newtonian form (2.56), then this equation becomes the standard non-relativistic Hamilton–Jacobi equation (2.18) of the original newtonian system, with hamiltonian H_o! But it can also take more general forms.

2.3.2 Partial observables

When treated on equal footing, the variables q and t are called *partial observables*. Configuration space variable and time are, by definition, partial observables. Physics is about the relative evolution of partial observables. The reason for the adjective "partial" is to stress the fact that the partial observables include (for a newtonian system) also the t variable; namely, to distinguish these quantities from the functions on the configuration or phase space, which are sometimes called "observables."

The expression "partial observables" is important because the word "observable" alone generates confusion, since it takes different meanings in quantum mechanics, non-relativistic physics, general-relativistic physics, Dirac theory of constrained systems, ... The main source of confusion comes from mixing up two notions:

- Quantities that can be predicted from the knowledge of the initial state. For instance: the position of a particle at some later time.
- Quantities that can be measured, namely, those for which we have measuring apparatus. For instance: the position variable q, *and* the time variable t.

The number of partial observables in a system is always larger than the number of (physical) degrees of freedom, because the number of degrees of freedom is given by the number of quantities whose evolution can be predicted by the theory, and these, in turn are given by *relations* among partial observables.

The space of the partial observables is called the extended configuration space $\mathcal{C}_{ext} = \mathcal{C} \times \mathbb{R}$. We shall denote by $x \in \mathcal{C}_{ext}$ a generic point in this space (not to be confused with spacetime coordinates, sometimes denoted with the same letter). In the examples discussed $x = (q, t)$. Before taking these tools to the quantum theory, we summarize the covariant formalism developed above, which leads to a generalized definition of dynamics. This is done in the next section.

2.3.3 Classical physics without time

Let C_{ext} be an extended configuration space and $x \in C_{ext}$. The action of a system is a functional of the motions $x(\tau)$ in C_{ext}

$$S[x] = \int d\tau\, \mathcal{L}(x, \dot{x}) \tag{2.71}$$

invariant under reparametrizations $x(\tau) \to x(\tau'(\tau))$. This invariance implies the vanishing of the canonical hamiltonian and the existence of a constraint $C(x,p) = 0$, where $p = \frac{\partial \mathcal{L}}{\partial \dot{x}}$. The Hamilton function $S(x,x')$ is a function of two points in C_{ext}, defined as the value of the action on a solution of the equations of motion bounded by the two points. The Hamilton function satisfies the constraint equation

$$C\left(\frac{\partial S}{\partial x}, x\right) = 0, \tag{2.72}$$

which is the covariant form of the Hamilton–Jacobi equation. Knowledge of the Hamilton function gives the general solution of the equations of motion as follows. The derivative

$$p = -\frac{\partial S(x,x')}{\partial x} = p(x,x') \tag{2.73}$$

defines the initial momenta. If the initial momenta p and partial observables x are fixed, Eq. (2.73) gives a relation between the partial observables x'. This relation is the predictive content of the dynamical theory.

Notice that the construction does not ever require us to mention the word "time" or to refer to a "time variable."

Example 2.4: Free particle In this case there are two partial observables: $x = (q, t)$. The Hamilton function is $S(x, x') = \frac{m(q'-q)^2}{2(t'-t)}$. The momenta are

$$p_q = -\frac{\partial S(x,x')}{\partial q} = m\frac{q'-q}{t'-t}, \qquad p_t = -\frac{\partial S(x,x')}{\partial t} = \frac{m(q'-q)^2}{2(t'-t)^2}. \tag{2.74}$$

From the first relation we derive

$$q'(t') = q + \frac{p_q}{m}(t'-t), \tag{2.75}$$

which is the general solution of the equation of motion, here derived as a relation between two equally treated partial observables.

In summary, a system invariant under reparametrizations of the lagrangian evolution parameter is a system where the evolution parameter is pure gauge. The canonical hamiltonian vanishes, the momenta satisfy a constraint $C(p, x) = 0$, and this constraint encodes the dynamical information. The Hamilton function is independent of the lagrangian evolution parameter and the Hamilton–Jacobi equation it satisfies is simply (2.72). Any standard newtonian system can be reformulated in this manner, but not every system formulated in this manner can be expressed as a newtonian system (an example of this is given in the Complements of this chapter).

2.4 Quantum physics without time

The quantum theory of a covariant system is defined by:

- A Hilbert space $\mathcal{K}$ called the *kinematical* Hilbert space, where self-adjoint operators x and p_x in $\mathcal{K}$ corresponding to classical variables $x \in \mathcal{C}_{ext}$ and their momenta are defined.
- A constraint operator C, whose classical limit is the constraint $C(x,p)$. Equivalently, we may work with the transition amplitudes $W(x,x')$ that it defines.

Notice the absence of hamiltonian, time variable, and Schrödinger equation.

Discrete spectrum

Assume for the moment that zero is in the discrete spectrum of C. Then the subspace of $\mathcal{K}$ formed by the states that satisfy the equation

$$C\psi = 0 \tag{2.76}$$

is a proper subspace of $\mathcal{K}$; therefore it is a Hilbert space. It is called the *physical* state space and denoted $\mathcal{H}$. Equation (2.76) generalizes the Schrödinger equation in the covariant case. This can be seen in many ways. First, this equation *is* precisely the Schrödinger equation for the parametrized form of the newtonian systems, where, as we have seen, the constraint has the form (2.56). (Write it explicitly!) Second, it is the wave equation whose classical limit is the Hamilton–Jacobi equation (2.72). Therefore this equation has the correct form for the newtonian systems and the correct classical limit. This is more than we need to take it seriously as the definition of the quantum dynamics of the covariant quantum systems. In doing so, we are making a genuine physical hypothesis.

Equation (2.76) is called the Wheeler–deWitt equation (DeWitt 1967). It was first derived by Bryce deWitt in the context of quantum general relativity starting from the Hamilton–Jacobi equation, interpreted as the geometrical-optics approximation of a wave equation, that is, precisely in the same manner as Schrödinger found his equation starting from the Hamilton–Jacobi equation of a particle in a central potential. John Wheeler understood the importance of the equation and took it seriously (Wheeler 1968).

Let us see how this equation defines the transition amplitudes, which are the actual things we need to derive predictions from the theory. There exists a map P that sends $\mathcal{K}$ to $\mathcal{H}$ given simply by the orthogonal projection. The transition amplitude is defined by the matrix elements of P in the basis that diagonalizes the operators x:

$$W(x,x') = \langle x'|P|x \rangle. \tag{2.77}$$

The transition amplitude determines transition probabilities. Notice that this is a function that treats all partial observables on the same ground. Formally, we can rewrite the definition as

$$W(x,x') = \langle x'|\delta(C)|x \rangle \sim \int_{-\infty}^{\infty} d\tau \ \langle x'|e^{i\tau C}|x \rangle. \tag{2.78}$$

(the integral of an exponential is a delta function) and since C generates evolution in the parameter τ, we can translate this into the Feynman language in the form

$$W(x,x') = \int_x^{x'} D[x(\tau)]\, e^{\frac{i}{\hbar}S[x]}. \tag{2.79}$$

where the integration is over all paths that start at x and end at x', for any parametrization of these. Since the action does not depend on the parametrization, the integration includes a large gauge redundancy that needs to be factored out. For a newtonian system, gauge fixing the parameter τ by $t = \tau$ reduces this definition of the transition function to the previous newtonian one, Eq. (2.39). In the next section we will come back to a more precise definition of the functional integral (2.79) as a limit.

Continuum spectrum

The same construction holds if zero is in the continuous spectrum of C, but in this context it requires more refined mathematics, analogous to that allowing us to treat continuous-spectrum operators and their generalized eigenstates in conventional quantum theory. The need for this refined mathematics is sometimes mistakenly taken as a sign of deep issues about the nature of time or probability, but it is not so. Let us sketch how to deal with them. A simple possibility is to pick a dense subspace S of $\mathcal{K}$ whose dual defines a space S^* of generalized states.[8] Then the space $\mathcal{H}$ of the solutions of (2.76) is interpreted as the subset of S^* formed by the states $\psi \in S^*$ such that $\psi(C\phi) = 0$ for any $\phi \in S$ and the map P is defined from S to $\mathcal{H}$ by $(P\phi)(\phi') = \int d\tau \langle \phi | e^{i\tau C} | \phi' \rangle$. The space $\mathcal{H}$ is still a Hilbert space, with the scalar product $\langle P\phi | P\phi' \rangle \equiv (P\phi)(\phi')$. The operator P is often called "the projector," by extension, even if it is a true projector operator only in the discrete-spectrum case.

As an example of a continuous spectrum, consider a free newtonian particle in one dimension. The kinematical Hilbert space is $\mathcal{K} = L_2[R^2, dq\,dt]$, the partial-observable operators and their momenta are the diagonal operators q and t and the momentum operators $-i\hbar\frac{\partial}{\partial q}$ and $-i\hbar\frac{\partial}{\partial t}$, and the constraint operator is

$$C = -i\hbar\frac{\partial}{\partial t} - \frac{\hbar^2}{2m}\frac{\partial^2}{\partial q^2}, \tag{2.80}$$

so that the Wheeler–deWitt equation is precisely the Schrödinger equation. We can take S to be the Schwartz space and S^* the space of of tempered distributions (Gel'fand and Shilov 1968). The physical state space $\mathcal{H}$ is the space of solutions in S^* and not in $\mathcal{K}$, because they are not square-integrable in $dq\,dt$. The transition amplitude is given by

$$W(x,x') = W(q,t,q',t') = \int d\tau \, \langle q',t' | e^{\frac{i}{\hbar}\tau C} | q,t \rangle. \tag{2.81}$$

This can be computed as before by inserting resolutions of the identity

$$W(q,t,q',t') = \frac{1}{2\pi} \int d\tau \int dp \int dp_t \, e^{ip(x'-x)} e^{ip_t(t'-t)} e^{\frac{i}{\hbar}\tau(p_t + \frac{p^2}{2m})}. \tag{2.82}$$

---

[8] Giving the Gel'fand triple $S \subset \mathcal{K} \subset S^*$. In fact, strictly speaking the Hilbert space structure on $\mathcal{K}$ is not needed.

The τ integration gives the delta function $\delta(p_t + \frac{p^2}{2m})$ and then a gaussian integration gives back the result (2.29). The operator P is simply given by $\delta(p_t + \frac{p^2}{2m})$ in Fourier transform. Notice again that we have obtained the physically interpreted transition function without ever referring to a physical hamiltonian, or selecting one of the partial observables as the time parameter.

Interpretation

The equation

$$W(q,t,q',t') = \langle q',t'|P|q,t \rangle \tag{2.83}$$

can be interpreted as follows. The unphysical state $|q,t\rangle$ in $\mathcal{K}$ is a delta-function concentrated on the spacetime point (q,t). It represents the system being in the configuration q at time t. For a particle, this is the event of a particle being at the spacetime point $x = (q,t)$. The operator P projects this kinematical state down to a solution of the Schrödinger equation, namely, to a wave function in spacetime that is a solution of the Schrödinger equation and is concentrated on the space point q at time t. The contraction with the state $\langle q',t'|$ gives the value of this wave function at the point (q',t'); namely, the amplitude for a particle that was at q at t to be at q' at t'. This is the *physical* overlap of the kinematical state representing the event "particle at (q,t)" and the kinematical state representing the event "particle at (q',t')." All this is just a covariant language for describing the conventional physics of a quantum particle. The formalism treats q and t on equal footing. The difference between the behavior of the theory with respect to one or the other is only in the fact that they enter differently in C, and therefore the solutions of the Wheeler–deWitt equation have different properties with respect to q or t.

While the hamiltonian language (time evolution, hamiltonian, Poincaré symmetry, ...) cannot be used in quantum gravity, the language described here remains effective.

2.4.1 Observability in quantum gravity

So far, the language we have developed above is just a language: all the examples considered admit a newtonian formulation where the time variable is used as an independent evolution variable. However, general relativity comes already formulated in this language, and there is no general way of selecting one of its variables and interpreting it as "the time variable." General relativity asks us to describe the world in terms of relative evolution of partial observables, rather than in terms of evolution of degrees of freedom in time.

As an emblematic case, consider the following physical situation. You keep in your hand a precise clock whose reading is T and throw upward a second clock, identical to the first and synchronized to the first in the past, whose reading is T'. When the second clock falls down and you grab it again in your hand, it will be late with respect to the first. Knowing sufficient initial data, GR predicts the value $T'(T)$ it will display when the first clock reads T. Or, which is the same, it predicts the value $T(T')$ the first will display when the second reads T'. Does this describe the evolution of T as a function of the time T' or the evolution of T' as a function of the time T? Clearly the question is silly.[9] T and T' are partial observables, and the theory treats them on equal footing in describing their relative evolution. In general relativity there is no preferred observable for time: time evolution is always measured with respect to some arbitrary variable. Windshield wipers slappin' time.

In his great book the *Principia*, Newton asked "What is time?" and answered that time is not something that we observe directly. It is a convenient function of observable quantities that we can usefully single out for playing the role of an independent evolution parameter. This works when we disregard the dynamics of the gravitational field, because the fixed configuration of this field itself provides a convenient time variable. But this does not work anymore when we consider the full quantum dynamics of spacetime. For this, we have to resort to a fully relational view of dynamics, where the notion of a specific "time" variable plays no role.

This does not mean that there is no evolution in the world or that there is no change. It means that evolution and change in the real world are too complicated to be well represented as evolution in a single variable.

We now have to extend this technique to field theory, and in particular to a general context. This is an important issue, and we ask the special attention of the reader.

We call *process* what happens to a system between an initial and a final interaction. Dynamics has been presented in the previous sections in terms of *finite* portions of the trajectory of a system, expressed in terms of relations between the values of physical variables at the *boundaries* of a process. These relations can be coded in the Hamilton function S. In the classical theory, dynamics establishes relations between initial and final coordinates and momenta. In quantum mechanics, trajectories between two interactions cannot be deduced from the interaction outcomes. The transition amplitudes W determine probabilities of alternative sets of *boundary* values.[10]

2.4.2 Boundary formalism

Taking the above idea to field theory requires us to move the boundary formalism, championed and developed by Robert Oeckl (Oeckl 2003, 2008). In field theory, we can still consider a *finite* portion of the trajectory of a system, but now "finite" means finite in time as well as in space. Thus, we focus on a compact region of spacetime M.

The transition amplitude is a function of the field values on initial and final spacelike surfaces, but also on the "sides" of M: eventual timelike surfaces that bound the box. In other words, the transition amplitude W is a function of the values of the field on the *entire* boundary of the spacetime region M. Formally, W can be expressed as the Feynman path

---

[9] By the way, the clock in free fall during the period they are separated, that is, the one "moving straight," is the one launched up.

[10] A common language for describing processes in quantum theory is in terms of "preparation" and "measurement." This anthropomorphic language is misleading, since it appears to involve human intervention. The boundary of a process can be *any* physical interaction of the system with another – generic – physical system. Quantum mechanics describes the manner in which physical systems affect one another in the course of these interactions (Rovelli 1996b). It computes the probabilities for the different possible effects of such interactions. The theory is characterized, and in fact its structure is largely determined, by the fact that this description is consistent with arbitrary displacements of what we decide to consider the boundary between processes.

integral of the field in M, with fixed values on the boundary $\Sigma = \partial M$. The quantum state of the field on the entire boundary is an element of a boundary Hilbert space $\mathcal{H}$. The transition amplitude W is a linear functional $\langle W |$ on this space.

In the non-relativistic case, the boundary Hilbert space can be identified with the tensor product of the initial and final Hilbert spaces

$$\mathcal{H} = \mathcal{H}_0 \otimes \mathcal{H}_t^*, \qquad (2.84)$$

and

$$W_t(\psi \otimes \phi^*) := \langle \phi | e^{-iHt} | \psi \rangle. \qquad (2.85)$$

For a field theory on a fixed spacetime, W depends on the shape and geometry of Σ, for instance, on the time elapsed between its initial and final sides, precisely as in the last equation it depends on t.

But no longer so in gravity.

In gravity, a transition amplitude $\langle W | \Psi \rangle$ depends on the state Ψ of the gravitational field (as well as any other field that is present) on the boundary Σ, and that is all. Formally, this will be given by the Feynman path integral in the internal region, at fixed boundary values of the gravitational (and other) fields on Σ. How do we know then the shape, namely the geometry, of Σ?

Here comes the magic of quantum gravity: the answer is that the shape, the size and the geometry, of Σ are already determined by Ψ!

In fact, the gravitational field on the boundary Σ *is* precisely the quantity that specifies the shape of Σ! It includes any relevant metric information that can be gathered on the surface itself! Therefore we expect that $\langle W | \Psi \rangle$ is a function of Ψ *and nothing else*.

This of course is nothing else but the field analog of the phenomenon observed in the two previous sections for the parametrized systems, and in particular Eq. (2.68): the temporal information is stored and mixed among the dynamical variables, instead of being singled out and separated from other variables, as in unparametrized newtonian mechanics. In the general-relativistic context, this holds for temporal as well as for spatial locations: W will not be a function of space and time variables, but simply a function of the gravitational field on the boundary Σ (up to diffeomorphisms of Σ), which includes the entire relevant geometrical information on the boundary. This determines any information that can be gathered by clocks or meters on the boundary, because clocks and meters measure the gravitation field ($T = \int \sqrt{g_{\mu\nu} \dot{x}^\mu \dot{x}^\nu} d\tau$).

Therefore, in quantum gravity, dynamics is captured by a transition amplitude W that is a function of the (quantum) state of the field on a surface Σ. Intuitively, W is the "sum over geometries" on a *finite* bulk region bounded by Σ.[11] The explicit

---

[11] A common prejudice is that in quantum gravity we can only rely on observables at infinity, as one often does dealing with scattering in particle theory. A source for this misleading prejudice is the difficulty of defining bulk observables in a generally covariant theory. But this can be resolved: we measure and describe the relativistic dynamics of our solar system, in spite of the fact that we are immersed in it. A second source for the prejudice is the consideration that local observables require infinite precision and this can only be achieved with infinitely long or infinitely extended measurements: in a region of size L it does not make sense to discuss time evolution with a time resolution better than $\delta t \sim L_{\text{Planck}}^2 / L$ (Arkani-Hamed *et al.* 2007).

construction of W is the main objective in this book. It will be given explicitly in Chapter 7.

2.4.3 Relational quanta, relational space

The interpretation of quantum mechanics developed in Rovelli (1996b), which can be used in the context of quantum gravity, emphasizes the relational aspect of quantum mechanics. The theory yields probability amplitudes for processes, where a process is what happens between interactions. Thus quantum theory describes the universe in terms of the way systems affect one another. States are descriptions of ways a system can affect another system. Quantum mechanics is therefore based on *relations* between systems, where the relation is instantiated by a physical interaction.

The structure of general relativity is also relational, because the localization of dynamical objects is not given with respect to a fixed background structure; instead, "things" are only localized with respect to one another, where "things" are all dynamical objects, including the gravitational field. The relevant relation that builds the spacetime structure is of course contiguity: the fact of being "next to one another" in spacetime. We can view a general relativistic theory as a dynamical patchwork of spacetime regions adjacent to one another at their boundaries.

Now, a fundamental discovery and universal ingredient of twentieth-century physics is *locality*. Interactions are local. That is, interactions require spacetime contiguity. But the converse is also true: the only way to ascertain that two objects are contiguous is to have them interacting. Therefore locality reveals the existence of a structural analogy between the relations on which quantum mechanics is based and those on which spacetime is woven.

Quantum gravity makes this connection explicit: a process is not *in* a spacetime region: a process *is* a spacetime region. A state is not somewhere in space: it *is* the description of the way two processes interact, or two spacetime regions passing information to one another. Conversely, a spacetime region *is* a process: because it is actually a Feynman sum of everything that can happen between its boundaries.

The argument, however, *assumes* continuous background spacetime, which is exactly what is not present in quantum gravity. Time resolution is limited by uncertainty relations, but this is consistent with standard uncertainly relations for the gravitational field at the boundary. What is fuzzy is the expected value of the gravitational field, therefore physical localization of the measurement, not the possibility itself of making a measurement somewhere else than infinity. The point is related to holography (see again Arkani-Hamed *et al.* 2007): the Bekenstein bound limits the number of states an apparatus with given area can resolve, therefore an apparatus localized in spacetime can only distinguish a finite number of states and cannot resolve arbitrarily small distances. This again is correct, but saved by the physical granularity of spacetime. All these arguments show that in the presence of gravity there are no local observables in the sense of local quantum field theory: localized in arbitrarily small regions. They are all versions of Bronstein's original argument on the fact that space and time are ill-defined in quantum gravity. The solution is not to take refuge at infinity. It is to accept observables that do not resolve space and time more finely than Planck-scale. It is conventional quantum field theory that needs to be upgraded, in order for us to deal with observables that are local in a more general sense.

Quantum mechanics	General relativity
Process	Spacetime region
$\longleftarrow$ Locality $\longrightarrow$	
State	Boundary, space region

As noticed, a remarkable aspect of quantum theory is that the boundary between processes can be moved at will. Final total amplitudes are not affected by displacing the boundary between "observed system" and "observing system." The same is true for spacetime: boundaries are arbitrarily drawn in spacetime. The physical theory is therefore a description of how arbitrary partitions of nature affect one another. Because of locality and because of gravity, these partitions are at the same time subdivided subsystems (in the sense of quantum theory) and partitions of spacetime. A spacetime region is a process; a state is what happens at its boundary.

These abstract considerations will become concrete with the construction of the theory, and its applications, in the final chapter.

2.5 Complements

2.5.1 Example of a timeless system

Problem

Study a system that can be formulated in the covariant language of this chapter but not in newtonian language. The system is given by the extended configuration space $\mathcal{C}_{ext} = \mathbb{R}^2$ with coordinates a and b, and the lagrangian

$$\mathcal{L}(a,b,\dot{a},\dot{b}) = \sqrt{(2E - a^2 - b^2)(\dot{a}^2 + \dot{b}^2)}, \qquad (2.86)$$

where E is a constant. Solve the equations of motion. (Hint: the motions that minimize this action are the geodesics of the Riemannian metric $ds^2 = (2E - a^2 - b^2)(da^2 + db^2)$. Gauge fix the norm of the velocity to 1, and separate variables) and show that this is formally similar to two harmonic oscillators with total energy equal to E. Show that the motions in the extended configuration space are ellipses and therefore it is impossible to deparametrize this system and cast it in newtonian form. Show that the hamiltonian vanishes and the dynamics is determined by the hamiltonian constraint

$$C = \frac{1}{2}\left(p_a^2 + a^2 + p_b^2 + b^2\right) - E = 0. \qquad (2.87)$$

Define $\mathcal{K}$ and the relevant operators and show that zero is in its discrete spectrum provided that .... (Hint: just harmonic oscillators math!) Show that the physical phase space $\mathcal{H}$ is finite-dimensional. Compute its dimension (what determines it?) and write an integral expression for the transition amplitudes $W(a,b,a',b')$. Discuss the physical interpretation of this system. What would "time" be here? See Figure 2.3.

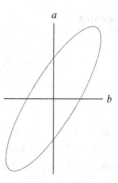

Figure 2.3   The motion in the space (a, b). Which of the two variables is the time variable?

Solution

The action $S[a, b] = \int d\tau\, \mathcal{L}(a, b, \dot{a}, \dot{b})$ is clearly invariant under reparametrization of τ. Therefore the canonical hamiltonian vanishes. The momenta are

$$p_a = \frac{\partial \mathcal{L}}{\partial \dot{a}} = \frac{(2E - a^2 - b^2)\dot{a}}{\mathcal{L}}, \tag{2.88}$$

$$p_b = \frac{\partial \mathcal{L}}{\partial \dot{b}} = \frac{(2E - a^2 - b^2)\dot{b}}{\mathcal{L}}, \tag{2.89}$$

and it is immediately seen that they satisfy (2.87).

The evolution in τ is generated by C. Since this is just the sum of two harmonic oscillator hamiltonians, the evolution in τ is given by

$$a(\tau) = A \sin(\tau + \phi_a), \tag{2.90}$$

$$b(\tau) = B \sin(\tau + \phi_b). \tag{2.91}$$

The constraint itself gives

$$A^2 + B^2 = 2E. \tag{2.92}$$

Therefore the motions are *closed* curves in the extended configuration space (a, b). They are ellipses. The constant ϕ_b can be set to zero by redefining τ. The space of the solution is therefore two-dimensional, parametrized by A and $\phi = \phi_a$. The relation between a and b, which is the physics predicted by the system, is

$$\arcsin \frac{a}{A} - \arcsin \frac{b}{\sqrt{E - A^2}} = \phi. \tag{2.93}$$

Since the motions are closed this is a system that does not admit a conventional newtonian description. In other words, in a conventional newtonian there is always one variable, t, going from $-\infty$ to $+\infty$, while here all partial observables are bounded.

The kinematical Hilbert space is $\mathcal{K} = L^2[R^2, da\, db]$. The constraint is the sum of two harmonic oscillator hamiltonians minus E. In the energy basis, C is diagonal and the Wheeler–de Witt equation reads

$$C|n_a, n_b\rangle = \left((n_a + \frac{1}{2}) + (n_b + \frac{1}{2}) - E \right) |n_a, n_b\rangle. \tag{2.94}$$

There are solutions only if $E = N + 1$ is an integer. The solutions are the linear combinations of the states $|n_a, n_b\rangle$ with $n_a + n_b = N$. Therefore $\mathcal{H}$ is the proper subspace of $\mathcal{K}$ formed by the states of the form

$$|\psi\rangle = \sum_{n=1}^{N} c_n |n, N-n\rangle, \tag{2.95}$$

and the project on this subspace is

$$P = \sum_{n=1}^{N} |n, N-n\rangle\langle n, N-n|. \tag{2.96}$$

The transition amplitude is

$$W(a,b,a',b') = \sum_{n=1}^{N} \langle a',b'|n,N-n\rangle\langle n,N-n|a,b\rangle. \tag{2.97}$$

This can be rewritten in the form

$$W(a,b,a',b') = \sum_{n=1}^{N} \langle a'|n\rangle\langle n|a\rangle\langle b'|N-n\rangle\langle N-n|b\rangle, \tag{2.98}$$

and the explicit form of the transition amplitude is therefore

$$W(a,b,a',b') = \sum_{n=1}^{N} \psi_n(a')\overline{\psi}_n(a)\psi_{N-n}(b')\overline{\psi}_{N-n}(b), \tag{2.99}$$

where $\psi_n(x)$ are the eigenstates of the harmonic oscillator hamiltonian.[12]

2.5.2 Symplectic structure and Hamilton function

Here we give a more precise mathematical picture of the formal structure of generalized hamiltonian mechanics, following Robert Littlejohn (Littlejohn 2013). If you like math, this is very pretty. Our starting point is a covariant system with *extended* configuration space $\mathcal{C}$ with dynamics defined by a constraint C (here single, for simplicity). We call $T^*\mathcal{C}$ the extended phase space. This space has a naturally symplectic structure $\omega = d\theta$, because it is a cotangent space.[13] If x are coordinates on $\mathcal{C}$, namely partial observables, then natural coordinates on the extended phase space are (x, p_x). For a newtonian system with n degrees of freedom, these include configurations variables and time $x = (q, t)$, as well as momenta and energy $p_x = (p, -E)$; the dimension of the extended phase space is $2(n+1)$ and $C = p_t + H(q, p, t)$. The constraint C defines the constraint surface Σ by $C = 0$ in $T^*\mathcal{C}$; the 2-form ω is degenerate (presymplectic) when restricted to this surface and its orbits are the motions. In this language the Hamilton equations read

$$\omega|_{\Sigma}(X) = 0. \tag{2.100}$$

X is the hamiltonian vector field of C, or the vector field that generates the motions. The space of these motions is the physical phase space Γ_{ph}. We call π the natural projection from Σ to Γ_{ph} which sends each point to the orbit to which it belongs.

Now consider the extended *boundary* phase space $D = T^*\mathcal{C} \times T^*\mathcal{C}$. We call P and P', respectively, the projection of this space onto its cartesian components. This carries the symplectic form $\omega_b =$

---

[12] For more details on this system and its interpretation, see Colosi and Rovelli (2003).

[13] It is defined by $\theta(v) := z(f_* v)$; here $z \in T^*\mathcal{C}$, $v \in T_z(T^*\mathcal{C})$, and f is the natural projection $T^*\mathcal{C} \to \mathcal{C}$; see for instance Arnold (1989). In local coordinates, $z = (x^a, p_a)$, $\theta = p_a dx^a$ and $\omega = dp_a \wedge dx^a$.

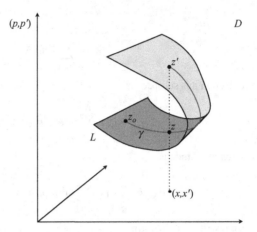

Figure 2.4   The Hamilton function $S(x,x')$ is the line integral of θ_b from a reference point z_o to z. If there is another point z' above (x,x'), then $S(x,x')$ is multivalued.

$\omega' - \omega = d\theta_b = d\theta' - d\theta$. For a newtonian system with n degrees of freedom, these include the initial and final configurations variables and times (x,x'), and the initial and final momenta and energy (p_x,p'_x); and the dimension of D is $2(2(n+1))$. The cartesian product of the constraint surface Σ with itself defines a subspace of D, with co-dimension 2, where the pull-back of the symplectic form has two degenerate directions. In this subspace, there is a surface L formed by the points $p \in D$ such that

$$\pi P p = \pi' P' p. \tag{2.101}$$

The surface L was introduced by Robert Littlejohn. For a newtonian system we can use C to solve for p_t as a function of the other variables and we can coordinatize Σ with (q,p,t), L is formed by the sets (q,p,t,q',p',t') such that the initial (q,p) at time t happen to evolve to (q',p') at time t'. It can be coordinatized by (q,p,t,t'). The map $\Sigma \times R \to L$ defined by $(q,p,t,t') \to (q,p,t,q',p',t')$ is unique and gives the full solution of the equations of motion, since (q',p') are precisely the evolved data from (q,p,t) at time t'.

The 2-form $\omega_b = d\theta_b$ vanishes when restricted to L; therefore the line integral of θ_b on L depends only on the initial and final points of the line. Fixing an initial point z_o (lying on the diagonal), this line integral defines a function on L (Figure 2.4):

$$S(z) := \int_{\gamma\,:\,z_o \to z} \theta_b, \tag{2.102}$$

where $\gamma \in L$. This is a clean definition of the Hamilton function. The function $S(z)$ does not depend on the path; it is single-valued and defined everywhere on L.

The surface L has the same dimension as $\mathcal{C} \times \mathcal{C}$. Consider the natural projection $f: (x,p,x',p') \to (x,x')$ from D to $\mathcal{C} \times \mathcal{C}$. This projection maps L to $\mathcal{C} \times \mathcal{C}$; the two spaces have the same dimension, but the map is not injective or surjective in general, therefore the function S defined on L is sent to a "function" on $\mathcal{C} \times \mathcal{C}$ which is not defined everywhere and may have "branches." This function on $\mathcal{C} \times \mathcal{C}$ is what is usually called the Hamilton function, that is,

$$S(x,x') \equiv S(x,p,x',p'), \quad \text{if } (x,p,x',p') \in L. \tag{2.103}$$

For a given (x,x'), if there is no point $(x,p,x',p') \in L$ then $S(x,x')$ has no value; if there is more than one point $(x,p,x',p') \in L$ then $S(x,x')$ has more than one branch. On the other hand, $S(x,p,x',p')$ is well defined, uniquely and everywhere, on L.

The interest of this construction is that it extends to general covariant field theory. In this case the space $D = T^*C_b$ can be taken to be the cotangent space of the space C_b of the fields on a surface Σ bounding a region R of spacetime. A point in D represents then a configuration of the field and its momentum on the boundary surface. The surface L is defined as the set of these configurations which bound a solution of the field equations in R. The line integral of the restriction of the boundary phase space 1-form θ_b to L defines the Hamilton function. We will see in the next chapter how this applies to general relativity.

3

Gravity

Let us come to what we know about gravity. General relativity, the "most beautiful of the physical theories,"[1] has received an impressive amount of empirical corroboration in the last decades. The theory deserves to be taken seriously: what it tells us about the world is likely to be important, and to be key to further progress. Here we give the basic formulation of general relativity in the form it will be used in the following chapters (tetrads, Holst action, linear simplicity constraint, . . .).[2]

3.1 Einstein's formulation

In Einstein's original formulation Einstein (1915), the gravitational field is a symmetric tensor field $g_{\mu\nu}(x)$, which can be interpreted as the pseudo-Riemannian *metric* of spacetime. The dynamics is coded in the Einstein equations, which follow from the Einstein–Hilbert action

$$S[g] = \frac{1}{16\pi G} \int d^4x \sqrt{-g}(R - 2\Lambda), \tag{3.1}$$

where g is the determinant of the metric $g_{\mu\nu}(x)$, R is the Ricci scalar, and Λ is the cosmological constant. Greek indices $\mu, \nu, \ldots = 0, 1, 2, 3$ are spacetime tangent indices. The Einstein equations, in the absence of matter, are

$$R_{\mu\nu} - \frac{1}{2}Rg_{\mu\nu} + \Lambda g_{\mu\nu} = 0. \tag{3.2}$$

The theory is defined by two constants, both of which have been measured. Their measured values are approximately $G \sim 6 \times 10^{-8}$ cm^3 g^{-1} s^{-1} and $\Lambda \sim 10^{-52}$ m^{-2}.

The cosmological constant is sometimes presented as something more mysterious than it actually is: it is just a constant of nature, like several others. In particular, it is often wrongly confused with vacuum energy. This is like confusing the charge of the electron with its radiative corrections.[3] The presence of the cosmological constant implies that there

---

[1] Landau and Lifshitz (1951), Sec. 82.

[2] For the mathematically inclined reader who desires a more precise treatment of the geometry used, here are some references. A good and simple introductory book which we recommend is Baez and Munian (1994). A standard reference for mathematical physics and in particular geometry is the two volumes Choquet-Bruhat and Dewitt-Morette (2000, 2004). Other good texts are Frankel (2003) and Chandia and Zanelli (1997). If you like math, go for the classic of Kobayashi and Nomizu (1969).

[3] Vacuum energy can be computed using quantum field theory on curved spacetime, and in this context it can be renormalized to its measured value (Hollands and Wald 2008). The argument often heard that the cosmological

is a dimensionless constant in the (quantum) theory. Its measured value is

$$\frac{\Lambda \hbar G}{c^3} \sim 10^{-120}, \tag{3.3}$$

which happens to be a very small number. The cosmological constant is close to the scale of the Hubble radius $R_{Universe} \sim \Lambda^{-1/2}$, therefore this number is more or less given by the ratio between the largest and the smallest things we are aware of in the universe (namely, the cosmological scale and the Planck scale). This is

$$\frac{c^3}{\Lambda \hbar G} \sim \frac{R_{Universe}^2}{L_{Planck}^2} \sim 10^{120}. \tag{3.4}$$

This is a very large number, or so it looks from our parochial human perspective; perhaps for the gods it is still a small number. This dimensionless number sits inside quantum gravity.[4]

In the following we drop Λ wherever it does not contribute conceptually. We come back to it in Chapter 6, where we shall see that the dimensionless number above plays an important role in the complete theory. Also, wherever convenient we use units where $\hbar = 8\pi G = c = 1$.

3.2 Tetrads and fermions

The formulation of the gravitational field as a pseudo-Riemannian metric cannot be fundamentally correct, because it does not allow coupling to fermions, and fermions exist in the world. For this, we need the tetrad formulation.

The dynamics of a fermion in flat space is governed by the Dirac equation,

$$i\gamma^I \partial_I \psi - m\psi = 0. \tag{3.5}$$

In order to couple this equation to gravity, we need a different description of the gravitational field. Instead of $g_{\mu\nu}(x)$, we must use the *tetrads* $e_\mu^I(x)$ (an introduction to tetrads and other mathematical tools can be found in Baez (1994a). Other references are Wald (1984), Misner *et al.* (1973), Hawking and Ellis (1973)). Here $I = 0,1,2,3$ are "internal" flat Minkowski indices. The relation with the metric is

$$g_{\mu\nu}(x) = e_\mu^I(x)e_\nu^J(x)\eta_{IJ}, \tag{3.6}$$

---

constant should naturally be at the cut-off scale shows that we do not yet have good control of quantum field theory at the Planck scale, not that the cosmological constant itself is mysterious (Bianchi and Rovelli 2010a,b).

[4] Much has been made about the lack of "naturalness" in the existence of a large number in fundamental physics. We do not think this is necessarily good thinking. The heliocentric model proposed by Aristarchus of Samos in antiquity was discarded because it required that the distance to the stars was a number "too large to be realistic," otherwise the Earth movement would have determined a stellar parallax, which was not observed. This "naturalness" argument is in Ptolemy, and is wrong. Similar "naturalness" arguments were used until the late nineteenth century against the atomic hypothesis, because the Avogadro number was "too large" to be realistic. When we learn more about the universe, we see farther away, and we learn that what looks "natural" to us may just be so because of our limited experience.

where η_{IJ} is the Minkowski metric, which we use (with its inverse) to lower and raise the flat Minkowsli internal indices $I, J = 0, \ldots, 3$. Geometrically, $e^I_\mu(x)$ is a map from the tangent space at x to Minkowski space. It directly captures Einstein's central intuition that spacetime is locally like Minkowski space. The metric $g_{\mu\nu}$ is the pull-back of the Minkowski metric to the tangent space: this is the meaning of the last equation. Fermions can then be defined locally in Minkowski space as usual.

The gravitational action can be written replacing the metric with its expression in terms of the tetrad: $S[e] = S[g[e]]$.

The tetrad formalism satisfies an additional local Lorentz $SO(3,1)$ gauge invariance under the transformations

$$e^I_\mu(x) \to \Lambda^I{}_J(x) e^J_\mu(x), \tag{3.7}$$

where $\Lambda^I{}_J(x)$ is a Lorentz matrix.[5] The metric is not affected by this transformation because the Lorentz matrices transform the Minkowski metric η_{IJ} into itself. Hence the action is invariant.

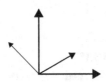

This local Lorentz gauge invariance has a geometrical interpretation: the inverse $e^\mu_I(x)$ of the matrix $e^I_\mu(x)$ defines four vector fields (one for each I) at each point x. These are orthonormal:

$$e^\mu_I(x) e^\nu_J(x) g_{\mu\nu} = \eta_{IJ}. \tag{3.8}$$

Given a Riemannian manifold, the four tetrad fields are such that everywhere they provide an orthonormal frame; conversely, if one has just the four vectors at each point, these define a Riemannian manifold by defining the frame that they determine to be orthonormal.

Related to this local gauge invariance, we can introduce a connection, as in Yang–Mills theories. The Lorentz connection is an object in the Lie algebra of the Lorentz group

$$\omega^{IJ}_\mu = -\omega^{JI}_\mu. \tag{3.9}$$

This connection defines a covariant derivative D_μ on any object that transforms under a finite-dimensional representation of the Lorentz group. In particular, this allows

---

[5] What is gauge invariance? If we consider a system in isolation, we can interpret gauge invariance just as a mathematical redundancy in its description. However, many gauge-invariant physical systems, such as gravity and electromagnetism, do not couple to other systems via a gauge-invariant quantity. Rather, the interaction couples a non-gauge-invariant quantity of the system A with a non-gauge-invariant quantity of the coupled-to system B. The coupled system is still gauge-invariant, but the total gauge invariance does not factor in the product of the gauge invariance of the two component-systems. Thus, it is not correct to regard non-gauge-invariant quantities of a system as purely mathematical artifacts; they are a handle for possible interactions of the system. In this sense, they represent physically meaningful quantities. See Rovelli (2013b) for a detailed discussion of this point.

us to define a general covariant Dirac equation, which couples the fermion field to the tetrads and the connection:

$$i\gamma^I e_I^\mu D_\mu \psi - m\psi = 0 .$$

(3.10)

Let us introduce a form notation[6] by writing

$$e^I = e_\mu^I dx^\mu \qquad \text{and} \qquad \omega^{IJ} = \omega_\mu^{IJ} dx^\mu .$$

(3.11)

Tetrad and connection define a "geometry" in the sense of Cartan, which is more general than Riemann geometry because it includes also torsion. The torsion 2-form is defined by

$$T^I = de^I + \omega^I{}_J \wedge e^J$$

(3.12)

(this is called the first Cartan equation) and the curvature 2-form is defined by

$$F^I{}_J = d\omega^I{}_J + \omega^I{}_K \wedge \omega^K{}_J$$

(3.13)

(the second Cartan equation). Given a tetrad, the condition that the torsion vanishes,

$$de^I + \omega^I{}_J \wedge e^J = 0,$$

(3.14)

can be shown to have a unique solution $\omega[e]$ for the connection. This solution is called the (torsionless) *spin connection*, or the Levi-Civita connection. Therefore if the connection is torsionless, it is uniquely determined by the tetrad and Cartan geometry reduces to Riemanniann geometry (plus the extra gauge invariance given by the tetrads). The curvature of the torsionless spin-connection is directly related to the *Riemann curvature tensor* by

$$R^\mu{}_{\nu\rho\sigma} = e_I^\mu e_\nu^J F^I{}_{J\rho\sigma} ,$$

(3.15)

or equivalently

$$F^{IJ} = e_\mu^I e_\nu^J R^{\mu\nu}{}_{\rho\sigma} dx^\rho \wedge dx^\sigma .$$

(3.16)

Using this, it is a simple exercise to rewrite the Einstein–Hilbert action in the form

$$\mathcal{S}[e] = \frac{1}{2} \int e^I \wedge e^J \wedge F^{KL} \varepsilon_{IJKL}$$

(3.17)

where F is the curvature (3.13) of the torsionless spin connection determined by the tetrad field, and ε_{IJKL} is the completely antisymmetric four-index object, normalized by $\epsilon^{0123} = 1 = -\epsilon_{0123}$. This action yields the Einstein equations (see the Complements to this chapter). We denote the Hodge dual in Minkowski space by a star, that is: $F^\star_{IJ} \equiv \star F_{IJ} := \frac{1}{2}\epsilon_{IJKL}F^{KL}$, or $T^\star_I \equiv T^{JKL}\epsilon_{IJKL}$ and we can write (3.17) as

$$\mathcal{S}[e] = \int e \wedge e \wedge F^\star,$$

(3.18)

here and subsequently we will frequently suppress contracted indices. The quantity

$$\Sigma^{IJ} = e^I \wedge e^J$$

(3.19)

---

[6] For an introduction to differential forms, see for instance Arnold (1989); Flanders (1963); Choquet-Bruhat and Dewitt-Morette (2000, 2004).

is called the Plebanski 2-form. It is a 2-form with antisymmetric Minkowski indices, and can therefore be interpreted as a 2-form with values in the Lorentz algebra.

Finally, the action of the fermion field interacting with gravity which gives the curved Dirac equation (3.10) is

$$S_f[\psi, e] = \int \bar{\psi} \gamma^I D \psi \wedge e^J \wedge e^K \wedge e^L \epsilon_{IJKL} \equiv \int \bar{\psi} \gamma D \psi \wedge (e \wedge e \wedge e)^\star, \tag{3.20}$$

where $D = D_\mu dx^\mu$.

3.2.1 An important sign

We wrote above that the tetrad formalism is equivalent to the metric one. That is not exactly true, and this is of importance for the quantum theory. There is a difference between the theory defined by the Einstein–Hilbert action,

$$S_{EH}[g] = \frac{1}{2} \int \sqrt{-\det g} \, R \, d^4x, \tag{3.21}$$

and that defined by the tetrad action,

$$S_T[e] = \int e^I \wedge e^J \wedge F^\star_{IJ}. \tag{3.22}$$

These two actions are not equivalent. This can be seen by performing an internal time-reversal operation

$$^{(i)}Te^0 := -e^0, \qquad ^{(i)}Te^i := e^i, \qquad i = 1,2,3. \tag{3.23}$$

Under this transformation, S_{EH} is clearly invariant as the metric $g = e^I e_I$ is not affected by this transformation, while S_T flips sign, $S_T[^{(i)}Te] = -S_T[e]$. The difference becomes manifest by writing both actions in tensor notation and in terms of tetrads:

$$S_{EH}[e] = \frac{1}{2} \int |\det e| R[e] d^4x, \tag{3.24}$$

$$S_T[e] = \frac{1}{2} \int (\det e) R[e] d^4x. \tag{3.25}$$

They differ by the sign factor

$$s \equiv \text{sgn}(\det e). \tag{3.26}$$

Hence the gravitational field defined by the triad $^{(i)}Te$ has a different action from the gravitational field e, in spite of the fact that they determine the same Riemannian metric. The difference between the two gravitational fields e and $^{(i)}Te$ has no effect on the matter that couples to the metric, but recall that fermions couple to the tetrad. In fact, the dynamics of the fermion, determined by the Dirac equation (3.10), is sensitive to the sign s: the phase of the fermion evolves in the opposite direction in a region where s has opposite sign. Consider for instance the tetrad field

$$e^i = dx^i, \qquad e^0 = N(t) dx^0, \tag{3.27}$$

where the "Lapse" function $N(t)$ is continuous but becomes negative in a finite interval in the time t. It is easy to see that this tetrad defines a flat metric. But in the negative Lapse region the time-derivative term of the Dirac equation flips sign, and therefore the phase of a Dirac particle rotates in the opposite direction to that in the positive Lapse region. Since interference experiments can in principle detect a relative phase shift, the sign s is in principle observable (Christodoulou *et al.* 2012).

In non-relativistic physics, the action (and the Hamilton function) always changes sign for a time-reversed trajectory. The Einstein–Hilbert action does not. Therefore, in this respect the tetrad action is the natural one. All this has not much effect on the classical theory, but it is important in the quantum theory: in defining a path integral for the gravitational field, it is not the same whether we integrate over metrics or over tetrads, because the integration over tetrads includes integrations over configurations with $s < 0$, which contribute to the Feynman integral with a term that is of the form

$$e^{-\frac{i}{\hbar} S_{EH}[g]},$$ (3.28)

in addition to the term

$$e^{+\frac{i}{\hbar} S_{EH}[g]}.$$ (3.29)

The two contributions are akin to the forward-propagating and the backward-propagating paths in a path integral for a relativistic particle. In this sense, one could call a negative s region an "antispacetime": a spacetime region that contributes to the action with a negative sign, as if it were time-reversed. We will see these two terms appearing over and over in the quantum theory.

3.2.2 First-order formulation

Consider a different action, which is now a function of a tetrad and a Lorentz connection considered as independent fields,

$$S[e, \omega] = \int e \wedge e \wedge F[\omega]^{\star}.$$ (3.30)

This is a simple *polynomial* action, sometimes denoted the Palatini action (or, better, the tetrad-Palatini action). As first noticed by Palatini in a similar context, this action yields general relativity: the variation of the connection gives precisely Eq. (3.14), which states that the torsion vanishes and therefore on-shell the connection is the Levi-Civita connection. Then the variation of the tetrad gives the Einstein equations:

$$\begin{cases} \delta\omega: & \omega = \omega[e], \\ \delta e: & \text{Einstein equations.} \end{cases}$$ (3.31)

This is called a *first-order formulation*, and is the analog of the free-particle action (2.23); while the formulations where the tetrad or the metric are the only variables are called *second-order formulations*.[7]

3.3 Holst action and Barbero–Immirzi coupling constant

The Palatini action is nicely polynomial, like the action of Yang–Mills theory. Furthermore, it admits a purely geometric formulation in terms of differential forms. One can ask which other terms with all these symmetries can be added to this action. In fact, up to boundary terms there is essentially only one other term that can be added, which is $\int e \wedge e \wedge F := \int e_I \wedge e_J \wedge F^{IJ}$. Let us add it to the action with a coupling constant $1/\gamma$:

$$S[e,\omega] = \int e \wedge e \wedge F^\star + \frac{1}{\gamma} \int e \wedge e \wedge F. \tag{3.35}$$

The equations of motion of this action are again the same as those of GR: the second term has no effect on the equations of motion. What happens is that the variation with respect to the connection again gives the torsionless condition, and when this is used, the second term becomes

$$\int e^I \wedge e^J \wedge F_{KL} = \int R_{\mu\nu\rho\sigma} \, \varepsilon^{\mu\nu\rho\sigma} \, d^4x = 0, \tag{3.36}$$

---

[7] The first-order and second-order formulations are equivalent for pure gravity, but not so with minimally coupled fermions. The difference between the second-order action

$$S[\psi,e] = \int e \wedge e \wedge F[\omega(e)]^\star + \int \bar{\psi}\gamma D[\omega[e]]\psi \wedge (e \wedge e \wedge e)^\star \tag{3.32}$$

and the first-order one

$$S[\psi,e,\omega] = \int e \wedge e \wedge F[\omega]^\star + \int \bar{\psi}\gamma D[\omega]\psi \wedge (e \wedge e \wedge e)^\star, \tag{3.33}$$

can be shown to be a four-fermion interaction term. This is because in the presence of fermions the variation of ω does not give the vanishing of the torsion, but rather a torsion term proportional to the fermion current. The spin connection that solves this equation has a term depending on the fermion, which couples back to the fermions via the covariant derivative. Thus, schematically

$$S[\psi,e] = S[\psi,e,\omega] + \int \bar{\psi}\psi\bar{\psi}\psi. \tag{3.34}$$

The minimally coupled theories are different, but in either theory we can add or subtract a four-fermion interaction, making them equivalent.

which vanishes because of the symmetry properties of the Riemann tensor (its totally antisymmetric part always vanishes). The action (3.35) can be written in the compact form

$$S[e, \omega] = \int e \wedge e \wedge \left(F^* + \frac{1}{\gamma} F \right)$$

$$\equiv \int e \wedge e \wedge \left(\star + \frac{1}{\gamma} \right) F$$

$$= \int \left(\star e \wedge e + \frac{1}{\gamma} e \wedge e \right) \wedge F \qquad (3.37)$$

and is known as the Holst action; the second term is called the Holst term.[8] The coupling constant γ is called the Barbero–Immirzi constant. This is the action with which we do 4-dimensional quantum gravity, as it is the most generic one: it is polynomial and has all the relevant symmetries. The Holst term has no effect on classical physics, but plays a substantial role in the quantum theory. The term in parentheses,

$$B = \left(\star e \wedge e + \frac{1}{\gamma} e \wedge e \right) = \left(\star + \frac{1}{\gamma} \right) \Sigma, \qquad (3.38)$$

plays an important role in what follows. On a $t = 0$ boundary, B is the derivative of the action with respect to $\partial \omega / \partial t$ (because the quadratic part of the action is $\sim B \wedge d\omega$), therefore B is the momentum conjugate to the connection. More precisely, if we reinstate the dimensionful constant $\frac{1}{8\pi G}$ in front of the full action, and we go to a time gauge where the restriction of $\star e \wedge e$ on the boundary vanishes, the momentum is the 2-form on the boundary taking values in the SL(2, $\mathbb{C}$) algebra

$$\Pi = \frac{1}{8\pi \gamma G} B. \qquad (3.39)$$

Notice that in QCD there is a similar term, which has no effect on the equations of motion but plays a role in the quantum theory as well:

$$\mathcal{S}_{\text{QCD}} = \int F \wedge F^* + \theta_{\text{QCD}} \int F \wedge F. \qquad (3.40)$$

The constant θ_{QCD} plays a role in quantum theory, for instance in the theory of instantons. The same happens with the Barbero–Immirzi constant γ.

3.3.1 Linear simplicity constraint

Consider a spacelike boundary surface Σ in the tetrad-connection formalism. The tetrad is a map from the tangent space to the spacetime manifold to Minkowski space. In particular, at each point σ of Σ it maps the tangent space at σ into a three-dimensional linear spacelike subspace of the Minkowski space. The subgroup of the Lorentz group that leaves this subspace invariant is the SO(3) *rotation* subgroup, and its existence breaks the local

---

[8] Holst was the second to discover it (Hojman *et al.* 1980; Holst 1996): it is common in science to call something after the second discoverer.

SO(3,1) invariance down to SO(3) at the boundary. That is, the boundary allows us to pick up a preferred Lorentz frame.

In coordinates, we can view this by writing the covector n_I normal to all vectors tangent to Σ, that is

$$n_I \sim \epsilon_{IJKL}\, e^J_\mu e^K_\nu e^L_\rho \, \frac{\partial x^\mu}{\partial \sigma^1} \frac{\partial x^\nu}{\partial \sigma^2} \frac{\partial x^\rho}{\partial \sigma^3}, \tag{3.41}$$

where $\{\sigma^i\}$, with $i = 1,2,3$, are the coordinates of the point $\sigma \in \Sigma$ and $x^\mu(\sigma)$ is the embedding of the boundary Σ into spacetime. We can use n_I to gauge-fix SO(3,1) down to the SO(3) subgroup that preserves it. That is, we can orient the local Lorentz frame in such a way that the boundary is locally a fixed-time surface, so that $n_I = (1,0,0,0)$.

The pull-back to Σ of the momentum 2-form B^{IJ}, defined in (3.38), can be decomposed into its electric $K^I = n_J B^{IJ}$ and magnetic $L^I = n_J(\star B)^{IJ}$ parts, in the same manner in which the electromagnetic tensor F^{IJ} can be decomposed into electric and magnetic parts once a Lorentz frame is chosen. Since B is antisymmetric, L^I and K^I do not have components normal to Σ, that is, $n_I K^I = n_I L^I = 0$ and are hence three-dimensional vectors in the space normal to n, which we can denote as $\vec{K}$ and $\vec{L}$. In the gauge $n_I = (1,0,0,0)$ these are simply

$$K^i = B^{i0}, \qquad L^i = \frac{1}{2}\epsilon^i{}_{jk}B^{jk}, \tag{3.42}$$

where we write $\vec{K} = \{K^i\}$ and $\vec{L} = \{L^i\}$. These are relations analogous to the ones that relate electric and magnetic fields to the Maxwell tensor.

Now, a simple result that has major importance for the quantum theory follows easily: from the definition (3.38) of B we have

$$n_I B^{IJ} = n_I(\star e \wedge e + \frac{1}{\gamma}e \wedge e)^{IJ} = n_I(\epsilon^{IJ}{}_{KL}e^K \wedge e^L + \frac{1}{\gamma}e^I \wedge e^J), \tag{3.43}$$

but on the boundary we have $n_I e^I|_\Sigma = 0$ by the very definition of n, therefore

$$n_I B^{IJ} = n_I(\star e \wedge e)^{IJ}. \tag{3.44}$$

For the same reason,

$$n_I(B^\star)^{IJ} = n_I((\frac{1}{\gamma}e^I \wedge e^J)^\star)^{IJ} = \frac{1}{\gamma}n_I(\star e \wedge e)^{IJ} = \frac{1}{\gamma}n_I B^{IJ}. \tag{3.45}$$

By definition of K^I and L^I, the last equation gives

$$\boxed{\vec{K} = \gamma \vec{L}.} \tag{3.46}$$

In words, the magnetic and electric part of the momentum 2-form B are proportional to one another, and the proportionality constant is the Barbero–Immirzi constant. This relation is called the "linear simplicity constraint" and is one of the most important equations in covariant loop quantum gravity, as we will see in Chapter 7.

3.3.2 Boundary term

Consider gravity defined in a compact region $\mathcal{R}$ of spacetime, with the topology of a four-dimensional ball. The boundary of $\mathcal{R}$ is a three-dimensional space with the topology of a 3-sphere, which we call Σ. In Section 2.1.1 we have seen that in order to have a well-defined variational principle and to have a well-defined Hamilton function, boundary terms may be required. These are known for general relativity (Gibbons and Hawking 1977; York 1972). In the metric formulation, the boundary term is

$$S_{EH\ boundary} = \int_\Sigma k^{ab}\, q_{ab}\, \sqrt{q}\, d^3\sigma, \tag{3.47}$$

where k^{ab} is the extrinsic curvature of the boundary (its explicit form is given below in (3.55)), q_{ab} is the three-metric induced on the boundary, q its determinant, and σ are coordinates on the boundary. In the case of pure gravity without cosmological constant, since the Ricci scalar vanishes on the solutions of the Einstein equations, the bulk action vanishes and the Hamilton function is just given by the boundary term[9]

$$S_{EH}[q] = \int_\Sigma k^{ab}[q]\, q_{ab}\, \sqrt{q}\, d^3\sigma. \tag{3.48}$$

This is a very non-trivial functional to compute, because the extrinsic curvature $k^{ab}[q]$ is determined by the bulk solution singled out by the boundary intrinsic geometry. Therefore $k^{ab}[q]$ is going to be non-local. Knowing the general dependence of k^{ab} from q is equivalent to knowing the general solution of the Einstein equations.

3.4 Hamiltonian general relativity

The task of finding a hamiltonian formulation for general relativity was first addressed by Dirac (Dirac 1958), with the aim of quantizing the theory. The task turned out to be difficult for two reasons. The first is that the theory has constraints. It is for this reason that Dirac developed his theory of constrained hamiltonian systems, which we briefly review below. The second is due to the horrendous intricacy of the algebra of the canonical analysis of the theory, when using the metric variable. Dirac succeeded also in this, but the result was cumbersome. A few years later, the task was greatly simplified by Arnowit, Deser, and Misner, who found a change of variables that simplifies all this, and has a clean geometric interpretation. Let us start from this.

---

[9] Note that here the Hamilton function is a *functional* of the boundary metric. Therefore one cannot rely on the difference of notation between $S(\ )$ and $S[\ ]$ to distinguish it from the action, as in finite-dimensional systems. Which is which, however, would be clear from the context: the action is a local functional of the bulk 4-metric, the Hamilton function is a non-local functional of the boundary 3-metric.

3.4.1 ADM variables

The change of variables introduced by Arnowit, Deser, and Misner (Arnowitt *et al.* 1962) is the following. From $g_{\mu\nu}$, define the three fields

$$q_{ab} = g_{ab}, \tag{3.49}$$

$$N_a = g_{ao}, \tag{3.50}$$

$$N = \sqrt{-g^{oo}}, \tag{3.51}$$

where $a, b = 1, 2, 3$. Notice that in the last equation N is defined by the time–time component of $g^{\mu\nu}$ with upper indices. N and N_a are called "Lapse" and "Shift" functions, q_{ab} is called the three-metric. This change of variables turns out to be extremely smart, for a number of reasons. The main one is that when writing the action in terms of these variables, one discovers that the lagrangian does not depend on the time derivatives $\dot{N}$ and $\dot{N}_a$ of Lapse and Shift, and this immediately simplifies the canonical analysis.

The second is the fact that these variables have a nice intrinsic geometrical interpretation. To see this, consider the hypersurfaces $t = constant$. Assume that these surfaces foliate spacetime. Then q_{ab} is clearly the three-dimensional metric induced on the surfaces by the spacetime metric. We use this metric to raise and lower three-dimensional indices $a, b = 1, 2, 3$. Next, consider the metric normal n^μ to this surface. It is not hard to see that the point $y^\mu = x^\mu + Nn^\mu dt$ lies on the surface $t + dt$. That is, the Lapse function is the proper-time elapsed between a point on the t surface and a point on the $t + dt$ surface, along the normal to the surface, or, equivalently, for a static observer on the surface. In a sense, the Lapse determines the rate at which physical time elapses in the coordinates chosen. Finally, the Shift $\vec{N} = \{N^a = q^{ab}N_b\}$ gives the separation (on the $t + dt$ surface) between the two points y^μ and $(t + dt, \vec{x})$. That is, the Shift function measures the shift of the spatial coordinates from one constant surface to the next, with respect to the coordinates that observers not moving on the initial surface would carry with them.

The Lapse can equivalently be defined by

$$N^2 \det q = \det g \tag{3.52}$$

or

$$g_{oo} = -N^2 + N_a N^a, \tag{3.53}$$

so that the line element in these variables reads

$$ds^2 = -(N^2 - N_a N^a)dt^2 + 2N_a dx^a dt + q_{ab} dx^a dx^b. \tag{3.54}$$

The extrinsic curvature of the $t = constant$ surfaces is given by

$$k_{ab} = \frac{1}{2N}(\dot{q}_{ab} - D_{(a}N_{b)}) \tag{3.55}$$

where the dot indicates the derivative with respect to t and D_a is the covariant derivative of the three-metric. Using this, the action in terms of these variables reads

$$S[N, \vec{N}, q] = \int dt \int d^3x \sqrt{q} N \left(k_{ab} k^{ab} - k^2 + R[q] \right), \tag{3.56}$$

where $k = k_a{}^a$ and $\sqrt{q} \equiv \sqrt{\det q}$. That is, in terms of the main variables, the lagrangian density reads

$$\mathcal{L}[N, \vec{N}, q] = \frac{\sqrt{q}(g^{ac}g^{bd} - g^{ab}g^{cd})(\dot{q}_{ab} - D_{(a}N_{b)})(\dot{q}_{cd} - D_{(c}N_{d)})}{4N} + \sqrt{q}NR[q]. \qquad (3.57)$$

In this form the hamiltonian analysis is easy. The canonical momenta of Lapse and Shift vanish because $\dot{N}$ and $\dot{N}_a$ do not appear in the action. The canonical momentum of the three-metric is

$$\pi^{ab} = \frac{\partial L}{\partial \dot{q}_{ab}} = \sqrt{q}G^{abcd}k_{cd} = \sqrt{\det q}\,(k^{ab} - kq^{ab}), \qquad (3.58)$$

and the action written in hamiltonian form reads

$$S[N, \vec{N}, q, \pi] = \int dt \int d^3x \left(\pi^{ab}\dot{q}_{ab} - NC(\pi, q) - 2N^a C_a(\pi, q) \right), \qquad (3.59)$$

where

$$C = G_{abcd}\pi^{ab}\pi^{cd} - \sqrt{q}\,R[q] \qquad (3.60)$$

is called the scalar constraint, or hamiltonian constraint, and

$$C_a = D_a\pi^{ab} \qquad (3.61)$$

is called the vector, or diffeomorphism constraint. Here

$$G_{abcd} = \frac{1}{2\sqrt{q}}(g_{ac}g_{bd} + g_{ad}g_{bc} - g_{ab}g_{cd}), \qquad (3.62)$$

which is called the DeWitt super metric. These must vanish because of the variation of Lapse and Shift. The hamiltonian theory is therefore entirely given by the vanishing of the momenta associated to Lapse and Shift, the vanishing of the hamiltonian and diffeomorphism constraints, and by a hamiltonian which vanishes when the constraints are satisfied.

Notice that this leads us immediately to the structure of the dynamics described in the previous chapter. In particular, the constraints (3.60) and (3.61) realize the constraint (2.61) for general relativity, and the hamitonian vanishes for the reasons explained in Chapter 2.

The formalism presented in Section 2.5.2 takes a particularly nice form for general relativity: The space $\mathcal{C}_b$ is the space of the intrinsic geometries $q_{ab}(x)$ of a surface with the topology of a 3-sphere. D is the space of the intrinsic and extrinsic geometries $[q_{ab}(x), k^{ab}(x)]$ of this surface. L is the set of boundary (extrinsic and extrinsic) geometries that bound a Ricci flat ball. The Hamilton function on L is the restriction to L of the expression

$$S[q, k] = \int_\Sigma k^{ab}q_{ab}\,\sqrt{q}d^3x, \qquad (3.63)$$

and its pull-back to $\mathcal{C}_b$

$$S[q] = \int_\Sigma k^{ab}[q]q_{ab}\,\sqrt{q}d^3x, \qquad (3.64)$$

is the usual Hamilton function, which codes the solutions of the theory.

Notice that in the case of gravity the region R and its boundary Σ have no metric geometry until the field is specified, because the metric is determined by the gravitational field. Therefore there is a limit in the space of the boundary fields where the geometry of R shrinks to a point. This is the natural reference point z_o.

3.4.2 What does this mean? Dynamics

The result above is clarified by comparing it with the hamiltonian structure of two well-known theories, which we review here briefly.

Electromagnetism

Maxwell theory can be written in terms of the potential A_μ. The Maxwell field is $F_{\mu\nu} = \partial_{[\mu} A_{\nu]}$ and the Maxwell equations are invariant under the gauge transformation $A_\mu \rightarrow A_\mu + \partial_\mu \lambda$. The action is

$$S[A] = \frac{1}{4} \int d^4x \, F_{\mu\nu} F^{\mu\nu} \qquad (3.65)$$

and it is immediately seen that it does not depend on $\dot{A}_0$. Therefore A_0 has the same role in electromagnetism as the Lapse and Shift functions. Notice that a gauge transformation allows us to choose A_0 as we want. For instance, we can choose it to be $A_0 = 0$. Analogously, Lapse and Shift can be chosen arbitrarily. They simply determine the position on the coordinates of the next $t = constant$ surface, after the first is chosen. The simplest choice is the choice $N = 1, \vec{N} = 0$, called the time gauge, where the line element has the form $ds^2 = -dt^2 + q_{ab}dx^a dx^b$. The momentum of A_0 vanishes as the momenta of Lapse and Shift and the momentum of $\vec{A}$ is

$$\pi^a = \frac{\partial L}{\partial \dot{A}_a} = F^{oa}, \qquad (3.66)$$

namely, the electric field $\vec{E}$. The action in hamiltonian terms then reads

$$S[A_0, \vec{A}, \vec{E}] = -\int dt \int d^3x \left(E^a \dot{A}_a - (E^2 + B^2) + A_o C(\vec{E}) \right), \qquad (3.67)$$

where

$$C(\vec{E}) = \partial_a E^a \qquad (3.68)$$

is called the Gauss constraint. This is the analog of the constraints of gravity. Thus the hamiltonian theory is determined by the vanishing of the momenta associated with A_0, by the vanishing of the Gauss constraints, and by a non-vanishing hamiltonian.

The physical interpretation of the Gauss constraint is related to the residual gauge symmetry left by the choice $A_0 = 0$. In fact, we can still gauge transform $A_a \rightarrow A_a + \partial_a \lambda$ with a time-independent function λ. This means that there is a residual gauge freedom in the initial data. It is a general fact of constrained systems that any first-class constraint[10] is related to a gauge freedom. In fact, it is easy to see that the Gauss constraint generates the gauge transformation. Defining $C[\lambda] = \int d^3x\,\lambda(x)C(x)$, we have

$$\{A, C[\lambda]\} = d\lambda = \delta_\lambda A; \tag{3.69}$$

that is: the Poisson bracket of the variable with the constraint gives its infinitesimal gauge transformation. Analogously, the scalar and vector constraints of gravity generate gauge transformations, which correspond, respectively, to deformations of the $t = $ constant surface, and to change of spatial coordinates on this surface.

There remains one feature of the hamiltonian formulation of gravity that is not reflected by the Maxwell analogy: the vanishing of the hamiltonian. To illustrate this, let us turn to the second example.

Special relativistic dynamics

The action of a single relativistic particle can be written in the form

$$S = m \int d\tau \sqrt{\dot{x}^\mu \dot{x}_\mu}. \tag{3.70}$$

The momenta are

$$p_\mu = \frac{\partial L}{\partial \dot{x}^\mu} = m\frac{\dot{x}_\mu}{|\dot{x}|} \tag{3.71}$$

and clearly satisfy the constraint

$$C = p^2 - m^2 = 0. \tag{3.72}$$

This constraint generates the gauge transformations of the theory, which are the reparametrizations in τ. To see the analogy with gravity, notice that the same dynamics can be obtained from the action

$$S = \frac{1}{2} \int d\tau \left(\frac{\dot{x}^\mu \dot{x}_\mu}{N} + Nm^2 \right). \tag{3.73}$$

Notice that this has precisely the same structure as (3.57), and that the time derivative of N does not appear. The momenta are now $p_\mu = \dot{x}_\mu/N$ and the action in hamiltonian form is

$$S = \frac{1}{2} \int d\tau \left(p_\mu \dot{x}^\mu + NC(p) \right). \tag{3.74}$$

That is, the hamiltonian vanishes when the constraint is satisfied. This clarifies that the vanishing of the gravitational hamiltonian is simply the consequence of the fact that the evolution parameter given by the coordinate time is not a physical quantity, but only a free parameter. The vector constraint reduces the six degrees of freedom in q_{ab} to three, and the hamiltonian constraint determines the dynamics in the form of a relation between these three variables. Therefore the theory has two degrees of freedom per space point.

---

[10] In Dirac terminology, a constraint is first class if its Poisson brackets, with all other constraints and with the hamiltonian, vanish when the constraints are satisfied.

3.4.3 Ashtekar connection and triads

Instead of introducing tetrads on spacetime, it is also possible to introduce triads on each $t = $ constant surface. These are defined by

$$q_{ab}(x) = e_a^i(x)e_b^j(x)\delta_{ij}, \tag{3.75}$$

like their spacetime counterparts. Here $i,j = 1,2,3$ are flat indices. The introduction of this variable adds a local SO(3) invariance to the theory, with the geometrical interpretations of spatial rotations on fixed-time surfaces. We can also define the triad version k_i^a of the extrinsic curvature by

$$k_i^a e_b^i \equiv k_{ab}. \tag{3.76}$$

We can consider these variables as a 9+9-dimensional canonical conjugate pair, namely, posit $\{k_i^a, e_b^j\} \sim \delta_i^j \delta_b^a$. Notice that this increases by three the number of variables. However, by definition the antisymmetric part of $k_i^a e_i^b$ must vanish (because it is k^{ab}), therefore if we want to recover the 6+6-dimensional (q_{ab}, k^{ab}) phase space we must impose the constraint

$$G_c = \epsilon_{cab} k_i^a e_i^b = 0. \tag{3.77}$$

It is immediately seen that the Poisson brackets of this constraint generate precisely the local SO(3) gauge rotations. In 1986, Abhay Ashtekar realized that a curious connection introduced not long before by Amitaba Sen (Sen 1982) gives rise to an extremely useful canonical transformation (Ashtekar 1986).[11] Consider the connection

$$A_a^i = \Gamma_a^i[e] + \beta k_a^i, \tag{3.78}$$

where $\Gamma[e]_a^i$ is the torsionless spin connection of the triad (the unique solution to the three-dimensional first Cartan equation $de^i + \epsilon^i{}_{jk}\Gamma^j \wedge e^k = 0$), and β is an arbitrary parameter, and the "Ashtekar electric field"

$$E_i^a(x) = \frac{1}{2}\epsilon_{ijk}\,\epsilon^{abc}\,e_b^j e_c^k, \tag{3.79}$$

namely, the inverse of the triad multiplied by its determinant. Ashtekar realized that the connection satisfies the Poisson brackets

$$\{A_a^i(x), A_b^j(y)\} = 0 \tag{3.80}$$

(which is highly non-trivial, since the connection depends on non-commuting quantities) and

$$\{A_a^i(x), E_j^b(y)\} = \beta \delta_a^b\, \delta_j^i\, \delta^3(x,y). \tag{3.81}$$

Therefore A_a^i and E_i^a are canonically conjugate variables.

If $\beta = i$, the connection A can be shown to be the pull-back to the $t = $ constant of the self-dual component of the spacetime spin connection and is called the complex Ashtekar

---

[11] Ashtekar–Barbero variables are derived in Ashtekar and Lewandowski (2004), with emphasis on their covariant four-dimensional character.

connection (or simply the Ashtekar connection). With this choice the scalar and vector constraints turn out to have an impressively simple form:

$$C = \epsilon_{ijk} F^i_{ab} E^{aj} E^{bk},$$ (3.82)

$$C_a = F^i_{ab} E^{bi},$$ (3.83)

where F is the curvature of the Ashtekar connection

$$F^i_{ab} = \partial_a A^i_b - \partial_b A^i_a + \epsilon^i_{jk} A^j_a A^k_b,$$ (3.84)

and the constraint that generates the additional SO(3) local rotations is the same as in Yang–Mills theory:

$$G^i = D_a E^{ai}.$$ (3.85)

This formulation of general relativity has long been the basis of loop quantum gravity. It still plays a central role in the canonical theory and a less evident but more subtle role in the covariant theory.

In more recent years, the interest has shifted to the case with real β, largely thanks to the work by Fernando Barbero (Barbero 1995). The connection with real β appears naturally in the hamiltonian analysis of the Holst action, with $\beta = \gamma$. But it is important to notice that γ (the coupling constant in the action) and β (the constant entering in the definition of the connection) do not necessarily need to be taken as equal.[12] With real $\gamma = \beta$, the hamiltonian constraints turns out to be more complicated. With cosmological constant, it reads

$$H = E^a_i E^b_j \left(\epsilon^{ij}_k F^k_{ab} - 2\left(1+\beta^2\right) k^i_{[a} k^j_{b]} + \frac{\Lambda}{3}\, \epsilon_{abc} \epsilon^{ijk} E^c_k \right).$$ (3.86)

The field E^a_i has a direct geometric interpretation as the area element. A simple calculation, indeed, shows that the area of a two-surface S in a $t = constant$ hypersurface is

$$A_S = \int_S d^2\sigma \, \sqrt{E^a_i n_a E^b_i n_b}.$$ (3.87)

To show this, choose coordinates where the surface is given by $x^3 = 0$. Then

$$A_S = \int_S d^2\sigma \, \sqrt{det^{(2)}q} = \int_S d^2\sigma \, \sqrt{q_{11}q_{22} - q^2_{12}}$$
$$= \int_S d^2\sigma \, \sqrt{det q \; q^{33}} = \int_S d^2\sigma \, \sqrt{E^3_i E^3_i} = \int_S d^2\sigma \, \sqrt{E^a_i n_a E^b_i n_b}.$$ (3.88)

By introducing the 2-form

$$E^i = \frac{1}{2}\epsilon_{abc} E^{ai} dx^b dx^c,$$ (3.89)

this can be written in the form

$$A_S = \int_S |E|.$$ (3.90)

[12] For instance an elegant and useful formulation of the hamiltonian dynamics is given by Wieland (2012), keeping γ real but using $\beta = i$, namely, using the complex Ashtekar connection as main variable.

Notice that in the limit in which the surface is small, the quantity $\vec{E}_S$, defined by

$$E_S^i = \int_S E^i, \tag{3.91}$$

is a vector normal to the surface, whose length is the area of the surface.[13] In terms of the triad, this reads

$$E_S^i = \frac{1}{2}\epsilon^i{}_{jk}\int_S e^j \wedge e^k, \tag{3.92}$$

a formula that we have already seen in (1.11) in the first chapter. Remember?

Now, the momentum (3.39) is conjugate to the connection and therefore it is the canonical generator of Lorentz transformation. In particular, the generator of a boost in, say, the z direction is

$$\mathcal{K}_z = \frac{1}{8\pi\gamma G}K_z. \tag{3.93}$$

Using the simplicity constraint (3.46), this gives

$$\mathcal{K}_z = \frac{1}{8\pi G}L_z. \tag{3.94}$$

But we have just seen that L_z is the area element A of a surface normal to the z direction. Therefore the area element A of a surface is related to the generator K of a boost in the direction normal to the surface by

$$K = \frac{A}{8\pi G}. \tag{3.95}$$

This important simple equation encloses, in a sense, the full dynamics of general relativity, as we shall later see.

3.5 Euclidean general relativity in three spacetime dimensions

In the following chapters we use the euclidean version of general relativity in three spacetime dimensions (3d) as a toy model to introduce techniques and ideas used in the physically relevant case of the four-dimensional (4d) lorentzian theory. Here, let us briefly define this theory.

In three-dimensional the Riemann tensor is fully determined by the Ricci tensor, therefore the vacuum Einstein equations *Ricci* $= 0$ imply *Riemann* $= 0$, namely, that spacetime is flat. Therefore the theory has no local degrees of freedom: spacetime is just flat everywhere. This does not mean that the theory is completely trivial, for two reasons. First, if space has non-trivial topology it can be locally flat, but have a global dynamics (change size in time, for instance). Second, even on a trivial topology, where there is a single physical state, this single state still determines non-trivial relations between boundary partial

---

[13] In electromagnetism, the quantity conjugate to the connection is the electric field. Here it is E^i. Hence "the length of the gravitational electric field is the area."

observables. Let us see how this happens. In the metric formalism, the action (including constants) is

$$S[g] = \frac{1}{16\pi G} \int_{\mathcal{M}} d^3x \sqrt{-g}\, R + \frac{1}{8\pi G} \int_{\Sigma} d^2\sigma\, k^{ab}\, q_{ab}\sqrt{q}, \tag{3.96}$$

where the second term lives on the boundary $\Sigma = \partial\mathcal{M}$ of a compact ("bulk") spacetime region $\mathcal{M}$ and is needed for the reasons examined in Section 2.1.1. The Hamilton function of the theory is a function of the restriction of the metric to the boundary, which is given by the two-dimensional metric q induced by g on Σ. The bulk term in the action does not contribute to the Hamilton function, because the Ricci scalar vanishes on the solutions of the equation of motion, namely on flat space, leaving

$$S[q] = \frac{1}{8\pi G} \int_{\Sigma} d^2\sigma\, k^{ab}[q]\, q_{ab}\sqrt{q}. \tag{3.97}$$

This of course depends only on the intrinsic metric of Σ, and not on the coordinates in which this is expressed, because of general covariance. The dependence of the extrinsic curvature k^{ab} on q_{ab}, namely, on the intrinsic metric, is what codes the dynamics. This is determined by embedding the two-dimensional Riemannian space defined by q into a *flat* three-dimensional space. The dependence of k^{ab} on q_{ab} is obviously highly non-local. The non-triviality of the theory is therefore in the (global!) dependence of the extrinsic curvature on the intrinsic metric.

The solution of the dynamics is expressed by giving the boundary extrinsic curvature $k^{ab}(\sigma)$ as a function of the boundary partial observable $q_{ab}(\sigma)$. This is analogous to giving the solution of the equations of motion of a newtonian system by giving p as a function of (q,t,q',t'), which in turn, as we have seen, is equivalent to giving the dependence of $q'(t')$ from the initial data (q,p,t).

In the triad-connection formalism, the gravitational field is described by a triad field $e^i = e^i_a dx^a$ and an SO(3) connection $\omega^i{}_j = \omega^i_{aj} dx^a$, where indices $a, b, \ldots = 1, 2, 3$ are spacetime indices and $i, j = 1, 2, 3$ are internal indices raised and lowered with the Kronecker delta δ_{ij}. In three-dimensional, the bulk action is

$$S[e, \omega] = \frac{1}{16\pi G} \int \epsilon_{ijk}\, e^i \wedge F^{jk}[\omega], \tag{3.98}$$

where the curvature is $F^i{}_j = d\omega^i{}_j + \omega^i{}_k \wedge \omega^k{}_j$. The equations of motion are the vanishing of torsion and curvature.

Exercise 3.1 *Derive these equations of motion from the action.*

The four-dimensional local Lorentz invariance of the tetrad formalism is here replaced by a local SO(3) gauge invariance under the transformations

$$e^i_a(x) \mapsto R^i{}_j(x) e^j_a(x), \qquad R \in \mathrm{SO}(3). \tag{3.99}$$

This is three-dimensional euclidean general relativity.

A word about notation. For the connection, and all other antisymmetric two-index tensors, we use also the single-index notation

$$\omega^i = \frac{1}{2}\epsilon^i{}_{jk}\omega^{jk}. \tag{3.100}$$

The index i labels in fact a basis in the so(3)=su(2) Lie algebra. It is convenient also to write the connection as the su(2) generator in the fundamental representation of su(2), by using the Pauli matrices $\sigma_i = (\sigma_i{}^A{}_B)$ basis. That is, to use the notation

$$\omega = \omega^i \tau_i \tag{3.101}$$

where $\tau = -\frac{i}{2}\sigma_i$ are the generators of su(2).

We consider this theory on a compact region $\mathcal{R}$ of spacetime, with trivial topology (a ball), bounded by a two-dimensional boundary with the topology of a sphere which we call Σ. The pull-back of the connection on the boundary is an SO(3) connection ω on Σ. The pull-back of the triad is a 1-form e on Σ with values in the su(2) algebra. It is easy to see from the action that these two variables are canonically conjugate to one another: in analogy with the $p\dot{q}$ term of particle mechanics, the derivative term ∂_0 normal to a boundary surface in the action has the structure $\sim \frac{1}{8\pi G}n_c\epsilon^{abc}e^i_a\partial_0\omega^i_b$, where n_a is the normal one-form to the boundary.[14] Therefore, using $\epsilon_{ab} \equiv n^c\epsilon_{cab}$ the Poisson bracket between ω and e is

$$\{e^i_a(\sigma), \omega^j_b(\sigma')\} = 8\pi G\, \delta^{ij}\, \epsilon_{ab}\, \delta^2(\sigma, \sigma') \tag{3.102}$$

where $\sigma, \sigma' \in \Sigma$.

3.6 Complements

3.6.1 Working with general covariant field theory

Problem

Relate the Einstein–Hilbert action to the tetrad action. How does the sign difference come about?

Solution

Let us introduce a bit of useful notation:

- The gauge exterior covariant derivative D acts on a p-form ϕ with Lorentz indices as

$$\begin{aligned}
D\phi &= d\phi, \\
D\phi^I &= d\phi^I + \omega^I{}_J \wedge \phi^J, \\
D\phi^{IJ} &= d\phi^{IJ} + \omega^I{}_K \wedge \phi^{KJ} + \omega^J{}_K \wedge \phi^{IK}.
\end{aligned} \tag{3.103}$$

---

[14] Defined by $n = \epsilon_{abc}\frac{\partial x^b}{\partial\sigma^1}\frac{\partial x^b}{\partial\sigma^2}dx^c$, where (σ^1, σ^2) are coordinates on the boundary.

It has the Leibnitz property. In this way we can write the first Cartan's equation as

$$De = T,\qquad(3.104)$$

where T is the torsion 2-form. Notice that (cf. (3.104)), $R = d\omega + \omega \wedge \omega \neq D\omega$ (because ω is not a tensor; the factor 2 is missing, cf. second relation in (3.103)). However,

$$\delta R = D\delta\omega.\qquad(3.105)$$

(Note that although the connection ω is not a tensor, $\delta\omega$ is.)

• For an object V^{IJKL} with four Lorentz indices the following trace notation is useful:

$$\text{Tr}(V) \equiv \epsilon_{IJKL} V^{IJKL}.\qquad(3.106)$$

If A^{IJ} and B^{IJ} are two antisymmetric tensors, in terms of the notation used in the chapter ($A \wedge B \equiv A^{IJ} \wedge B_{IJ}$), this reads

$$\text{Tr}(A \wedge B) = \text{Tr}(B \wedge A) = 2A \wedge *B.\qquad(3.107)$$

This trace has the cyclic property, as a normal matrix trace, for example $\text{Tr}(a \wedge b \wedge B) = \text{Tr}(b \wedge B \wedge a) = \text{Tr}(B \wedge a \wedge b)$, but also $\text{Tr}(a \wedge b \wedge B) = \text{Tr}(b \wedge a \wedge B)$.

1. For integration of a function on a manifold we can write

$$e^0 \wedge e^1 \wedge e^2 \wedge e^3 = e^0{}_\mu e^1{}_\nu e^2{}_\kappa e^3{}_\lambda dx^\mu \wedge dx^\nu \wedge dx^\kappa \wedge dx^\lambda \qquad(3.108)$$
$$= e^0{}_\mu e^1{}_\nu e^2{}_\kappa e^3{}_\lambda \epsilon^{\mu\nu\kappa\lambda} dx^0 \wedge dx^1 \wedge dx^2 \wedge dx^3$$
$$= \det(e)\, dx^0 \wedge dx^1 \wedge dx^2 \wedge dx^3 = s|\det(e)|\, d^4x = s\sqrt{-g}\, d^4x = s\, dV,$$

where s is the sign of the determinant of e and dV is the invariant volume on the manifold. The appearance of s gives the sign difference between the metric and tetrad formalism. Let us drop it in what follows, assuming $s = +1$. This implies that

$$e^I \wedge e^J \wedge e^K \wedge e^L = \epsilon^{IJKL}|e|\, d^4x.\qquad(3.109)$$

We have the following identities for ϵ:

$$\epsilon^{IJKL}\epsilon_{IMNO} = -\left(\delta^J_M\delta^K_N\delta^L_O + \delta^L_M\delta^J_N\delta^K_O + \delta^K_M\delta^L_N\delta^J_O - \delta^K_M\delta^J_N\delta^L_O - \delta^L_M\delta^K_N\delta^J_O - \delta^J_M\delta^L_N\delta^K_O\right),$$

$$\epsilon^{IJKL}\epsilon_{IJMN} = -2\left(\delta^K_M\delta^L_N - \delta^K_N\delta^L_M\right),$$

$$\epsilon^{IJKL}\epsilon_{IJKN} = -6\,\delta^L_N,\qquad(3.110)$$

$$\epsilon^{IJKL}\epsilon_{IJKL} = -24.$$

We compute

$$\text{Tr}(e \wedge e \wedge e \wedge e) = \epsilon_{IJKL}e^I \wedge e^J \wedge e^K \wedge e^L = \epsilon_{IJKL}\epsilon^{IJKL}|e|d^4x = -24|e|d^4x,\qquad(3.111)$$

where we have used (3.109) and the last of the relations (3.110). This is important when we want to add to the action the cosmological term. Similarly, we get

$$\text{Tr}(e \wedge e \wedge F) = \epsilon_{IJKL}e^I \wedge e^J \wedge F^{KL} = \frac{1}{2}\epsilon_{IJKL}F^{KL}{}_{MN}e^I \wedge e^J \wedge e^M \wedge e^N$$

$$= \frac{1}{2}\epsilon_{IJKL}\epsilon^{IJMN}F^{KL}{}_{MN}|e|d^4x$$

$$= -\left(\delta^M_K\delta^N_L - \delta^M_L\delta^N_K\right)F^{KL}{}_{MN}|e|d^4x = -2|e|Rd^4x,\qquad(3.112)$$

where we have used (3.16) in the last step. Hence the gravitational action (3.1) can be written as

$$S[g[e]] = -\frac{1}{32\pi G} \int \mathrm{Tr}[e \wedge e \wedge F] = -\frac{1}{16\pi G} \int e \wedge e \wedge F^\star. \tag{3.113}$$

Problem

Show that treating e and ω as independent variables in the tetrad action gives the same physics.

Solution

Varying (3.113), we get

$$-32\pi G\,\delta S = \int \mathrm{Tr}[\delta e \wedge e \wedge F(\omega)] + \int \mathrm{Tr}[e \wedge \delta e \wedge F(\omega)] + \int \mathrm{Tr}[e \wedge e \wedge \delta F(\omega)]$$

$$= 2\int \mathrm{Tr}(e \wedge F \wedge \delta e) + \int \mathrm{Tr}(e \wedge e \wedge D\delta\omega), \tag{3.114}$$

where we have used the cyclic property and Eq. (3.105). Using the identity

$$\int d[\mathrm{Tr}(e \wedge e \wedge \delta\omega)] = \int D[\mathrm{Tr}(e \wedge e \wedge \delta\omega)] = 2\int \mathrm{Tr}(De \wedge e \wedge \delta\omega) + \int \mathrm{Tr}(e \wedge e \wedge D\delta\omega), \tag{3.115}$$

Eq. (3.104), and the fact that the term on the l.h.s. is a total derivative, we can write the previous equation as

$$\delta S = -\frac{1}{16G}\left[\int \mathrm{Tr}(e \wedge F \wedge \delta e) - \int \mathrm{Tr}(T \wedge e \wedge \delta\omega)\right]. \tag{3.116}$$

The vanishing of the first term implies

$$\epsilon_{IJKL} e^J \wedge F^{JK} = 0. \tag{3.117}$$

To show that this is equivalent to Einstein's equations, expand the curvature 2-form F^{JK} in the basis, $F^{JK} = \frac{1}{2} R^{JK}{}_{MO} e^M \wedge e^O$. Hence we have

$$\epsilon_{IJKL} R^{JK}{}_{MO} e^J \wedge e^M \wedge e^O = 0 \quad \Leftrightarrow \quad \epsilon_{IJKL} \epsilon^{IMOP} R^{JK}{}_{MO} = 0, \tag{3.118}$$

where we have taken the Hodge dual. Applying (3.110), we find that this is equivalent to

$$G^P{}_L = \mathrm{Ric}^P{}_L - \frac{1}{2}\delta^P_L R = 0, \tag{3.119}$$

which are the vacuum Einstein equations.

Notice that we have three possibilities, which are called x-order formalism:

- *2nd-order formalism.* Assumes $\omega = \omega(e)$, i.e., $T = 0$. Then (3.114) gives the Einstein equations.
- *1st-order formalism.* Treats ω and e as independent. The variation with respect to e gives the Einstein equations, whereas the variation with respect to ω implies $T \wedge e = 0 \Rightarrow T = 0$. That is, in the absence of matter the torsion T vanishes as a consequence of the variational principle. This then establishes the relation $\omega = \omega(e)$.
- *1.5-order formalism.* Treats e and ω as independent. However, we know that variation with respect to ω just establishes the relation $\omega = \omega(e)$, whatever this is. To obtain the equation of motion, it is therefore enough to vary S with respect to e only, i.e.,

$$EOM = \frac{\delta S[e,\omega]}{\delta e} = 0. \qquad (3.120)$$

Problem

Show that the Holst term does not modify the equations of motion.

Solution

Consider the Holst term

$$S_\gamma[e,\omega] \sim \frac{1}{\gamma} \int \mathrm{Tr}(e \wedge e \wedge *F) = \frac{2}{\gamma} \int e \wedge e \wedge F = \frac{2}{\gamma} \int e^I \wedge e^J \wedge F_{IJ}, \qquad (3.121)$$

Varying with respect to e gives

$$\delta_e S_\gamma \sim \frac{1}{\gamma} \int \mathrm{Tr}(e \wedge *F \wedge \delta e), \qquad (3.122)$$

which implies that

$$\epsilon_{IJKL} e^J \wedge (*F)^{KL} = 0. \qquad (3.123)$$

That is, we get the following equations of motion:

$$0 = \epsilon^{KLMN}\epsilon_{KLIJ} e^J \wedge F_{MN} = -2(\delta_I^M \delta_J^N - \delta_I^N \delta_J^M) e^J \wedge F_{MN} = -4e^J \wedge F_{IJ}. \qquad (3.124)$$

In the absence of torsion, the last expression is precisely the Bianchi identity and vanishes. This proves that the Holst action gives classical Einstein equations and hence it is equally good action for GR.

3.6.2 Problems

1. Derive the Dirac equation from the Dirac action on a curved spacetime.
2. Develop the formalism discussed in Section 2.5.2 for the tetrad-connection formulation of general relativity.

4

Classical discretization

To define a quantum theory for the gravitational field, we apply the general structure devised in Chapter 2 to the theory defined in Chapter 3. For this, we need to study how to discretize general relativity. In this chapter, we review two classic discretizations: lattice Yang–Mills theory and Regge calculus, and then we introduce the discretization of general relativity which we use in the following.

Lattice Yang–Mills theory is the basis for defining QCD in the strong coupling regime, and the best tool for computing physical quantities such as hadron masses and comparing them with measurements (Durr *et al.* 2008). The theory is the result of some beautiful intuitions by Kenneth Wilson (Wilson 1974), important also for quantum gravity. But there is a crucial difference between QCD and general relativity, as we discuss in detail below. The difference is illustrated by Tullio Regge's natural discretization of general relativity, called Regge calculus (Regge 1961), also important for the following. The discretization of general relativity that we use in quantum gravity is introduced at the end of this chapter; it is a mixture of lattice Yang–Mills theory and Regge calculus.

A discretization is an approximation: a truncation in the number of degrees of freedom where we disregard those likely to be irrelevant for a given problem. In quantum field theory, truncations play also a constructive role. For describing the weak field regime we utilize Fock-space methods and perturbation theory. Fock space is constructed by defining the N-particle state spaces. The theory is formally given by the $N \rightarrow \infty$ limit. The same is true on the lattice. The theory is formally defined in the limit where the number N of lattice sites goes to infinity. The same is true in quantum gravity: the theory can be defined via a discretization/truncation.

4.1 Lattice QCD

We consider for simplicity an SU(2) Yang–Mills theory in four dimensions. The field variable in the continuous theory is an SU(2) connection $A^i_\mu(x)$, where i is an index in the Lie algebra of SU(2). We also write this field as a 1-form with value in the Lie algebra

$$A(x) = A^i_\mu(x)\, \tau_i\, dx^\mu, \qquad (4.1)$$

where τ_i are the su(2) generators, namely, the Pauli matrices multiplied by $i/2$. The central idea of Wilson, on which loop quantum gravity also relies, is that we must view the algebra as the tangent space to the group, and A as the log of a group variable.

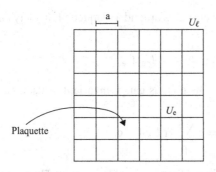

Plaquette

For this, fix a cubic lattice with N vertices connected by E edges in spacetime. This of course breaks the rotation and Lorentz invariance of the theory, which will only be recovered in a suitable limit. Let a be the length of the lattice edges, which is determined by the (here fixed) spacetime flat metric. Associate a *group element* $U_e \in SU(2)$ to each (oriented) edge e of the lattice (and the inverse U_e^{-1} of U_e to the same edge with opposite orientation). Then the set of group elements U_e, for all the edges of the lattice, provides a natural *discretization* of the continuous field A. Wilson's idea is that the quantum theory is better defined starting from these group variables than from the algebra variables. Physical quantities must then be studied in the limit where $N \to \infty$ and $a \to 0$. The different manner in which this limit enters gravity and Yang–Mills theory determines the structural difference between the two theories.

The formal relation between U_e and A is given by the equation

$$U_e = P\, e^{\int_e A}, \tag{4.2}$$

that is: the group element can be seen as the path ordered exponential of the connection along the edge. In quantum gravity parlance, the path ordered exponential is commonly called "holonomy," a term used with slightly different meaning in mathematics. The definition of this object is recalled in the Complements to this chapter.

Expanding in the length a of the edge, this gives, to first order

$$U_e = \mathbb{1} + a A_\mu(s_e)\, \dot{e}^\mu, \tag{4.3}$$

where $\dot{e}^\mu$ is the unit tangent to the edge and s_e is the initial point of e ("source"). The holonomy is invariant under all local gauge transformations $A \to A + D\lambda$ except those at the boundary points of the edge. Therefore, when truncating the theory to the group variables, the local SU(2) gauge symmetry is reduced to a symmetry under SU(2) rotations at the vertices of the lattice only. The gauge group of the lattice theory is therefore $SU(2)^V$, where V is the number of vertices. The group variables transform as

$$U_e \to \lambda_{s_e}\, U_e\, \lambda_{t_e}^{-1}, \qquad \lambda_v \in SU(2) \tag{4.4}$$

under such a gauge transformation. Here s_e and t_e are the initial and final vertices of the edge e (source and target).

The ordered product of four group elements around a plaquette f, namely an elementary square in the lattice

$$U_{\mathsf{f}} = U_{\mathsf{e}_1} U_{\mathsf{e}_2} U_{\mathsf{e}_3} U_{\mathsf{e}_4} \tag{4.5}$$

is a discrete version of the curvature. Its trace is gauge-invariant. Wilson has shown that the discrete action

$$S = \beta \sum_{\mathsf{f}} \operatorname{Tr} U_{\mathsf{f}} + \text{c.c.} \tag{4.6}$$

approximates the continuous action in the limit in which a is small. The coupling constant β is a number that depends on a and goes to zero when a goes to zero.

Consider now a boundary of the lattice. Let l be the boundary edges or "links." If we fix the value of these and integrate the exponential of the action over the bulk groups elements,

$$W(U_\ell) = \int dU_{\mathsf{e}} \, e^{\frac{i}{\hbar} S[U]}, \tag{4.7}$$

we obtain the transition amplitude of the truncated theory. This integral fully defines the theory. The analytic continuation of this in imaginary time can be computed numerically using Monte Carlo methods, since the integrand becomes positive definite. To get the continuous transition amplitudes, one must study the $\beta \to 0, N \to \infty$ continuum limit.

4.1.1 Hamiltonian lattice theory

Of particular interest for us is the hamiltonian formulation of this theory, which was early developed by Kogut and Susskind (1975). The hamiltonian formulation lives on a boundary of the lattice, which we assume to be spacelike. The hamiltonian coordinates are the group elements U_ℓ on the boundary edges. We call the boundary edges "links" ℓ, and the boundary vertices "nodes" n, for reasons that will become clear later on. These group elements correspond to the space components of the connection on the boundary, namely (in the time gauge) they code the magnetic field. In lagrangian language its time component, or electric field, is coded by the edges normal to the boundary, while in hamiltonian language it is coded in the momentum conjugate to the boundary group elements.

The canonical configuration space is therefore $\mathrm{SU}(2)^L$, which is a group, and the reduced gauge is given by the gauge transformations at the nodes on the boundary. Here L is the number of links. The corresponding phase space is the cotangent space $T^*\mathrm{SU}(2)^L$. There is one conjugate momentum L_ℓ^i in the algebra of $\mathrm{SU}(2)$ for each link ℓ, and it is identified with the electric field. A cotangent space T^*Q carries a natural symplectic structure (see footnote in Section 2.5.2). The symplectic structure of a space cotangent to a group is well studied. The corresponding Poisson brackets are

$$\{U_\ell, U_{\ell'}\} = 0 \,, \tag{4.8}$$

$$\{U_{\ell'}, L_\ell^i\} = \delta_{\ell\ell'} \, U_\ell \tau^i \,, \tag{4.9}$$

$$\{L_\ell^i, L_{\ell'}^j\} = \delta_{\ell\ell'} \, \epsilon^{ij}{}_k L_\ell^k \tag{4.10}$$

(no summation over ℓ). The Hilbert space of the discrete theory can therefore be represented by states $\psi(U_\ell)$, functions on the configuration space. The space of these states carries a natural scalar product, which is invariant under the gauge transformations on the boundary: the one defined by the SU(2) Haar measure

$$(\phi, \psi) = \int_{SU(2)} dU_\ell \, \overline{\phi(U_\ell)} \psi(U_\ell). \tag{4.11}$$

The boundary gauge transformations act at the nodes of the boundary and transform the states as follows

$$\psi(U_\ell) \to \psi(\lambda_{s_\ell} U_\ell \lambda_{t_\ell}), \qquad\qquad \lambda_n \in SU(2). \tag{4.12}$$

The states invariant under this transformation form the Hilbert space of the gauge-invariant states, which has the structure $L_2[SU(2)^L / SU(2)^N]$, where L is the number of links in the boundary and N the number of nodes on the boundary.

The operator corresponding to the group elements themselves is diagonal in this basis. The operator that corresponds to the quantity conjugate to U_ℓ is the natural derivative operator on functions of SU(2), namely, the left-invariant vector field $\vec{L}_\ell$. In the classical theory, the conjugate momentum to the connection is the electric field, namely the time–space component of the field strength. Therefore $\vec{L}_\ell$ can be identified with this quantity (more precisely, with the flux of the electric field across an elementary surface dual to link ℓ). These operators realize the Poisson algebra above.

All these structures will reappear in loop quantum gravity.

4.2 Discretization of covariant systems

Before moving to gravity, let us study what happens to the classical and the quantum theory when we discretize a *covariant* system (in the sense of Chapter 2).

In Section 2.2.1 we have seen that a quantum theory can be defined by a functional integral, and this can be defined as the limit of a multiple integral, namely, the limit of a discretization of the theory. See Eq. (2.39).

$$W(q, t, q', t') = \lim_{N \to \infty} \int dq_n \, e^{\frac{i}{\hbar} S_N(q_n)}, \tag{4.13}$$

where the discretized action at level N is

$$S_N(q_n) = \sum_{n=1}^{N} \frac{m(q_{n+1} - q_n)^2}{2\epsilon} - \epsilon V(q_n). \tag{4.14}$$

Since $\epsilon = (t' - t)/N$, the parameter N appears in this expression twice: in the number of steps and in the size of the time step. The second can be in part traded by a change of integration variables, but not completely. Redefining $q_n \to q_n/\sqrt{\epsilon}$, we can replace

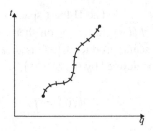

Figure 4.1 Discretization of the parameter time.

$\mathcal{S}_N(q_n)$ by

$$\mathcal{S}_N(q_n) = \sum_{n=1}^{N} \frac{m(q_{n+1} - q_n)^2}{2} - \epsilon V(\sqrt{\epsilon} q_n). \tag{4.15}$$

For concreteness, for instance, consider the simple case where $V(q) = \frac{1}{2}\omega^2 q^2$, namely, a harmonic oscillator. Then

$$\mathcal{S}_N(q_n) = \sum_{n=1}^{N} \frac{m(q_{n+1} - q_n)^2}{2} - \frac{\Omega^2}{2} q_n^2, \tag{4.16}$$

where $\Omega = \epsilon\omega$. In this manner ϵ has formally disappeared from the discrete action, but in reality it is hidden into Ω. To take the $N \to \infty$ limit we have to do two things: send the upper limit of the sum to ∞, as well as sending Ω to its critical value $\Omega_c = 0$.

This is well known to anybody who deals with discretized systems: the continuum limit is recovered by increasing the number of "lattice sites" and *also* scaling down the lattice spacing to zero, or, equivalently, sending appropriate constants (here Ω) to a critical limit value (here zero). This seems to be a universal feature of discretization and it is often presented as such.

But it is not.

Something remarkable happens on (appropriately) discretizing covariant systems.

Consider a discretization of a parametric system, like the one defined by the action described in Section 2.3,

$$S = \int d\tau \left(\frac{m}{2}\frac{\dot{q}^2}{\dot{t}} - iV(q) \right), \tag{4.17}$$

which is physically equivalent to the system above. Discretize this action in the natural manner. This gives

$$S = \sum_{n=1}^{N} \epsilon \left(\frac{m}{2} \frac{\frac{(q_{n+1}-q_n)^2}{\epsilon^2}}{\frac{(t_{n+1}-t_n)}{\epsilon}} - \frac{(t_{n+1} - t_n)}{\epsilon} V(q_n) \right). \tag{4.18}$$

A look at this is sufficient to show that

$$S = \sum_{n=1}^{N} \frac{m}{2} \frac{(q_{n+1} - q_n)^2}{(t_{n+1} - t_n)} - (t_{n+1} - t_n)V(q_n). \tag{4.19}$$

The parameter ϵ cancels!

At first this looks like magic: the discretization of the lagrangian does not depend on the "lattice spacing"! The number N appears *only* in the upper limit of the sum. But a moment of reflection can convince the dubious reader that it has to be so. The original action did not depend on the parametrization, so its discretized version will not either.

Notice that the path integral (Eq. (2.79)) is now both on the q_n's and the t_n's. In a sense, from the newtonian perspective, we are "integrating over the positions of the time steps of the discretization."

As we have seen in Chapter 2, the equations of motion of the continuous theory are not independent. It is easy to check that the equations of motion of the discretized theory lose this dependence: they become independent. The equation for the variable t, which gives energy conservation in the continuum, fixes the values of the t_n in such a way that the energy remains conserved also in the discretization. It is only in the $N \to \infty$ limit that reparametrization invariance is restored.

Restoring of invariances in the limit in which a discretization is removed is a general feature of discretizations. For instance, the discretization of a rotationally invariant system typically loses rotational invariance, and this is recovered in the limit.

The reparametrization-invariant systems are characterized by the fact that there is no parameter to rescale in the discrete action in the limit: the only place where N appears is in the number of steps.

This peculiar behavior of the discretization of reparametrization-invariant systems can be found in the "good" discretizations of GR. While the Wilson action used in lattice QCD depends on a parameter that has to be scaled to its critical value in order to gain the continuous limit, there is no similar parameter in Regge calculus, illustrated below, which is a natural discretization of GR. It is only the number of steps N (or cells) that has to be increased, to get a better approximation of the theory, without any other parameter to tune.

This is a profound difference between the discretization of theories on a metric space and (good) discretizations of general covariant theories. It is at the root of the peculiarity of the covariant form of quantum gravity.

4.3 Regge calculus

Tullio Regge introduced a very elegant discretization of general relativity, which is called "Regge calculus." It can be defined as follows. A d-simplex is a generalization of a triangle or a tetrahedron to arbitrary dimensions: formally, the convex hull of its $d + 1$ vertices. These vertices are connected by $d(d + 1)/2$ line segments[1] whose lengths L_s fully specify the shape of the simplex – that is to say its metric geometry. For example, a simplex of dimension $d = 3$ (a tetrahedron) has $d + 1 = 4$ vertices connected by 6 line segments whose lengths $L_1, \ldots, L_6$ fully determine its shape.

---

[1] The segments are sometimes called "edges" in the Regge-calculus literature. Here we prefer to use "segment," in order to avoid confusion with the edges of the dual complex, introduced below.

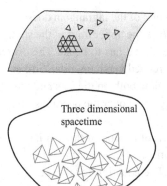

A Regge space (M, L_s) in d dimensions is a d-dimensional metric space obtained by gluing d-simplices along matching boundary $(d-1)$-simplices.

For instance, in two-dimensional we can obtain a surface by gluing *triangles*, bounded by *segments*, which meet at *points*.

In three-dimensional we chop space into *tetrahedra*, bounded by *triangles*, in turn bounded by *segments*, which meet at *points*.

In four-dimensional we chop spacetime into *4-simplices*, bounded by *tetrahedra*, in turn bounded by *triangles*, in turn bounded by *segments*, which meet at *points*.

These structures are called triangulations (see Table 4.1). We assume always that the triangulation is oriented. This means that we conventionally assign a direction to each segment, a preferred side to each triangle and a preferred cyclic ordering to the vertices of each tetrahedron, and so on; this is just to simplify the construction and the notation.

A moment of reflection shows that gluing flat d-simplices can generate curvature on the $(d-2)$ simplices (sometimes called "hinges"). For instance, in $d = 2$ dimensions, we can glue four equilateral triangles as in the boundary of a tetrahedron, and there is clearly curvature on the vertices of the tetrahedron. In $d = 3$ dimensions, we can glue several tetrahedra all around a common segment, and obtain a manifold flat everywhere except at this segment. In $d = 4$ dimensions, curvature is on the triangles. The metric of the resulting space is uniquely determined by the length L_s of all its segments s. This follows immediately from the fact that the metric of any d-simplex is uniquely determined by the length of its sides.

Now, it is easy to see that a Riemannian manifold (M, g) can be approximated arbitrarily well by a Regge manifold. This means that for any (M, g) and any ϵ, we can find an (M, L_s) (with sufficiently many simplices) such that for any two points x and y in M the difference between the Riemaniann distance and the Regge distance is less than ϵ.

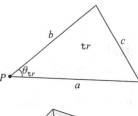

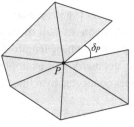

This is the basis for the approximation of curved surfaces by triangulations used for instance in industry. Regge's idea is to approximate general relativity by a theory of Regge manifolds.

For this, we need a notion of curvature for a Regge manifold, which converges appropriately to Riemann's curvature. Regge's notion of curvature is very beautiful. Consider first the case $d = 2$, which is simple. Consider a point P of the triangulation. Around it there are a certain number of triangles tr. Let the angles at P of these triangles be θ_{tr}.

These angles can be immediately computed from the lengths, using the elementary-geometry formula

$$\cos(\theta_{tr}) = \frac{c^2 - a^2 - b^2}{2ab}, \tag{4.20}$$

which gives the angle opposite to c of a triangle with sides $a, b,$ and c.

Table 4.1 Triangulations

Two-dimensional			Triangle	Segment	**Point**
Three-dimensional		Tetrahedron	Triangle	**Segment**	Point
Four-dimensional	4-Simplex	Tetrahedron	**Triangle**	Segment	Point

Now, if the angles around P sum up to 2π, clearly the manifold is flat at P. If not, there is curvature. Therefore the Regge curvature at P can be defined as the angle

$$\delta_P(L_s) = 2\pi - \sum_t \theta_t(L_s), \tag{4.21}$$

called the "deficit angle" at P. In this formula, the sum is over the triangles around P and θ_t is the angle at P of the triangle t.

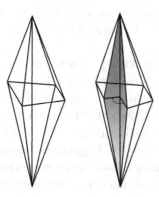

The same logic can be used in higher dimensions. In d dimensions, the Regge curvature is still given by a single deficit angle, but now P is replaced by a $(d-2)$ simplex (a segment in three-dimensional and a triangle in four-dimensional; see Table 4.1), the sum is over the $(d-1)$ simplices around it (triangles in three-dimensional, tetrahedra in four-dimensional) and the angles become the dihedral angles of the flat d-simplices, which can be computed from the sides using formulas of elementary geometry. For instance, the dihedral angle on the ab side of a tetrahedron with vertices (a,b,c,d) is

$$\cos\theta_{ab} = \frac{\cos\theta_{acd} - \cos\theta_{abc}\cos\theta_{abd}}{\sin\theta_{abc}\sin\theta_{abd}} \tag{4.22}$$

where θ_{abc} is the angle at the vertex a of the triangle of vertices a, b, c.

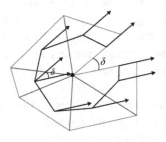

The geometrical interpretation of the deficit angle is simple. If we parallel transport a vector in a loop around a $d-2$ simplex, the vector gets back-rotated by the deficit angle (loops wrap points in two-dimensional, segments in three-dimensional, surfaces in four-dimensional). Using this, Regge defines the Regge action of a Regge manifold (M, L_s) to be

$$S_M(L_s) = \sum_h A_h(L_s)\delta_h(L_s), \tag{4.23}$$

where the sum is over the $(d-2)$ simplices of the triangulation, or "hinges" (points P in two-dimensional, segments in three-dimensional, triangles in four-dimensional) and A_h is the $(d-2)$-volume of h (length in three-dimensional, area in four-dimensional). The remarkable result obtained by Regge is then that the Regge action (4.23) converges to the Einstein–Hilbert action $S[g]$ when the Regge manifold (M, L_s) converges to the Riemann manifold (M,g). This tells us that the Regge theory is a good discretization of general relativity.

The Regge equations of motion are obtained by varying the action with respect to the length. This gives two terms: the first from the variation of the $A's$ and the second from the variation of the angles. The second term vanishes. This is a general feature of gravity theories: it is the analog of the fact that the variation of the Riemann tensor in the Einstein–Hilbert action vanishes. Thus, the Regge equations are

$$\sum_h \frac{\partial A_h}{\partial L_s} \delta_h(L_s) = 0, \qquad (4.24)$$

where the sum is over the hinges adjacent to the segment s, and there is one equation per segment. We can identify $\delta_h(L_s)$ with a measure of the discrete Riemann curvature and the left-hand side of the equation of motion as a measure of the discrete Ricci tensor.

In three spacetime dimensions the hinges are the same things as the segments and therefore the Regge equations reduce to $\delta_h(L_s) = 0$, that is: flatness, as in the continuum case.

There is a peculiarity of the Regge discretization that plays a role in what follows. The Regge curvature is concentrated at the hinges. If we parallel transport a vector around a hinge in a Regge manifold, the vector comes back to its original position rotated: this is the standard manifestation of curvature. However, a moment of reflection will convince the reader that the only rotation that can result is around an axis parallel to the hinge. That is, the rotation never rotates the hinge itself. This is why the Regge curvature is captured by a single number, the deficit angle: because it is restricted to rotations around the hinge itself. Therefore the Riemannian curvature of a Regge manifold is a curvature concentrated on the plane normal to the hinge and generating a rotation on this same plane. If the plane normal to the hinge is the (x^1, x^2) plane, for instance, the only non-vanishing component of the Riemann tensor would be $R^{12}{}_{12}$.

A curvature with these algebraic restrictions is very special. But since the physical curvature in a region is obtained by averaging the Regge curvature over the region, and since the special feature of the Regge curvature is not a linear property, the average curvature is not going to inherit this special property. Thus a Regge geometry can still approximate a generic Riemannian space. But it is important to remember that it does so with a rather restricted kind of curvature.

The action can also be rewritten as a sum over the d-simplices v of the triangulation. From (4.23) and the definition (4.21) of deficit angle, we have in fact

$$S_M(L_s) = 2\pi \sum_h A_h(L_s) - \sum_v S_v(L_s), \qquad (4.25)$$

where the action of a d-simplex is

$$S_v(L_s) = \sum_h A_h(L_s)\theta_h(L_s). \qquad (4.26)$$

This becomes particularly useful if the quantities A_h are integers (as will happen later), in which case $e^{2\pi i \sum_h A_h} = 1$, or half integers, in which case this term gives only a sign factor in a path integral.

If we fix the triangulation Δ, we obtain only a finite approximation to Riemannian manifolds and to general relativity. This defines a *truncation* of general relativity. It is the analog of replacing a wave equation with a finite difference equation, or QCD with lattice QCD.

Observe the difference between the lattice discretization of Yang–Mills theory and the Regge discretization of general relativity: in the first (Section 4.1), the continuum limit is obtained by taking the number of vertices N to infinite *as well as* the coupling constant β (or, equivalently, the lattice spacing a) to zero. Not so for the Regge discretization. Here the continuum limit is obtained by refining the triangulation, with no parameter to be taken to zero. This is a consequence of the covariant character of the theory as explained in Section 4.2. The difference between lattice QCD and Regge calculus is therefore deeply rooted in the general covariance of general relativity. The physical size (length) of the segments of the Regge triangulation are the dynamical variables themselves, and not a fixed external quantity that we can scale in a specific limit. The continuum limit of a proper discretization of a general-covariant theory does not require any parameter, besides the number of lattice steps, to be taken to a limit.

4.4 Discretization of general relativity on a two-complex

The Regge discretization is not very good for the quantum theory, for two reasons. First, it is based on the metric variables. The existence of fermions indicates that we need tetrads at the fundamental level. More importantly, the segments of a Regge triangulation are subjected to triangular inequalities: a segment connecting two points P and Q cannot be longer than the sum of two segments connecting P and R and Q and R. Therefore the configuration space of the theory, which is the set of segment's lengths that satisfies these inequalities, is a complicated space with boundaries, and this makes the search of a quantum theory with this classical limit far more complicated.

The tetrad-connection formulation of general relativity offers an alternative, and leads to a discretization of the theory which is closer to that of Yang–Mills theory, while at the same time retaining the specific feature of a discrete covariant theory. This discretization of general relativity is better described in terms of 2-complexes, which bridge between a Yang–Mills lattice and a Regge triangulation. We introduce here this discretization of general relativity, which we use extensively in the following.

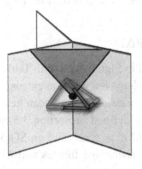

In this chapter we give the basic construction in three space-time dimensions, and we restrict to the euclidean theory for simplicity, namely the theory presented in Section 3.5. The discretization of the physical four-dimensional lorentzian theory (which is a straightforward generalization) is given later in Chapter 7.

The key notion we need is the *dual* of a triangulation, illustrated in the diagram, where the tetrahedron belongs to the original triangulation, while the grey faces meeting along edges, in turn meeting at points, form the dual. More precisely,

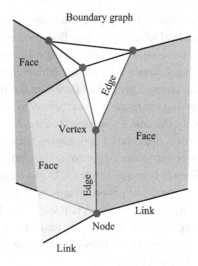

Figure 4.2   Triangulation and 2-complex terminology and notation in three-dimensional.

let Δ be a three-dimensional triangulation. The dual Δ^* of the triangulation Δ is defined as follows. It is obtained by placing a *vertex* within each tetrahedron, joining the vertices of two adjacent tetrahedra by an *edge* "dual" to the triangle that separates the two tetrahedra, and associating a *face* to each segment of the triangulation, bounded by the edges that surround the segment. These objects inherit an orientation from their dual.

The set of vertices, edges and faces, with their boundary relations, is called a *2-complex*. A 2-complex can be visualized by a set of faces meeting at edges, which in turn meet at vertices. What defines the 2-complex is the combinatorial structure of the boundary relations between these elements.

Boundary

The discretization of the bulk of a spacetime region $\mathcal{R}$ induces a discretization of the boundary $\Sigma = \partial \mathcal{R}$ of this region. The boundary Σ is discretized by the boundary triangles of Δ, separated by the boundary segments of Δ. The end points of the edges dual to these triangles are called *nodes*, or boundary vertices, and the boundaries of the faces dual to these segments are called *links*, or boundary edges, and together they form the *graph* Γ of the boundary. The boundary graph is at the same time the boundary of the 2-complex and the dual of the boundary of the triangulation:

$$\Gamma = \partial(\Delta^*) = (\partial \Delta)^*. \tag{4.27}$$

This terminology is important. It is summarized in Figure 4.2 and in Tables 4.2 and 4.3. Study it carefully because we will use this terminology and notation extensively.

We now use this construction to define a discretization of euclidean general relativity in three euclidean dimensions, namely, the theory defined in Section 3.5, where, recall, the gravitational field is described by a triad field $e^i = e^i_a dx^a$ and an SO(3) connection $\omega^i{}_j = \omega^i_{aj} dx^a$, where indices $a, b, \ldots = 1, 2, 3$ are spacetime indices and $i, j = 1, 2, 3$ are

Table 4.2 Bulk terminology and notation	
Bulk triangulation Δ	Two-complex Δ^*
Tetrahedron (τ)	Vertex (v)
Triangle (t)	Edge (e)
Segment (s)	Face (f)
Point (p)	

Table 4.3 Boundary terminology and notation	
Boundary triangulation $\partial\Delta$	Boundary graph Γ
Triangle (t)	Node, (n) (boundary vertex)
Segment (s)	Link, (ℓ) (boundary edge)

internal indices raised and lowered with the Kronecker delta δ_{ij}. To do so, we introduce discrete variables on the 2-complex.

We discretize the connection, as in Yang–Mills theory, by using it to assign an SU(2) group element U_e to each edge e of the 2-complex. We discretize the triad by associating a vector L_s^i of $\mathbb{R}^3$ to each segment s of the original triangulation (Figure 4.3).

$$\omega \longrightarrow U_e \qquad (4.28)$$

$$e \longrightarrow L_s^i . \qquad (4.29)$$

The formal relation between the continuous and the discrete variables can be taken to be the following:

$$U_e = P \exp \int_e \omega \in SU(2) , \qquad (4.30)$$

$$L_s^i = \int_s e^i \in \mathbb{R}^3 . \qquad (4.31)$$

U_e is the holonomy of the connection along the edge, namely, the matrix of the parallel transport generated by the connection along the edge, in the fundamental representation of SU(2). This association of a group variable to an edge is the same as in lattice gauge theory. L_s^i is the line integral of the 1-form e^i along the segment. The action (3.98) can be approximated in terms of these objects.

The definition (4.31) is somewhat imprecise, because of the gauge. Under a gauge transformation (3.99) the group elements U defined in (4.30) transform "well," namely, as

$$U_e \mapsto R_{s_e} U_e R_{t_e}^{-1} \qquad (4.32)$$

where s_e and t_e are the initial ("source") and final ("target") vertices of the edge e (recall that all the discrete structures are

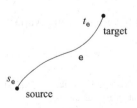

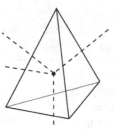

Figure 4.3 Group elements are on the edges (dotted lines). Algebra elements on the segments (continuous lines), or, equivalently, in their dual faces

oriented), and $R_v \in SO(3)$. Therefore in the discrete theory the continuous local SO(3) invariance is just reduced to rotations at the vertices. Not so for the algebra variables L^i defined in (4.31). To correct this,[2] assume that this definition is taken in a gauge where the connection is constant along the segment itself (at the possible price of being distributional at the boundaries of the segment), as well as along the first half of each edge (edges are oriented). In this way, also the L^i_s variables are invariant under all gauge transformations except those at the vertices; and they transform covariantly, in fact, in the adjoint representation, under rotation R_v at the vertex v.

The discretization approximates well the continuum theory when the curvature is small at the scale of the triangulation and the segments are straight lines. Then notice that the *norm* of the $\mathbb{R}^3$ element L^i_s associated with the segment s is the length L_s of the segment

$$L^2_s = |\vec{L}_s|^2 \, . \tag{4.33}$$

Since each face f of the 2-complex corresponds to a segment $s = s_f$ of the triangulation, we can view L^i_s as associated with the face: $L^i_f = L^i_{s_f}$. Also, it is convenient to view this as an element of the su(2) algebra. That is, define

$$L_f = L^i_f \tau_i. \tag{4.34}$$

Therefore the variables of the discretized theory are

- An SU(2) group element U_e for each edge e of the 2-complex
- An su(2) algebra element L_f for each face f of the 2-complex

In the four-dimensional lorentzian theory we will have these same variables, associated with edges and faces.

---

[2] There is a more precise mathematical way of doing this, which is to give a gauge equivalent definition of the quantities L^i by parallel transporting the triad to the same point; see Thiemann (2007).

Action

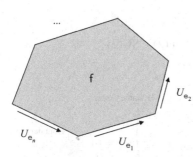

Given a face f bounded by the edges $e_1, \ldots, e_n$, we can multiply the group elements U_e around it and obtain a single group element U_f associated with the face itself.

$$U_f = U_{e_1} \ldots U_{e_n} . \tag{4.35}$$

This is the holonomy of the connection going around the segment s_f dual to the face f. If U_f is different from the identity $U_f \neq \mathbb{1}$, then there is curvature, and we can associate this curvature to the segment, as in Regge calculus.

Then the discretized action reads

$$\mathcal{S} = \frac{1}{8\pi G} \sum_f \mathrm{Tr}(L_f U_f), \tag{4.36}$$

where $L_f \in su(2)$ and $U_f \in SU(2)$. On the boundary, we must close the perimeter of the faces in order to write the quantity U_f for the faces that end on the boundary, and write the action. Therefore there must also be group quantities U_ℓ associated with the links of the boundary. In other words, the links of the boundaries, which are the boundary edges, are on the same footing as the bulk edges, and are assigned group variables as well. Using the relation (1.39) it is easy to verify that the variation of L_f in the action (4.36) gives $U_f = \mathbb{1}$; that is, flatness, which is equivalent to the continuous Einstein equations in three-dimensional.

The discretization of general relativity described in this section is similar, but not exactly equivalent, to the Regge discretization. One difference is that it can be coupled to fermions. A second difference is given by the sign discussed in Section 3.2.1.

Boundary variables

Let us study in more detail the boundary variables of the discrete theory, which will play an important role in the quantum theory. (See Figure 4.4.) These are of two kinds: the group elements U_ℓ of the boundary edges, namely, the links; and the algebra elements L_s of the boundary segment s. Notice that there is precisely one boundary segment s per each link ℓ, and the two cross. We can therefore rename L_s as L_ℓ, whenever ℓ is the link crossing the

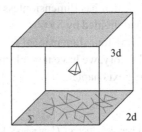

Figure 4.4 A triangulated region of three-dimensional spacetime and its two-dimensional triangulated boundary.

boundary segment s. Therefore the boundary variables are formed by one couple

$$(L_\ell, U_\ell) \in \mathrm{su}(2) \times \mathrm{SU}(2) \tag{4.37}$$

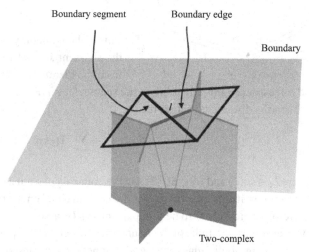

for each link ℓ of the graph. Since the algebra is the cotangent space of the group, we can write $\mathrm{su}(2) \times \mathrm{SU}(2) \sim T^*\mathrm{SU}(2)$. Thus the classical boundary phase space of the discretized theory is $T^*\mathrm{SU}(2)^L$, where L is the number of links of the graph. This is precisely the same boundary phase space as for an $\mathrm{SU}(2)$ Yang–Mills theory on a lattice. The Poisson brackets are the same as in the previous section:

$$\{U_\ell, U_{\ell'}\} = 0 \,,$$
$$\{U_{\ell'}, L_\ell^i\} = (8\pi G) \,\delta_{\ell\ell'} \, U_\ell \tau^i \,,$$
$$\{L_\ell^i, L_{\ell'}^j\} = (8\pi G) \,\delta_{\ell\ell'} \,\epsilon^{ij}{}_k L_\ell^k \,, \tag{4.38}$$

namely, the Poisson brackets (4.9) of an $\mathrm{SU}(2)$ Yang–Mills theory on the lattice.[3] The dimensional constant $8\pi G$ is determined by the action, because the momentum conjugate to the connection (which has dimensionless Poisson brackets with the connection) is not the triad but the triad divided by $8\pi G$.

We shall discuss the four-dimensional version of this construction in Chapter 7. For the three-dimensional theory, we have now all the ingredients to move to the quantum theory, which we do in the next chapter.

---

[3] With the gauge-invariant definition of L_ℓ^i mentioned in the previous footnote, these Poisson brackets can be derived from the continuous ones (3.102).

4.5 Complements

4.5.1 Holonomy

Given a connection A in a group G on a manifold, and a path γ in the manifold, the path ordered exponential

$$U_\gamma = P \, e^{\int_\gamma A} \tag{4.39}$$

is defined as the value for $s = 1$ of the solution of the differential equation

$$\frac{d}{ds} U(s) = \dot{\gamma}^a(s) \, A_a(\gamma(s)) \tag{4.40}$$

with initial condition $U(0) = \mathbb{1}$. Here $s \in [0, 1]$ is an arbitrary parameter along the path γ and $\dot{\gamma}^a = d\gamma^a/ds$ is the tangent of γ at s.

This is also the limit of large N of

$$U_N = \prod_n (1 + \dot{\gamma}^a(x_n) A_a(x_n) \Delta\tau), \tag{4.41}$$

where x_n are N ordered points along γ, at distance $\Delta\tau$.

Geometrically, $U_\gamma \in G$ is the group transformation obtained by following the path γ and rotating as dictated by the connection A. In other words, it is the rotation that undergoes a vector which is covariantly constant along γ.

The reason for the notation with P (path ordered) is that if we expand the exponential in the right-hand side of (4.39) in a power series in A and order each multiple integral by the order along γ, we obtain the left-hand side. That is, we can also write

$$U_\gamma = \mathbb{1} + \int_\gamma A + \frac{1}{2} P \int_\gamma A \int_\gamma A + \cdots, \tag{4.42}$$

where

$$P \int_\gamma A \int_\gamma A = 2 \int_0^1 ds \int_0^s dt \, \dot{\gamma}^b(s) A_b(\gamma(s)) \dot{\gamma}^a(t) A_a(\gamma(t)) \tag{4.43}$$

(A does not commute with itself) and so on. The group element U_γ is often called "holonomy" in the quantum gravity literature, although the relation with other uses of "holonomy" is only partial. If A is the Levi-Civita connection, its holonomy gives the standard parallel transport of vectors in general relativity. If A is the self-dual Ashtekar connection, its holonomy gives the parallel transport of a left-handed neutrino along γ.

The key property of the holonomy is that it transforms "nicely" under gauge transformations: it is insensitive to gauge transformations localized in the bulk of γ and transforms homogeneously for gauge transformations on the end points of γ. That is, if we gauge transform the connection by a gauge transformation with parameter $\Lambda(x)$, the holonomy transforms as

$$U_\gamma \to \Lambda(\gamma(0)) \, U_\gamma \, \Lambda^{-1}(\gamma(1)). \tag{4.44}$$

Finally, important properties of the holonomy are its variations under arbitrary variation of the connection or γ. A straightforward calculation gives for the first

$$\frac{\delta}{\delta A_a^i(x)} U_\gamma = \int_\gamma ds \, \delta^3(x, \gamma(s)) \, \dot{\gamma}^a(s) \, U_{\gamma_1} \tau^i U_{\gamma_2}, \tag{4.45}$$

where γ_1 and γ_2 are the two portions in which x cuts γ; and for the second

$$\frac{\delta}{\delta\gamma^a(s)}U_\gamma = U_{\gamma_1}F_{ab}(\gamma(s))\dot{\gamma}^b(s)U_{\gamma_2}. \tag{4.46}$$

In particular, an infinitesimal change of parametrization of $s \to f(s)$ where $f(s)$ vanishes at the end points gives

$$\int ds\, f(s)\dot{\gamma}^a(s)\frac{\delta}{\delta\gamma^a(s)}U_\gamma = \int ds\, f(s)\, U_{\gamma_1}F_{ab}(\gamma(s))\dot{\gamma}^a(s)\dot{\gamma}^b(s)U_{\gamma_2}, \tag{4.47}$$

which vanishes because of the contraction between the antisymmetric F_{ab} with the symmetric $\dot{\gamma}^a\dot{\gamma}^b$.

4.5.2 Problems

1. **Equations of motion.** Compute the equations of motion of the discretization of the covariant formulation of a newtonian system. Is energy conserved?

2. **Important: do this well.** Consider a tetrahedron. Place a point P in its interior and connect P with the four vertices of the tetrahedron. Let the resulting ten segments determine a Regge space. Carefully describe the corresponding triangulation Δ and its dual Δ^*, listing all vertices, edges, and faces, distinguishing those that are similar or are of different kinds. Write the boundary graph. What is the relation between the boundary graph and the original tetrahedron? Consider then the group elements associated with all the edges of Δ^*. Find a good notation to label these. How many conditions must these satisfy in order for the connection to be flat? (Careful: there is a bubble...)

3. **Gluing two complexes.** Consider two spacetime regions connected by a common portion of their boundary. Let both of them be triangulated, matching at the boundary. Describe how the dual triangulations match at the boundary. What happens to nodes and links? Build a concrete example, giving all the vertices, edges, faces, nodes, and links explicitly.

4. **Holonomy.** Show that the various definitions of the holonomy given above are equivalent. Derive (4.45), (4.46), and (4.47) from the formal definition of the holonomy.

5. **Curvature** (difficult). What is the relation between the Regge curvature defined by the deficit angle δ_h and the curvature defined by U_f? How are these two quantities related?

PART II

THREE-DIMENSIONAL THEORY

5 Three-dimensional euclidean theory

In order to find a quantum theory for the gravitational field, we can apply the general structure devised in Chapter 2 to the theory defined in Chapter 3, discretized as explained in Chapter 4. Since the resulting theory includes a number of technical complications, in this chapter we first complete a simpler exercise: apply this same strategy to euclidean general relativity in three spacetime dimensions, recalled in Section 3.5. This exercise allows us to introduce a number of ideas and techniques that will then be used in the physical case, which is the lorentzian theory in four spacetime dimensions.

There are major differences between general relativity in three dimensions and four dimensions. The foremost is that the 3d theory does not have local degrees of freedom: it is therefore infinitely simpler than the 4d theory. This is reflected in properties of the quantum theory that do not hold in the 4d theory. In spite of these differences, the exercise with the 3d theory is illuminating and provides insight into the working of quantum gravity, because in spite of its great simplicity the 3d theory *is* a generally covariant quantum theory of geometry.

5.1 Quantization strategy

To define the quantum theory we need two ingredients:

- A boundary Hilbert space that describes the quantum states of the boundary geometry.
- The transition amplitude for these boundary states. In the small $\hbar$ limit the transition amplitude must reproduce the exponential of the Hamilton function.

To construct the Hilbert space and the transition amplitude, we proceed in steps. First we discretize the classical theory. Then we study the quantum theory that corresponds to the discretized theory. Finally we discuss the continuum limit. We use the discretization of the theory both to construct its Hilbert space, following the ideas sketched in Chapter 1, and to define the transition amplitudes, following the strategy sketched in Chapter 2.

Recall from the previous chapter that three-dimensional general relativity can be truncated on a 2-complex by associating an SU(2) group element U_e to each edge e and an su(2) algebra element L_f to each face f of the dual triangulation.

$$\omega \longrightarrow U_e, \tag{5.1}$$

$$e \longrightarrow L_f. \tag{5.2}$$

The first is interpreted as the holonomy of gravitational connection along the edge, the second as the line integral of the triad along the segment dual to the face f. The action is given in (4.36), namely,

$$ S = \frac{1}{8\pi G} \sum_f \text{Tr}(L_f U_f), \tag{5.3} $$

where U_f is the product of the U_e of the edges around f. Gauge transformations act as

$$ U_e \rightarrow \lambda_{s_e} U_e \lambda_{t_e}^{-1}. \tag{5.4} $$

The 2-complex is bounded by a graph Γ where the boundary variables are a group element and an algebra element

$$ (U_\ell, L_\ell) \tag{5.5} $$

for each link ℓ of the graph Γ. These coordinatize the phase space of the discrete theory, which is the cotangent space to $\text{SU}(2)^L$.

This is the discrete theory. Let us move to the quantum theory.

5.2 Quantum kinematics: Hilbert space

The quantization of a phase space that is the cotangent space to a group is standard. We are seeking operators U_ℓ and L_ℓ realizing the quantum version of (4.38), and in particular

$$ [U_\ell, L_{\ell'}^i] = i\,(8\pi\hbar G)\,\delta_{\ell\ell'}\, U_\ell \tau^i \tag{5.6} $$

on a Hilbert space. This is a basis requirement for the theory in order to have the correct classical limit. Consider the group elements U_ℓ to be the "coordinates" and the algebra elements L_ℓ to be "momenta." Let the Hilbert space be the space of square integrable functions of the coordinates:

$$ \mathcal{H}_\Gamma = L_2\left[\text{SU}(2)^L\right]. \tag{5.7} $$

The index Γ reminds that this comes from a discretization that induces the graph Γ on the boundary. States are wavefunctions $\psi(U_\ell)$ of L group elements U_ℓ. The scalar product compatible with the SU(2) structure is given by the group-invariant measure, which is the Haar measure dU (see Complements to Chapter 1),

$$ \langle\psi|\phi\rangle = \int_{\text{SU}(2)^L} dU_\ell\,\overline{\psi(U_\ell)}\phi(U_\ell)\,. \tag{5.8} $$

The (self-adjoint) natural SU(2)-covariant derivative operator on the functions $\psi(U_\ell)$ is the left invariant vector field. This is defined as

$$ (J^i\psi)(U) \equiv -i\frac{d}{dt}\psi(Ue^{t\tau_i})\bigg|_{t=0}. \tag{5.9} $$

To get a triad operator satisfying (5.6) it sufficient to scale this with the appropriate dimensionful factor:

$$\vec{L}_\ell \equiv (8\pi\hbar G)\,\vec{J}_\ell, \tag{5.10}$$

where $\vec{J}_\ell$ is the left invariant vector field acting on the argument U_ℓ. This realizes the commutation relations (5.6).

Summarizing: we have the Hilbert space $\mathcal{H}_\Gamma = L_2\left[\mathrm{SU}(2)^L\right]$, where L is the number of links of Γ, and operators associated with the discrete variables U_ℓ and L_ℓ: U_ℓ is a multiplicative operator and the triad is $\vec{L}_\ell = 8\pi\hbar G\vec{J}_\ell$.

This is enough to start doing physics!

5.2.1 Length quantization

Recall that the length L_s of a segment is given by the norm $L_s = |\vec{L}_f|$ where f is the face dual to the segment. Therefore length of a boundary segment s is given by the operator $L_\ell = |\vec{L}_\ell|$ where ℓ is the link crossing s. Let us study the spectral properties of this operator. Since $\vec{J}$ is a generator of SU(2), its square $|\vec{J}|^2$ is the SU(2) Casimir, namely the same mathematical quantity as the total angular-momentum operator; its eigenvalues are $j(j+1)$ for half-integer j's. Therefore the spectrum of the length of a segment ℓ is[1]

$$L_{j_\ell} = 8\pi\,\hbar G\,\sqrt{j_\ell(j_\ell + 1)} \tag{5.11}$$

for half-integers j_ℓ. This is the first physical consequence of the three-dimensional theory: *length* is quantized.

Do not confuse the classical discreteness introduced by the triangulation with the quantum discreteness determined by the fact that the length operators have discrete spectrum. The first is classical, and is just a truncation of the degrees of freedom. It is the analog of expanding an electromagnetic field in a box in modes and truncating the theory to a finite number of modes. The second is quantum mechanical. It is the analog of the fact that the energy of each mode of the electromagnetic field is quantized: namely the field is made up by quanta.

In particular, the quantum discreteness that we find is not related to the discretization introduced in other tentative approaches to quantum gravity, such as quantum Regge calculus or dynamical triangulations. In that context, spacetime is discretized by means of a triangulation of fixed size, but this size is considered unphysical and scaled down to zero in the continuum limit. Not so here. Here the physical minimal (non-vanishing) size of a triangle is determined by $\hbar$ and G.

The quantum discreteness we find is a consequence of the compactness of SU(2): the length is conjugate to the group variables and the corresponding operator is a Laplace operator on a compact space, SU(2) and therefore has a discrete spectrum. It is *this* discreteness that will reappear also in four spacetime dimensions and is the core of the physical granularity of space.

---

[1] In 2d, $\hbar G$ has the dimensions of a length. This follows easily from the form of the action.

Since length operators associated with different (boundary) links commute (because each acts on different U_ℓ's), we can diagonalize all of them together. Therefore the j_ℓ's are good quantum numbers for labeling the kinematical boundary states of the theory. Are there other quantum numbers? It is time to introduce the basic structure of loop quantum gravity: the spin-network basis. This is done in the following section.

5.2.2 Spin networks

Gauge

We are not done with the definition of the boundary Hilbert space. The theory must be invariant under the SU(2) gauge transformations (5.4). We have not yet taken this local invariance into account. (In Maxwell's theory, if we fail to take into account gauge invariance and the Gauss law, we define three photon polarizations instead of two.) The gauge-invariant states must satisfy

$$\psi(U_\ell) = \psi(\Lambda_{s_\ell} U_\ell \Lambda_{t_\ell}^{-1}), \quad \text{where } \Lambda_n \in \text{SU}(2). \tag{5.12}$$

Equivalently,

$$\vec{C}_n \psi = 0 \tag{5.13}$$

for every node n of the boundary graph, where $\vec{C}_n$ is the generator of SU(2) transformations at the node n:

$$\vec{C}_n = \vec{L}_{\ell_1} + \vec{L}_{\ell_2} + \vec{L}_{\ell_3} = 0; \tag{5.14}$$

ℓ_1, ℓ_2 and ℓ_3 are the three links emerging from the node n. This relation is called the *closure constraint* or *gauge constraint* or *Gauss constraint*.

This equation has an important direct geometrical interpretation: consider a triangle of the boundary, bounded by the three segments s_1, s_2, s_3. These are dual to the three links ℓ_1, ℓ_2, ℓ_3 that meet at the node n at the center of the triangle. We have seen that $\vec{L}_\ell$ can be interpreted as the vector describing the side of the triangle. Then (5.14) is precisely the relation indicating that the triangle closes!

The subspace of $\mathcal{H}_\Gamma$ where (5.13) is verified is a proper subspace (because SU(2) is compact), which we call $\mathcal{K}_\Gamma$. We write it as

$$\mathcal{K}_\Gamma = L_2 \left[\text{SU}(2)^L / \text{SU}(2)^N \right]_\Gamma. \tag{5.15}$$

Here L is the number of links and N the number of nodes of Γ and the subscript Γ indicates that the pattern of the $\text{SU}(2)^N$ gauge transformations (5.4) on the $\text{SU}(2)^L$ variables is dictated by the structure of the graph. Let us study the structure of this space.

The length operators L_ℓ are gauge-invariant ($\vec{L}_\ell$ transforms as a vector). Therefore they are well defined on the gauge-invariant Hilbert space $\mathcal{K}_\Gamma$. Remarkably, they form a complete commuting set, as we show below. That is, a basis of $\mathcal{K}_\Gamma$ is given by the normalized eigenvectors of these operators, which we indicate as $|j_\ell\rangle$. An element of this basis is therefore determined by assigning a spin j_ℓ to each link l of the graph. A graph with a spin assigned to each link is called a "spin network." We say that the links of the graph of a spin network are "colored" by the spin. The spin network states $|j_\ell\rangle$ form a basis of $\mathcal{K}_\Gamma$. This is called a *spin-network basis* and spans the quantum states of the geometry. Let us show this in detail.

Spin-network basis

Let us show that the quantum numbers j_ℓ are sufficient to label a basis of $\mathcal{K}_\Gamma$. We do so in this paragraph in a rather abstract manner. The following paragraph will give a more concrete, explicit construction of these states.

The Peter–Weyl theorem states that the matrix elements $D^j_{mn}(U)$ of the Wigner matrices (see Complements of Chapter 1), seen as functions of SU(2), are orthogonal with respect to the scalar product defined by the Haar measure. Specifically:

$$\int dU\, \overline{D^{j'}_{m'n'}(U)} D^j_{mn}(U) = \frac{1}{d_j}\delta^{jj'}\delta_{mm'}\delta_{nn'}\, . \tag{5.16}$$

where $d_j = 2j + 1$ is the dimension of the representation j. Therefore the Wigner matrices form an *orthogonal basis* (up to the factor $1/d_j$ this is an ortho*normal* basis), in the Hilbert space $\mathcal{H} = L_2[\mathrm{SU}(2), dU]$, where dU is the invariant measure.

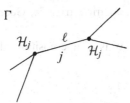

In other words, the Hilbert space $L_2[\mathrm{SU}(2)]$ can be decomposed into a sum of finite dimensional subspaces of spin j, spanned by the basis states formed by the matrix elements of the Wigner matrices $D^j(U)$. The dimension of these spaces is $(2j + 1)^2$, which is the number of entries of the matrix $D^j(U)$. This matrix is a map from the Hilbert space $\mathcal{H}_j$ of the spin-j representation to itself. Any such map can be viewed as an element of $\mathcal{H}_j \otimes \mathcal{H}_j$. Therefore the Peter–Weyl theorem states that

$$L_2[\mathrm{SU}(2)] = \oplus_j(\mathcal{H}_j \otimes \mathcal{H}_j). \tag{5.17}$$

This is nothing else than another way of saying that there is a basis labeled by spins and magnetic numbers whose range is determined by the spins.

In our case we have many such spaces, therefore the boundary state space has the structure

$$L_2[\mathrm{SU}(2)^L] = \otimes_\ell \left[\oplus_j(\mathcal{H}_j \otimes \mathcal{H}_j)\right] = \oplus_{j_\ell} \otimes_\ell (\mathcal{H}_j \otimes \mathcal{H}_j). \tag{5.18}$$

The two Hilbert spaces associated with a link naturally belong to the two ends of the link, because each transforms according to the gauge transformations at one end.

Let us therefore group together the Hilbert spaces next to the same node, which transform all together under a gauge transformation. We obtain

$$L_2[\mathrm{SU}(2)^L] = \oplus_{j_\ell} \otimes_n (\mathcal{H}_j \otimes \mathcal{H}_{j'} \otimes \mathcal{H}_{j''}), \tag{5.19}$$

where j, j', j'' are the spins coming out from the node n. Next, we want the space of gauge-invariant states. For this we should restrict to the invariant part of any set of spaces transforming at the same node. We obtain

$$L_2[\mathrm{SU}(2)^L/\mathrm{SU}(2)^N] = \oplus_{j_\ell} \otimes_n \mathrm{Inv}_{\mathrm{SU}(2)}\big(\mathcal{H}_{j_1} \otimes \mathcal{H}_{j_2} \otimes \mathcal{H}_{j_3}\big)\, . \tag{5.20}$$

But $\mathrm{Inv}_{\mathrm{SU}(2)}\big(\mathcal{H}_{j_1} \otimes \mathcal{H}_{j_2} \otimes \mathcal{H}_{j_3}\big)$ does not exist unless the sum of three spins is an integer and the three spins satisfy[2]

$$|j_1 - j_2| < j_3 < j_1 + j_2. \tag{5.21}$$

This relation is called the *triangular inequality* or the Mandelstam inequality. This is a condition for the existence of the state. If it satisfied, the invariant space is one-dimensional:

$$\mathrm{Inv}_{\mathrm{SU}(2)}\big(\mathcal{H}_{j_1} \otimes \mathcal{H}_{j_2} \otimes \mathcal{H}_{j_3}\big) = \mathbb{C}. \tag{5.22}$$

Therefore

$$L_2[\mathrm{SU}(2)^L / \mathrm{SU}(2)^N] = \oplus_{j_\ell} \mathbb{C}, \tag{5.23}$$

where the sum is restricted to the j_ℓ that satisfy the triangular relations. This is the same as saying that there is a basis $|j_\ell\rangle$ labeled by j_ℓ satisfying triangular inequalities.

Recall that the spins are the lengths of the triangle surrounding the node: the triangular identities state that the lengths L_1, L_2, L_3 of the side of the triangle satisfy the triangular inequalities

$$j_3 \geq |j_1 - j_2|$$
$$j_3 \leq j_1 + j_2$$

$$|L_1 - L_2| < L_3 < L_1 + L_2. \tag{5.24}$$

These are precisely the relations satisfied by a geometrical triangle. Geometry emerges nicely from quantum geometry.

A generic quantum state in loop quantum gravity is a superposition of spin-network states

$$|\psi\rangle = \sum_{j_\ell} C_{j_\ell} |j_\ell\rangle. \tag{5.25}$$

Summarizing, the spin-network states $|j_\ell\rangle$:

- Are an eigenbasis of all length operators.
- Span the gauge-invariant Hilbert space.
- Have a simple geometric interpretation: they just say how long the boundary links are.

Harmonic analysis on the group

Let us now write the states $|j_\ell\rangle$ in the $\psi(U_\ell)$ representation: namely, compute the spin-network wavefunctions

$$\psi_{j_\ell}(U_\ell) = \langle U_\ell | j_\ell \rangle. \tag{5.26}$$

In other words, we solve explicitly the eigenvalue equation for the length operators L_ℓ

$$L_\ell \, \psi_{j_\ell}(U_\ell) = L_{j_\ell} \, \psi_{j_\ell}(U_\ell) \tag{5.27}$$

---

[2] Recall that in the quantum theory of the composition of the angular momenta we write the tensor product of two representations as a sum over representations in the form $\mathcal{H}_{j_1} \otimes \mathcal{H}_{j_2} = \oplus_{j=|j_1-j_2|}^{j_1+j_2} \mathcal{H}_j$. Therefore $\mathcal{H}_{j_1} \otimes \mathcal{H}_{j_2} \otimes \mathcal{H}_{j_3} = \big(\oplus_{j=|j_1-j_2|}^{j_1+j_2} \mathcal{H}_j\big) \otimes \mathcal{H}_{j_3} = \oplus_{k=|j-j_3|}^{j+j_3} \mathcal{H}_k$. This shows that the invariant space, namely the spin zero representation, can enter only once in the tensor product of three representations, and it does if the triangular inequalities are satisfied.

in the group representation.

The Peter–Weyl theorem provides a notion of Fourier transform on the group. A state $\psi(U)$ can be written as a linear combination of states in the Wigner-matrix basis:

$$\psi(U) = \sum_{jmn} C_{jmn} D^j_{mn}(U). \qquad (5.28)$$

In our case the states are functions of L group elements $\psi(U_\ell)$, therefore we can write them as

$$\psi(U_\ell) = \sum_{\substack{j_1 \dots j_L \\ m_1 \dots m_L \\ n_1 \dots n_L}} C_{j_1 \dots j_L m_1 \dots m_L n_1 \dots n_L} D^{j_1}_{m_1 n_1}(U_{\ell_1}) \dots D^{j_L}_{m_L n_L}(U_{\ell_L}) \qquad (5.29)$$

A state invariant under SU(2) at the nodes must be invariant if we act with a transformation Λ_n at the node n. This acts on the three group elements of the three links that meet at the node. Since the Wigner matrices are representation matrices, the gauge transformation acts on the three corresponding indices. Therefore for the state to be invariant $C_{j_1 \dots j_L m_1 \dots m_L n_1 \dots n_L}$ must be invariant when acted upon by a group transformation on the three indices corresponding to the same node. (Up to normalization) there exists only one invariant object with three indices in three SU(2) representations: it is called the Wigner $3j$-symbol and written as

$$\iota^{m_1 m_2 m_3} = \begin{pmatrix} j_1 & j_2 & j_3 \\ m_1 & m_2 & m_3 \end{pmatrix}. \qquad (5.30)$$

See Haggard (2011) for a good introduction. That is, any invariant state in the triple tensor product of the three representations is proportional to the normalized state

$$|i\rangle = \sum_{m_1 m_2 m_3} \begin{pmatrix} j_1 & j_2 & j_3 \\ m_1 & m_2 & m_3 \end{pmatrix} |j_1, m_1\rangle \otimes |j_2, m_2\rangle \otimes |j_3, m_3\rangle. \qquad (5.31)$$

The $3j$-symbols are the symmetric form of the Clebsh–Gordon coefficients; they are described for instance in Chapter 14 of Landau and Lifshitz (1959); they are explicitly given by Mathematica (ThreeJSymbol$[\{j_1, m_1\}, \{j_2, m_2\}, \{j_3, m_3\}]$). In the quantum gravity literature the Wigner $3j$-symbols are usually called trivalent "intertwiners."

Therefore the gauge-invariant states must have the form

$$\psi(U_\ell) = \sum_{j_1 \dots j_L} C_{j_1 \dots j_L} \iota_1^{m_1 m_2 m_3} \dots \iota_N^{m_L - 2 m_L - 1 m_L} D^{j_1}_{m_1 n_1}(U_{\ell_1}) \dots D^{j_L}_{m_L n_L}(U_{\ell_L}), \qquad (5.32)$$

where all indices are contracted between the intertwiner ι and the Wigner matrices D. How these indices are contracted is dictated by the structure of the graph Γ. There is one matrix D for each link ℓ and a $3j$-symbol ι for each node n. The indices of a node are contracted with the indices of the adjacent links.

Therefore a generic gauge-invariant state is a linear combination

$$\psi(U_\ell) = \sum_{j_\ell} C_{j_\ell} \, \psi_{j_\ell}(U_\ell) \qquad (5.33)$$

of the orthogonal states

$$\psi_{j_\ell}(U_\ell) = \iota_1^{m_1 m_2 m_3} \ldots \iota_N^{m_{L-2} m_{L-1} m_L} D_{m_1 n_1}^{j_1}(U_{\ell_1}) \ldots D_{m_L n_L}^{j_L}(U_{\ell_L}) \tag{5.34}$$

labeled by a spin associated with each link (do not confuse the indices n the nodes n). These are the spin-network wavefunctions. We write them also in the simpler form

$$\psi_{j_\ell}(U_\ell) = \langle U_\ell | j_\ell \rangle = \bigotimes_{\text{n}} \iota_{\text{n}} \cdot \bigotimes_\ell D^{j_\ell}(U_\ell). \tag{5.35}$$

These are the three-dimensional spin-network states in the group representation.

5.3 Quantum dynamics: transition amplitudes

The transition amplitude is a function of the boundary states. We can consider a compact region of spacetime bounded by a boundary formed by two disconnected components, taken as the past and the future boundaries. But we can also consider a general boundary state (Oeckl 2003, 2008), possibly connected, and define the transition amplitude associated with any state on this boundary. This defines an amplitude for a process characterized by the given state on the boundary. That is, an amplitude associated with a given set of boundary values of partial observables.

Fix a triangulation Δ of spacetime. The transition amplitude is a function of the states defined on $\Gamma = (\partial \Delta)^*$, the boundary graph. Like the Hamilton function, the transition amplitude is a function either of the coordinates or the momenta. We write the two forms as

$$W_\Delta(U_\ell) = \langle W_\Delta | U_\ell \rangle \tag{5.36}$$

and

$$W_\Delta(j_\ell) = \langle W_\Delta | j_\ell \rangle. \tag{5.37}$$

The subscript Δ indicates that this is the amplitude computed on the discretization Δ. Notice that in the second case, where the momenta L_ℓ have discrete spectra, W is a function of the quantum numbers, not the classical variables. This has been discussed in detail in Section 2.2.2. The relation between the two expressions of the transition amplitude is easy to find, because we know how to express the states $|j_\ell\rangle$ in the (generalized) basis $|U_\ell\rangle$ and vice versa: the transition matrix is given by the spin-network states (5.34).

To compute the transition amplitude W_Δ of the theory discretized on the 2-complex dual to a triangulation Δ (see Figure 5.1), we use the Feynman path integral. The amplitude is given by the integral over all classical configurations weighted by the exponential of the classical action. That is,

$$W_\Delta(U_\ell) = \mathcal{N} \int dU_e \int dL_f \, e^{\frac{i}{8\pi\hbar G} \sum_f Tr[U_f L_f]}. \tag{5.38}$$

$\mathcal{N}$ is a normalization factor, where we will absorb various constant contributions. The integral over the momenta is easy to perform, since it is an integral of an exponential

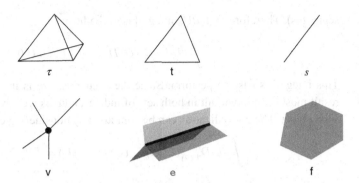

Figure 5.1 Tetrahedra, vertices, edges, faces, and links, and their relations.

which gives a delta function. Therefore we obtain

$$W_\Delta(U_\ell) = \mathcal{N} \int dU_e \prod_f \delta(U_f) \tag{5.39}$$

where the delta function is over SU(2). To compute this integral, we expand the delta function over the group in representations, using[3]

$$\delta(U) = \sum_j d_j \operatorname{Tr} D^{(j)}(U). \tag{5.40}$$

Recall that d_j is the dimension of the representation $d_j = 2j + 1$. A group element $U \in \mathrm{SU}(2)$ determines a rotation in SO(3): the quantity $\operatorname{Tr} D^{(j)}(U)$ is the character of the representation and depends on the angle but not on the rotation axis.

Using this, we can write

$$W_\Delta(U_\ell) = \mathcal{N} \int dU_e \prod_f \left(\sum_j d_j \operatorname{Tr} D^j(U_f) \right) \tag{5.41}$$

$$= \mathcal{N} \sum_{j_f} \left(\prod_f d_{j_f} \right) \int dU_e \prod_f \operatorname{Tr} \left(D^{j_f}(U_{1f}) \dots D^{j_f}(U_{nf}) \right). \tag{5.42}$$

In the expression above we have made use of the fact that the representation of a product is the product of representation, i.e., $D(UV) = D(U)D(V)$ and U_f is the product of the group elements associated with the links $1f, \dots, nf$ around the face f.

Consider one edge e and the corresponding integral dU_e. Observe that each edge e bounds precisely three faces (because the edge is dual to a triangle which is bounded by three

---

[3] This is the SU(2) analog of the well-known expansion of the delta over the circle U(1): $\delta(\phi) = \frac{1}{2\pi} \sum_n e^{in\phi}$, where n labels the irreducible representations of U(1), which are all one-dimensional and given by the exponential. Similar formulas hold for any compact group.

segments). Therefore each dU_e integral is of the form

$$\int dU\, D^{j_{e_1}}_{m_1 n_1}(U)\, D^{j_{e_2}}_{m_2 n_2}(U)\, D^{j_{e_3}}_{m_3 n_3}(U)\,. \tag{5.43}$$

This integral is easy to perform. Since the Haar measure is invariant on both sides, the result must be an invariant in both sets of indices, and, as we have seen, there is only one such object. The normalization can be computed by considering contractions. This gives

$$\int dU\, D^{j_1}_{m_1 n_1}(U)\, D^{j_2}_{m_2 n_2}(U)\, D^{j_3}_{m_3 n_3}(U) = \iota^{m_1 m_2 m_3}\, \iota^{n_1 n_2 n_3}$$

$$= \begin{pmatrix} j_1 & j_2 & j_3 \\ m_1 & m_2 & m_3 \end{pmatrix} \begin{pmatrix} j_1 & j_2 & j_3 \\ n_1 & n_2 & n_3 \end{pmatrix}\,. \tag{5.44}$$

Therefore the result of the integrals in (5.41) is a bunch of $3j$-symbols contracted among themselves. Let us see what the precise pattern of the contractions is. Each edge produces two $3j$-symbols, which we can view as located at the two ends of the edge, because they have indices that are contracted at that end. The contractions happen therefore at each vertex. At each vertex, there are four edges, and therefore four $3j$-symbols, contracted among themselves. The contraction must be SU(2) invariant, because the entire expression is invariant and therefore must be the one obtained by the invariant contraction of magnetic indices, given by the matrices computed for instance in Chapter 14 of Landau and Lifshitz (1959), or in Haggard (2011):

$$g_{mn} = \sqrt{2j+1}\, \begin{pmatrix} j & j & 0 \\ m & n & 0 \end{pmatrix} = \delta_{m,-n}(-1)^{j-m}\,. \tag{5.45}$$

(The $\delta_{m,-n}$ is easily understood as conservation of the third component of the angular momentum. The sign originates from tensoring the invariant object ϵ^{AB}.) The invariant contraction of four $3j$-symbols gives an object called Wigner $6j$-symbol (or Racah W-coefficient), indicated as follows:

$$\begin{Bmatrix} j_1 & j_2 & j_3 \\ j_4 & j_5 & j_6 \end{Bmatrix} \equiv \sum_{m_a, n_a} \prod_{a=1}^{6} g_{m_a n_a} \begin{pmatrix} j_1 & j_2 & j_3 \\ m_1 & m_2 & m_3 \end{pmatrix} \begin{pmatrix} j_1 & j_4 & j_5 \\ n_1 & m_4 & m_5 \end{pmatrix}$$

$$\times \begin{pmatrix} j_2 & j_4 & j_6 \\ n_2 & n_4 & m_6 \end{pmatrix} \begin{pmatrix} j_3 & j_5 & j_6 \\ n_3 & n_5 & n_6 \end{pmatrix} \tag{5.46}$$

or, more explicitly

$$\begin{Bmatrix} j_1 & j_2 & j_3 \\ j_4 & j_5 & j_6 \end{Bmatrix} = (-1)^{J-M} \sum_{m_a} \begin{pmatrix} j_1 & j_2 & j_3 \\ m_1 & m_2 & m_3 \end{pmatrix} \begin{pmatrix} j_1 & j_4 & j_5 \\ -m_1 & m_4 & m_5 \end{pmatrix}$$

$$\times \begin{pmatrix} j_2 & j_4 & j_6 \\ -m_2 & -m_4 & m_6 \end{pmatrix} \begin{pmatrix} j_3 & j_5 & j_6 \\ -m_3 & -m_5 & -m_6 \end{pmatrix}\,, \tag{5.47}$$

where $M = \sum_{a=1}^{6} m_a$ and $J = \sum_{a=1}^{6} j_a$. A good detailed introduction to this algebra, with all conventions carefully fixed is Aquilanti *et al.* (2010). The Wigner $6j$-symbols are widely used in nuclear physics, molecular physics, quantum information,

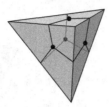

The tetrahedron τ associated to the $6j$-symbol is the boundary graph of a tetrahedron v^* of the triangulation.

etc. See Chapter 14 of Landau and Lifshitz (1959). They are given in Mathematica as SixJSymbol[$\{j_1,j_2,j_3\},\{j_4,j_5,j_6\}$].

Notice that the path of contractions reproduces the structure of a tetrahedron τ, where the $3j$-symbols are the four vertices of the tetrahedron and the spins are on the six sides. The three spins in the upper row (j_1,j_2,j_3) form a triangle; the three spins in the lower row (j_4,j_5,j_6) join at the vertex opposite to this triangle; and each column is formed by spins that do not meet.

Do not confuse this tetrahedron τ with the tetrahedron v^* of the triangulation Δ dual to the vertex v we are considering. τ and v^* are dual to one another: the vertices of τ are in the center of the faces of v^* and the sides of τ cross the sides of v^*. Rather, notice, since this is important, that the tetrahedron τ is the boundary graph of the triangulation formed by the single tetrahedron v^*. See Figure 5.2.

After integrating over all internal-edge group variables, the group variables of the boundary edges remain. We can integrate these as well contracting with a boundary spin network state, obtaining (carefully keeping trace of the signs (Barrett and Naish-Guzman 2009))

$$W_\Delta(j_\ell) = \mathcal{N}_\Delta \sum_{j_f} \prod_f (-1)^{j_f} d_{j_f} \prod_v (-1)^{J_v}\{6j\}, \qquad (5.48)$$

where the sum is over the association of a spin to each face, respecting the triangular inequalities at all edges, $J_v = \sum_{a=1}^6 j_a$, and j_a are the spins of the faces adjacent to the vertex v. (A vertex of the 2-complex is adjacent to six faces; see Figure 5.3.) We have indicated that the normalization factor may depend on the triangulation. This is the transition amplitude in the spin representation.

The expression (5.48) was first written by Ponzano and Regge in the 1960s (Ponzano and Regge 1968). Remarkably, Ponzano and Regge did not know that the spectrum of the length is discrete, but simply "guessed" that the length had to be discrete and determined by spins. The physical meaning of their ansatz became clear only with loop quantum gravity. The connection between loop quantum gravity and the Ponzano–Regge model was pointed out in Rovelli (1993b).

5.3.1 Properties of the amplitude

The following basic properties characterize the amplitude.

1. **Superposition principle.** This is the basic principle of quantum mechanics. The amplitudes are to be given by the sum of elementary amplitudes (in standard quantum theory,

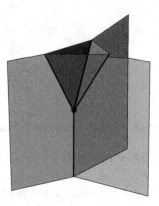

Figure 5.3 A vertex of the 2-complex and its six adjacent faces.

this sum is expressed by Feynman's sum over paths σ):

$$\langle W | \psi \rangle = \sum_{\sigma} W(\sigma) \tag{5.49}$$

2. **Locality.** This a fundamental discovery of nineteenth- and twentieth-century physics: interactions are local in spacetime. Therefore the elementary amplitudes can be seen as products of amplitudes associated with spacetime points (in standard quantum field theory, this product is expressed as the exponential of an integral over spacetime points):

$$W(\sigma) \sim \prod_{\mathsf{v}} W_{\mathsf{v}}. \tag{5.50}$$

3. **Local euclidean invariance.** In the physical four-dimensional theory, this criterion will translate into the local Lorentz invariance of general relativity. Here, it fixes the SU(2) invariance of the amplitude. We show in Section 6.1 at the beginning of the next chapter that the vertex amplitude, namely, the $6j$-symbol, can be written simply as the projection on the SU(2) invariant part of the state on the boundary graph of the vertex, that is,

$$W_{\mathsf{v}} = (P_{\mathrm{SU}(2)} \psi_{\mathsf{v}})(\mathbb{1}), \tag{5.51}$$

where ψ is a boundary state, $P_{\mathrm{SU}(2)}$ is the projector on its locally invariant SU(2) component, and ($\mathbb{1}$) indicates the evaluation of the spin-network state on the identity $U_{\ell} = \mathbb{1}$.

These fundamental properties will guide us to write the amplitudes in the 4d theory.

5.3.2 Ponzano–Regge model

The expression (5.48) seems a bit like magic. It is a very simple expression written in terms of basic objects in the representation theory of SU(2). How can it possibly have anything to do with general relativity?

The connection with the classical theory must be in a classical limit. In quantum mechanics, a classical limit is given by large quantum numbers, where the quantum discreteness becomes irrelevant. Let us therefore study this expression for large j's.

In their original paper, Ponzano and Regge prove the following remarkable result. Consider a flat geometrical tetrahedron. Let its geometry be determined by the six lengths of its sides: $L_1, \ldots, L_6$. Associate six spins $j_1, \ldots, j_6$ to these lengths so that $L_a = j_a + \frac{1}{2}$. Let V be the volume of this tetrahedron, and S be the Regge action of this tetrahedron. Then Ponzano and Regge have given evidence that in the large-spin limit

$$\{6j\} \underset{j \to \infty}{\sim} \frac{1}{\sqrt{12\pi V}} \cos\left(S + \frac{\pi}{4}\right). \tag{5.52}$$

This was proved rigorously only in 1998 by Roberts (Roberts 1999). The consequences of this asymptotic relation are important. The cosine is the sum of two terms and the $\pi/4$ term can be taken out of the cosine, therefore the formula above can be written as

$$\{6j\} \underset{j \to \infty}{\sim} \frac{1}{2\sqrt{-12i\pi V}} e^{iS} + \frac{1}{2\sqrt{12i\pi V}} e^{-iS}. \tag{5.53}$$

If we consider only large spins, we can disregard quantum discreteness and the sum over spins is approximated by an integral over lengths in a Regge geometry. The integrand is given by exponentials of the action. This is a discretization of a path integral over geometries of the exponential of the Einstein–Hilbert action. Therefore (5.48) is precisely a concrete implementation of the path-integral "sum over geometries" formal definition of quantum gravity:

$$Z \sim \int D[g] \, e^{\frac{i}{\hbar} \int \sqrt{-g}R}. \tag{5.54}$$

The fact that the sum has two terms with opposite signs of the action follows from the discussion in Section 3.2.1: since we are quantizing the triad theory and not the metric theory (recall that this is necessary because of the existence of fermions) there are *two* distinct triad configurations for each metric configuration, at every vertex. They have opposite signs and they give rise to the two terms above. The $\pi/4$ term, which gives rise to an extra phase difference between the two terms, is also well understood in saddle-point approximations: this phase is called the Maslov index and always appears when there is a flip of sign in the momentum between two saddle points, namely, a folding of the surface L of Littlejohn, discussed in Section 2.5.2, over configuration space (Figure 5.4). This is discussed in detail in Littlejohn (1992).

There is one last point to be discussed to define the theory: the continuum limit. What happens if we refine the discretization? Let us discuss this in two steps. First, let us refine the triangulation by keeping the boundary graph fixed. Ponzano and Regge proved in their paper that choosing the normalization factor to be

$$N_\Delta = w^p, \tag{5.55}$$

where w is a number and p is the number of points in the triangulation, the transition amplitude does not change under a change of triangulation, up to possible divergent terms that may appear in the refinement. Formally, divergent terms can be reabsorbed into w as

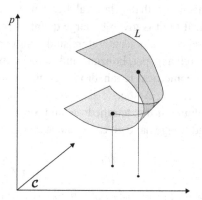

Figure 5.4   If surface L folds over configuration space $\mathcal{C}$, the relative phase of the saddle-point approximation of the path integral between two extrema includes a $\pi/4$, from a non-trivial Maslov index.

follows. First, correct the theory with a cut-off, by limiting the sum over each spin to a maximum value Λ. Next, take w of the form

$$w = \frac{w_o}{\Lambda}. \tag{5.56}$$

Then it is possible to prove that the resulting expression does not depend on the triangulation in the limit $\Lambda \to \infty$. That is, the amplitude is defined by

$$W = \lim_{\Lambda \to \infty} \left(\frac{w_0}{\Lambda} \right)^p \sum_{j_f=0}^{\Lambda} \prod_f (-1)^{j_f} d_{j_f} \prod_v (-1)^J \{6j\} . \tag{5.57}$$

Second, what happens if we refine the boundary graph? Also in this case the refinement has no physical effect. This can be seen as follows. Consider the transition amplitude for a boundary formed by two disconnected equal graphs Γ. The transition amplitude defines a map from the Hilbert space $\mathcal{H}_\Gamma$ to itself. This map is a projector that projects out from $\mathcal{H}_\Gamma$ the physical Hilbert space. It is possible to prove that the resulting physical Hilbert space does not depend on Γ. Both these properties are strictly connected to the fact that the corresponding classical theory has no local degrees of freedom and should not be expected in four-dimensional. Therefore we will not insist on this here, since they are of limited interest for the physical theory of quantum gravity.

When Regge and Ponzano found this model for quantum gravity in three-dimensional, there was great excitement in Princeton, where Regge was at the time. The asymptotic result (5.52) is truly amazing: general relativity is coded in simple SU(2) representation theory combinatorics. The action of GR is just hidden in a Wigner $6j$-symbol! The idea of trying to repeat the construction in four-dimensional was put forward immediately; but the project turned out to be more difficult than expected. The four-dimensional version of the model had to wait almost half a century. It will be discussed in Chapter 7.

5.4 Complements

5.4.1 Elementary harmonic analysis

We collect here a few basic facts about the space $L_2[\mathrm{SU}(2)]$, which plays a major role in quantum gravity. For a mathematically complete presentation of harmonic analysis on compact groups, see for instance Folland (1995).

It is good to keep in mind as reference a simple and well-known example of functions on a compact group: the space $L_2[U(1)]$ of the square integrable functions $\psi(\theta)$ of an angular variable. This is for instance the space of the quantum states of a particle on a circle. The scalar product is

$$(\psi, \phi) = \frac{1}{2\pi} \int_0^{2\pi} d\theta \, \overline{\psi(\theta)} \phi(\theta). \tag{5.58}$$

The space of these functions carries a representation of the group $U(1)$ which acts simply as $\psi(\theta) \to \psi(\theta + \alpha)$, for $\alpha \in U(1)$. A basis of functions orthogonal with respect to this scalar product is given by the functions $\psi_n(\theta) = e^{in\theta}$ with integer n. This is the (discrete) Fourier theorem. Notice that each of these functions defines a unitary irreducible (one-dimensional) representation of the abelian group $U(1)$, because $\psi_n(\theta + \phi) = \psi_n(\theta)\psi_n(\phi)$. Therefore the Fourier theorem states that $L_2[U(1)]$ decomposes into the orthogonal sum of irreducible representations of $U(1)$. This can be written in the form

$$L_2[U(1)] = \oplus_n C_n, \tag{5.59}$$

where C_n is the complex plane, viewed as a representation of $U(1)$ under the exponential map $\psi_n(\theta)$. Also, note that the delta distribution on the circle can be expanded in terms of this basis simply as

$$\delta(\theta) = \frac{1}{2\pi} \sum_n e^{in\theta}. \tag{5.60}$$

The main features of this well-known example are reproduced by the space $L_2[\mathrm{SU}(2)]$ of functions $\psi(U)$ over the group $\mathrm{SU}(2)$, square integrable with respect to the Haar measure. The scalar product is

$$(\psi, \phi) = \int_{\mathrm{SU}(2)} dU \, \overline{\psi(U)} \phi(U). \tag{5.61}$$

The space of these functions carries a representation of the group $\mathrm{SU}(2)$ which acts simply as $\psi(U) \to \psi(\Lambda^{-1}U)$, for $\Lambda \in \mathrm{SU}(2)$. A basis of functions orthogonal with respect to this scalar product is given by the functions $\psi_{jnm}(U) = D^j_{nm}(U)$, where the $D^j_{nm}(U)$ are the Wigner matrices. This is the Peter–Weyl theorem. Each of these functions is a matrix element of a unitary irreducible representation of the non-abelian group $\mathrm{SU}(2)$. Therefore the Peter–Weyl theorem states that $L_2[\mathrm{SU}(2)]$ decomposes in the orthogonal sum of irreducible representations of $\mathrm{SU}(2)$. This can be written in the form

$$L_2[\mathrm{SU}(2)] = \oplus_j V_j \tag{5.62}$$

where the sum is over the half integers j, which label the irreducible representations of $\mathrm{SU}(2)$. The space V_j is the space where a D^j matrix lives; it has dimension $(2j + 1)^2$ and transforms under the spin-j representation on either index. It can therefore be written as $\mathcal{H}_j \otimes \mathcal{H}_j$ where $\mathcal{H}_j$ is the spin-j representation of $\mathrm{SU}(2)$. Thus

$$L_2[\mathrm{SU}(2)] = \oplus_j(\mathcal{H}_j \otimes \mathcal{H}_j). \tag{5.63}$$

Finally, the delta distribution on the group can be expanded in terms of this basis simply as

$$\delta(U) = \sum_j (2j+1)\, Tr[D^j(U)]. \tag{5.64}$$

5.4.2 Alternative form of the transition amplitude

Here we derive an alternative form of the amplitude, which will turn out to be useful for generalizing the theory to four-dimensional. Let us start from

$$Z = \int dU_e \prod_f \delta(U_{e_1}\dots U_{e_n}). \tag{5.65}$$

where we consider the partition function of a triangulation without boundaries for simplicity. Now introduce *two* variables per each link l, namely $U_e = g_{ve}g_{ev'}$ where $g_{ev} = g_{ve}^{-1}$ is a variable associated with each couple vertex-edge (Figure 5.5). Then we can write

$$Z = \int dg_{ve} \prod_f \delta(g_{ve}g_{ev'}g_{v'e'}g_{e'v''}\dots). \tag{5.66}$$

Now imagine regrouping the g_{ev} variables differently, using $h_{vf} = g_{ev}g_{ve'}$, where e and e' are the two edges coming out from the vertex v and bounding the face f. Then clearly we can rewrite the amplitude as

$$Z = \int dh_{vf}\, dg_{ve} \prod_f \delta(g_{ve}g_{ev'}g_{v'e'}g_{e'v'}\dots) \prod_{vf} \delta(g_{e'v}g_{ve}h_{vf}). \tag{5.67}$$

Or equivalently

$$Z = \int dh_{vf}\, dg_{ve} \prod_f \delta(h_{vf}h_{v'f}\dots) \prod_{vf} \delta(g_{e'v}g_{ve}h_{vf}). \tag{5.68}$$

This can be reorganized as a transition amplitude where a delta function glues the group element around each face,

$$Z = \int dh_{vf} \prod_f \delta(h_f) \prod_v A_v(h_{vf}), \tag{5.69}$$

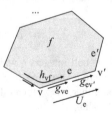

Figure 5.5   Splitting the group elements U_ℓ on the edges in two.

where we have defined h_f as the product of internal faces

$$h_f = \qquad\qquad\qquad = \prod_{v \in \partial f} h_{vf} \, . \qquad (5.70)$$

If a face f is bounded by a link ℓ of the boundary:

$$h_f = \qquad\qquad\qquad = \left(\prod_{v \in \partial f} h_{vf} \right) h_\ell . \qquad (5.71)$$

The vertex amplitude is then

$$A_v(h_{vf}) = \int dg_{ve} \prod_f \delta(g_{e'v}g_{ve}h_{vf}). \qquad (5.72)$$

The SU(2) integrals in this vertex amplitude are $n = 4$: one group element for *each* of the n (here four) edges coming out of the vertex. This is somewhat redundant, because a moment of reflection shows that if we perform $n - 1$ of these integrals the result is invariant under the last integration variable. Therefore we can drop one of the integrations without affecting the result. We write this by putting a prime on the measure

$$\int_{SU(2)^n} dg'_{ve} \equiv \int_{SU(2)^{n-1}} dg_{ve_1} \dots dg_{ve_{n-1}}. \qquad (5.73)$$

This is just cosmetic in 3d, because the volume of SU(2) is just unity, but it will be very useful in 4d, where we will have to deal with a non-compact group with infinite volume. Expanding the delta function in representations gives

$$A_v(h_{vf}) = \sum_{j_f} \int dg'_{ve} \prod_f (2j_f + 1) \mathrm{Tr}_{j_f}[g_{e'v}g_{ve}h_{vf}]. \qquad (5.74)$$

Notice that the vertex amplitude is a function of one SU(2) variable per face around the vertex. Imagine drawing a small sphere surrounding the vertex: the intersection between this sphere and the 2-complex is a graph Γ_v. The vertex amplitude is a function of the states in

$$\mathcal{H}_{\Gamma_v} = L_2[SU(2)^6/SU(2)^4]_{\Gamma_v}. \qquad (5.75)$$

Γ_v is the complete graph with four nodes, the boundary graph of the vertex (Figure 5.6).

In conclusion, and going back in an obvious way to the transition amplitudes, these are given by

$$W(h_\ell) = \int dh_{vf} \prod_f \delta(h_f) \prod_v A_v(h_{vf}), \qquad (5.76)$$

where the vertex amplitude is

$$A_v(h_{vf}) = \mathcal{N} \sum_{j_f} \int dg'_{ve} \prod_f (2j_f + 1) \mathrm{Tr}_{j_f}[g_{e'v}g_{ve}h_{vf}]. \qquad (5.77)$$

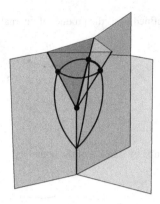

Figure 5.6   The vertex-graph Γ_v

This form of the amplitude will turn out to be the easiest to generalize to four-dimensional.

5.4.3 Poisson brackets

Problem

Compute the Poisson brackets (4.9) of L and U from their definition and the continuous Poisson bracket (3.102).

Solution

From the definitions, we have

$$\{U_\ell, L^i_{\ell'}\} = \left\{ P e^{\int_\ell \omega}, \int_{s_{\ell'}} e^i \right\}. \tag{5.78}$$

This is non-vanishing only if ℓ intersects $s_{\ell'}$, that is, if $\ell = \ell'$. In this case (dropping the link subscript for clarity),

$$\{U, L^i\} = \int_{s_{\ell'}} ds \left\{ P e^{\int_\ell dt \omega_b \frac{d\ell^b}{dt}}, e^i_a \frac{d\ell^a}{ds} \right\}. \tag{5.79}$$

Let us compute this using the Poisson brackets (3.102):

$$\{U, L^i\} = \int_{s_{\ell'}} ds \frac{ds^a_\ell}{ds} \int_\ell dt \frac{d\ell^b}{dt} U_{\ell_+} \{\omega^j_b(l(t))\tau_j, e^i_a(s_\ell(s))\} U_{\ell_-}$$

$$= 8\pi G \int_{s_{\ell'}} ds \frac{ds^a_\ell}{ds} \int_\ell dt \frac{d\ell^b}{dt} \epsilon_{ab} \delta^2(\ell(s), \ell(t)) U_{\ell_+} \delta_{ij} \tau_j U_{\ell_-}. \tag{5.80}$$

Here $U_{\ell_\pm}$ are the holonomies of the two halves of the edge cut by the segment. Since this is invariant under reparametrization and choice of coordinates, we can use coordinates where the link ℓ, the segment s_ℓ, and the normal n are orthogonal, and we have immediately that the integral gives precisely unity. Therefore

$$\{U_\ell, L^i_{\ell'}\} = 8\pi G \delta_{\ell\ell'} U_{\ell_+} \tau^i U_{\ell_-} \tag{5.81}$$

This is a bit disturbing because the right-hand side is not written in terms of the original variables. But recall that the definition of L^i was in a particular gauge, where $U_- = 1$. Using this,

$$\{U_\ell, L^i_{\ell'}\} = 8\pi G \delta_{\ell\ell'} U_\ell \tau^i \tag{5.82}$$

5.4.4 Perimeter of the universe

Problem

Consider euclidean GR in three spacetime dimensions and assume spacetime has the topology of a sphere $\mathcal{S}^2$ ("space") times $\mathbb{R}$ ("time)." Approximate this theory by means of a discretization, where space (namely, the 2-sphere) is discretized by two triangles joined by their boundaries. Call the "equator" of the sphere the (common) perimeter of the (two) triangles. The graph dual to this triangulation is the the graph with two nodes joined by three links. Therefore the Hilbert space is the space of functions $\psi(U, V, W)$, where U, V, and W are SU(2) group elements (with a given orientation of the links). Now, suppose the quantum state happens to be $\psi(U, V, W) = \text{Tr}(UV)$. How long is the equator? What is its expectation value and what is its quantum spread?

Hint: The quantum spread is zero because this state is an eigenstate of the perimeter $P = C_1 + C_2 + C_3$, where C_i are the three Casimirs. The mean value is the eigenvalue, which is the sum of the three spins (one happens to be zero) times $8\pi\hbar G$. . . .

6 Bubbles and the cosmological constant

If the reader is in a hurry to get to the physical four-dimensional theory, this chapter can be skipped at first reading without compromising understanding of what follows.

In this chapter we analyze the quantum theory of three-dimensional euclidean general relativity constructed in Chapter 5, focusing on the divergences that appear in the amplitudes and the way to correct them.

These divergences cannot be identified with the standard ultraviolet (UV) divergences of perturbative quantum general relativity, because UV divergences originate from arbitrary small spacetime regions, while here the theory is cut off at the Planck scale. Instead, the divergences come from *large* spacetime regions, and are, in a sense, infrared (IR), although the distinction is tricky in gravity, where a large portion of spacetime can be enclosed within a small boundary. Do these IR divergences spoil the theory?

A remarkable result, which we briefly describe in this chapter is that these divergences disappear in the theory with a positive cosmological constant. The cosmological constant happens to function in the theory as a natural IR cut-off, next to the UV cut-off provided by the Planck length.

This is beautiful, because two open issues in fundamental physics – the presence of a small cosmological constant, and the existence of quantum field theoretical divergences – resolve one another: the cosmological constant yields to and is required for the finiteness of the theory.

6.1 Vertex amplitude as gauge-invariant identity

Let us start the analysis of the divergences by better studying the amplitude of a single vertex. Consider a triangulation formed by a single tetrahedron τ. The boundary graph has again the shape of a tetrahedron as in Figure 5.2. Therefore the amplitude is a function of the variables of the links on this graph. Label with $a, b = 1, 2, 3, 4$ the nodes of the graph (the faces of the tetrahedron) and denote $U_{ab} = U_{ba}^{-1}$ the boundary group elements, which live on the links. The transition amplitude is a function $W(U_{ab})$.

To write it, we must construct the the 2-complex. We put a vertex inside the tetrahedron, connect it by four edges to the four boundary nodes, and draw the six faces that connect the vertex to the six boundary links. See Figure 6.1. Using (5.39) (dropping the normalization for simplicity), this is given by

$$W(U_{ab}) = \int dU_a \prod_{ab} \delta(U_a U_{ab} U_b^{-1}). \tag{6.1}$$

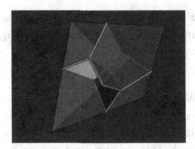

The spinfoam of a single tetrahedron.

The integrals are easily performed. One of them turns out to be redundant, and we obtain

$$W(U_{ab}) = \delta(U_{12}U_{23}U_{31})\delta(U_{13}U_{34}U_{41})\delta(U_{23}U_{34}U_{42}). \tag{6.2}$$

Notice that each sequence of U_{ab} inside the deltas corresponds to an independent closed loop in the boundary graph. The interpretation of this amplitude is therefore straightforward: the amplitude forces the connection to be flat on the boundary. Notice that what is flat is the three-dimensional connection, the one that generates three-dimensional rotations, not the two-dimensional rotations. That is, we can imagine that there is a *spacetime* reference frame in each face, and these are parallel transported along the boundary in such a way that any closed loop gives unity. In other words, $W(U_{ab})$ is just the gauge-invariant version of $\prod_{ab}\delta(U_{ab})$. And in fact, this it what the formula (6.1) says: it only makes the product of delta functions gauge-invariant. The reader may wonder why the fourth independent loop $U_{12}U_{24}U_{41}$ is missing. But it is immediately seen that if the connection is flat on the other three loops, it is also flat on this loop. In fact, it is easy to see that

$$\langle W|\psi\rangle = \int dU_{ab}\, W(U_{ab})\,\psi(U_{ab}) = \int dU_a\, \psi(U_a U_b^{-1}). \tag{6.3}$$

That is, W projects on the flat connections, averaged over the gauge orbits.

Let us now look at the same amplitude in the spin representation. This is given by

$$W(j_{ab}) = \left\{ \begin{array}{ccc} j_1 & j_2 & j_3 \\ j_4 & j_5 & j_6 \end{array} \right\}. \tag{6.4}$$

If all is consistent, this should be just the transform of the above, where the integral kernel is given by the spin-network states. That is,

$$W(j_{ab}) = \int dU_{ab}\psi_{j_{ab}}(U_{ab})W(U_{ab}). \tag{6.5}$$

Let us check this. Inserting the definitions,

$$W(j_{ab}) = \int dU_{ab}\int dU_a \prod_{ab}\delta(U_a U_{ab}U_b^{-1}) \otimes_a i_a \cdot \prod_{ab}D^{j_{ab}}(U_{ab}). \tag{6.6}$$

Performing the integral gives

$$W(j_{ab}) = \int dU_a \prod_{ab}\otimes_a i_a \cdot \prod_{ab}D^{j_{ab}}(U_a)D^{j_{ab}}(U_b^{-1}). \tag{6.7}$$

But now the representation's matrix elements act precisely on the invariant tensor, and since these are invariant, they do not do anything, so that what remains is the contraction of the four invariant tensors,

$$W(j_{ab}) = \text{Tr}\,[\otimes_a i_a] \tag{6.8}$$

namely the Wigner $6j$-symbol (6.4).

This shows that the $6j$-symbol is nothing other than a Fourier transform of the gauge-invariant delta functions on flat connections, in the Hilbert space associated with the tetrahedral graph. This can be written in the notation

$$W(j_{ab}) = \psi_{j_{ab}}(\mathbb{1}). \tag{6.9}$$

or, using the the projector $P_{\text{SU}(2)}$ on the SU(2) invariant part of a function, the vertex amplitude W_v can be written as

$$\langle \psi_v \,|\, W_v \rangle = (P_{\text{SU}(2)}\psi_v)(\mathbb{1}), \tag{6.10}$$

where ψ_v is a state in the boundary of the vertex. This is the notation that we anticipated in the previous chapter.

6.2 Bubbles and spikes

Consider again the single tetrahedron τ, but now add a point P inside it, and join P to the four points of the original tetrahedron. Consider the six triangles determined by P and one of the six sides of the original tetrahedron. These triangles split the original tetrahedron into four smaller tetrahedra $\tau_1, \ldots, \tau_4$. This defines a triangulation Δ_4 formed by four connected tetrahedra (Figure 6.2). (This step from one tetrahedron to four tetrahedra is called the "1-4 Pachner move".) Let us compute the amplitude W_{Δ_4}. The boundary of this triangulation is the same as the boundary of a single tetrahedron. Therefore the boundary variables can still be taken to be the six variables U_{ab} as in the previous section.

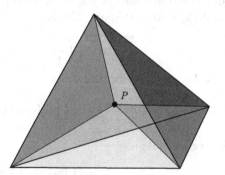

Figure 6.2 Δ_4.

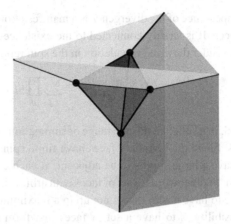

Figure 6.3 Δ_4^*. The bubble is the inside tetrahedron spanned by the four internal vertices.

To compute the amplitude, we need to understand the 2-complex Δ_4^*. This has four vertices, each joined to the other vertices and to one boundary node. It has two kinds of internal faces: six boundary faces, and four internal faces, each bounded by their vertices. A moment of reflection shows that these four internal faces surround P. Together they have the topology of a sphere. They form an example of a "bubble." A bubble is a collection of internal faces that form a surface with no boundaries. Bubbles are important. See Figure 6.3.

The amplitude is then easy to write. We call U_{ab} the boundary group elements, U_a and V_{ab} the internal ones, and we have

$$W_{\Delta_4}(U_{ab}) = \int dU_a dV_{ab} \prod_{ab} \delta(U_{ab}U_b^{-1}V_{ba}U_a) \prod_{abc} \delta(V_{ab}V_{bc}V_{ca}). \qquad (6.11)$$

Integrating, we have easily

$$W_{\Delta_4}(U_{ab}) = \delta(U_{12}U_{23}U_{31})\delta(U_{13}U_{34}U_{41})\delta(U_{23}U_{34}U_{42})\delta(U_{12}U_{24}U_{41}). \qquad (6.12)$$

At first sight, this may seem quite the same thing as the amplitude of a single tetrahedron that we have computed above:

$$W_{\Delta_1}(U_{ab}) = \delta(U_{12}U_{23}U_{31})\delta(U_{13}U_{34}U_{41})\delta(U_{23}U_{34}U_{42}) ; \qquad (6.13)$$

but there is a crucial difference: the last delta function. As already noticed, this is redundant. To see this in detail, let us integrate the amplitude against an arbitrary boundary function $\psi(U_{ab})$. We obtain

$$\langle W_{\Delta_4}|\psi\rangle = \int dU_{ab}W_{\Delta_4}(U_{ab})\psi(U_{ab}) = \delta(1)\int dU_a\psi(U_aU_b^{-1}). \qquad (6.14)$$

That is

$$W_{\Delta_4} = \delta(1)W_{\Delta_1}, \qquad (6.15)$$

and $\delta(1)$ is of course an infinite factor.

The appearance of the divergence is a manifestation of the standard quantum field theory divergences. It is strictly connected to the existence of the bubble. To see that this is the case, reconsider the same calculation in the spin representation. In this case,

$$W_{\Delta_4}(j_{ab}) = \sum_{j_{ab}} \prod_{ab} d_{j_{ab}} \prod_a \{6j\}. \tag{6.16}$$

In general, in a sum like this the range of summation of the j_{ab} is restricted by the triangular identities. Since the boundary faces have finite spins, the only possibility for an internal face to have a large spin is to be adjacent, at each edge, to at least one other face with a large spin. In other words, a set of faces with arbitrarily large spins cannot have boundaries. Therefore to have a sum that is not up to a maximum spin by the triangular identities the only possibility is to have a set of faces that form a surface without boundaries in the 2-complex. That is, a bubble.

All this is very similar to the ultraviolet divergences in the Feynman expansion of a normal quantum field theory, where divergences are associated with loops, because the momentum is conserved at the vertices. Here, divergences are associated with bubbles, because angular momentum is conserved on the edges. A Feynman loop is a closed set of lines where arbitrarily high momentum can circulate. A spinfoam divergence is a closed set of faces, that can have arbitrarily high spin.

Notice, however, that in spite of the formal similarity there is an important difference in the physical interpretation of the two kinds of divergence. The Feynman divergences regard what happens at very small scale. On the contrary, the spinfoam divergences concern large spins, namely, large geometries. Therefore they are not ultraviolet divergences: they are infrared. Let us see this more in detail.

Spikes

Consider a term contributing to the divergence of $W_{\Delta_4}(j_{ab})$. This is a configuration of the spins on the 2-complex where very large spins sit on the internal faces. What is the corresponding geometry? The spins of the internal faces are the lengths of the four bones that connect P to the boundary. Therefore the geometries creating the divergence are geometries where these lengths are very large.

To get an intuition about these, consider the analogous situation in two-dimensional instead of in three-dimensional. Imagine triangulating a plane having small curvature. But imagine that somewhere on this plane there is a "spike": an extremely thin and tall mountain (Figure 6.4). Choose a triangulation where a point P is on the top of this mountain and is surrounded by a triangle that surrounds the bottom of the mountain. This is a situation where the base triangle is small, but the distance of P from this triangle is large. It is important to emphasize that all this concerns the *intrinsic* geometry of the plane, and not its *extrinsic* geometry, which is irrelevant in this context. Therefore P is still *inside* the triangle, even if its distance from the boundary of the triangle is arbitrarily large. There is no sense for it to be "outside" the triangle, since what is relevant here is only the geometry of the two-dimensional surface itself. The possibility of having a small triangle including a

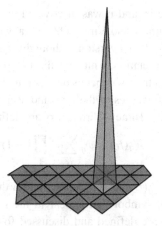

Figure 6.4   A spike in a two-dimensional triangulation. What matters is the intrinsic geometry of the two-dimensional surface.

point at very large distance from its boundary is just a normal feature of Riemannian space. It is because the space can be strongly curved.

This example provides the right intuition for the spikes, and therefore the divergences of spinfoams. These are geometries that are strongly curved, and occur where there is a region with a small boundary but a large internal volume.[1]

6.3 Turaev–Viro amplitude

How to deal with the divergences? In the three-dimensional theory, this is easy: just compute the amplitude using a 2-complex without bubbles. This is possible in general, and therefore the theory is well defined as it is. Since there are no local degrees of freedom, in general choosing a more complicated triangulation does not improve the approximation.

But in four-dimensional we will not have such a luxury; because the theory will have local degrees of freedom, we will *have* to refine the triangulation to capture all degrees of freedom of the theory, and the bubble amplitudes will describe genuine large radiative corrections. With this in mind, let us see whether there is a way to make sense of bubble divergences also in three-dimensional.

---

[1] The astute reader may raise an objection at this point. The classical configurations of three-dimensional gravity have vanishing curvature. Presumably, in the large-quantum-number regime, configurations that are very far from solutions of the equations of motion are suppressed, as one always expects in quantum theory. Why then is the weight of the spikes not suppressed, if these have large curvature? The answer is subtle: it is because of the second term in the semiclassical expansion (5.53) of the vertex amplitude. If only the first term was present, the divergences would probably not be present, because the large curvature would produce a rapidly oscillating phase in the integral, which might suppress the spike contribution. But the second term contributes with the opposite sign to the action. Since there are four vertices (four tetrahedra), there are four couples of such terms in the amplitude. A detailed analysis (Christodoulou *et al.* 2013) shows that divergences are associated with the mixed terms, where the positive contribution of one tetrahedron cancel the negative contribution of another.

In fact, there is, and it was discovered by the mathematicians Turaev and Viro in the early 1990s (Turaev and Viro 1992). Turaev and Viro found that it is possible to modify the theory and obtain a version without divergences. Remarkably, the modified theory has a physical interpretation, which we discuss in the next section.

The modified theory depends on a parameter $q = e^{\frac{i\pi}{r}}$ for an integer r (so that $q^r = -1$) and formally converges to the Ponzano–Regge model when $r \to \infty$, that is $q \to 1$. The amplitudes of the Turaev–Viro theory are defined by modification of (5.48), given by

$$W_q(j_\ell) = w_q^p \sum_{j_f} \prod_f (-1)^{j_f} d_q(j_f) \prod_v (-1)^{j_v} \{6j\}_q . \tag{6.17}$$

The quantities $d_q(j)$ and $\{6j\}_q$ appearing in this formula are the quantum dimension and the quantum $6j$-symbol that appear in the representation theory of the quantum group $SU(2)_q$. These are defined and discussed for instance in Carter et $al.$ (1995) or Kauffman and Lins (1994), and we do not discuss them here, apart from a few remarks below. The normalization factor is

$$w_q = -\frac{(q - q^{-1})^2}{2r}. \tag{6.18}$$

The reason we have not labeled the amplitude with the triangulation is that Turaev and Viro proved that the amplitude is in fact rigorously independent of the triangulation. It depends only on the global topology of the manifold on which it is defined. Most importantly, it is finite.

For large r, all the quantities entering into the amplitude converge to the corresponding $SU(2)$ quantities:

$$\left\{ \begin{array}{ccc} j_1 & j_2 & j_3 \\ j_4 & j_5 & j_6 \end{array} \right\}_q = \left\{ \begin{array}{ccc} j_1 & j_2 & j_3 \\ j_4 & j_5 & j_6 \end{array} \right\} + O(r^{-2}), \tag{6.19}$$

$$d_j^q = (2j + 1) + O(r^{-2}), \tag{6.20}$$

$$w_q = \frac{2\pi^2}{r^3} \left(1 + O(r^{-2}) \right). \tag{6.21}$$

In this sense the amplitude converges to the Ponzano–Regge one for large r. But the sum over representations is finite, because the q-deformed quantities vanish for large j. In particular, the quantum dimension is given by

$$d_j^q = (-1)^{2j} \frac{q^{2j+1} - q^{-2j-1}}{q - q^{-1}} = (-1)^{2j} \frac{\sin\left(\frac{\pi}{r}(2j+1)\right)}{\sin\frac{\pi}{r}} \tag{6.22}$$

and is plotted in Figure 6.5.

Notice from the plot that there is a maximum value of j, which increases as $q \to 1$. Therefore only spins up to a maximum value, which is

$$j_{max} \sim \frac{r - 2}{2} \tag{6.23}$$

enter this theory. The finiteness of the j' makes the amplitude manifestly finite. The theory predicts a maximum length for each individual segment. This works as an infrared cut-off

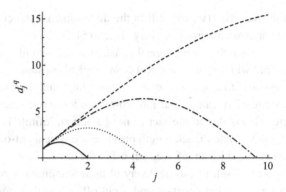

Figure 6.5 The quantum dimension as a function of the representation, for different values of q approaching $q = 1$.

that suppresses the infrared spike divergences. A completely explicit expression for the quantum $6j$-symbol $\{6j\}_q$ is given in the appendix of Noui and Roche (2003).

The triangular inequalities are also modified in this setting: the q-deformed $6j$-symbols vanish unless q-deformed versions of the triangular inequalities hold. These do not correspond to the inequalities satisfied by a triangle in *flat* space, but instead to the inequalities satisfied by a triangle on a sphere of finite radius.

The physical interpretation of this mathematics is discussed in the next section.

6.3.1 Cosmological constant

Ponzano and Regge proved that the asymptotic of the $\{6j\}$ symbol for large spins gives the Regge action of three-dimensional general relativity. This indicates that the classical limit of the theory is indeed general relativity. What does the asymptotic of the q-deformed symbol $\{6j\}_q$ give?

Remarkably, it was shown by Mizoguchi and Tada (1992) that it gives the Regge action for the theory *with cosmological constant* Λ, where the cosmological constant is determined by the deformation parameter by

$$q = e^{i\sqrt{\Lambda}\hbar G} \tag{6.24}$$

or

$$\Lambda = \frac{\pi^2}{(r\hbar G)^2}. \tag{6.25}$$

Therefore the Turaev–Viro amplitude defines a quantum theory whose classical limit is simply related to quantum general relativity with a positive cosmological constant. The presence of the cosmological constant is related to the finiteness of the theory. In the last few years, of course, we have learned from cosmological observations that in Nature the cosmological constant is not zero and is positive. The quantum theory of gravity appears to be consistent only by taking this fact into account.

In Planck units the measured cosmological constant is a very small number. This means that the deformation parameter r is very large, or $q \sim 1$, the maximum spin is very large in the "realistic" (we are still in the three-dimensional euclidean context) theory, and the relevant quantum group is "very close to SU(2)."

Notice that the cosmological constant Λ can be only *positive* in this theory, and this is in accord with the observed real cosmological constant.[2] A positive cosmological constant corresponds to a space of constant curvature, and therefore suggest that space is *closed*: the value of Λ can be put into correspondence with the size of space, which has to be finite. This is due to the fact of not having an infinite resolution for small things (minimal length). The Planck length provides a physical cut-off in the small; the cosmological constant provides an infrared cut-off in the large.

There is a simple physical way of understanding the connection between the presence of a cosmological constant and a cut-off in angular momentum. Let us do this in 3+1 dimensions, where $\hbar G$ has the dimension of a length squared and the relation between the cosmological constant and q is

$$q = e^{i\Lambda\hbar G} . \tag{6.26}$$

Consider an observer in a world where there is a minimal size ℓ determined by the Planck scale and a cosmological constant that determines a large distance scale L. In a maximally symmetric geometry, the observer in this world cannot see farther than the scale $L = 1/(2\Lambda)$. In the euclidean context, this is because the universe is a sphere; in the lorentzian context, because of the cosmological horizon. Now consider the angle ϕ under which such an observer sees an object. To make this angle smaller, we should either move the object farther away, or make it smaller. But there is a limitation in the smallness of the size of the object, given by ℓ, as well as a limitation in the distance, given by L. Therefore the minimal angle under which an object is seen (Figure 6.6) is

$$\phi_{min} \sim \frac{\ell}{L}. \tag{6.27}$$

Now suppose the observer is observing around him. What he sees can be decomposed in spherical harmonics in a sphere surrounding him. To capture smaller objects you need higher spherical harmonics. Since the number of dimensions of the spin-j spherical harmonics is $2j + 1$ and the full solid angle is 4π, the spherical harmonics of spin j capture solid angles $\phi^2 \sim 4\pi/(2j + 1)$, which together with the previous equation gives

$$j_{max} = \frac{2\pi}{\phi_{min}^2} \sim 2\pi \frac{L^2}{\ell^2} = \frac{\pi}{\Lambda\hbar G}. \tag{6.28}$$

Using the definition $q^r = 1$ and (6.23) we get

$$1 = q^r = q^{2j_{max}+1} = q^{2j_{max}+1} \sim q^{\frac{2\pi}{\Lambda\hbar G}} . \tag{6.29}$$

which is solved precisely by (6.26). In words, the maximum spin is the one for which the maximal angular resolution of the corresponding spherical harmonics is needed to capture the smallest object at the largest distance.

[2] Quite the opposite happens in string theory, where the theory seems to like the unphysical *negative* cosmological constants, as in the standard AdS/CFT scenario.

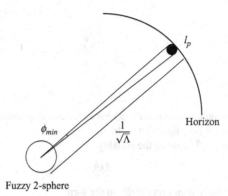

Fuzzy 2-sphere

Figure 6.6   The minimal angle in a four-dimensional universe with minimal size and maximal distance.

The geometry described by the Turaev–Viro theory can be visualized as follows. Consider a 3-sphere, with the radius given by r. Fix the north pole as the point where the observer is, and foliate the 3-sphere by 2-spheres corresponding to parallels. Consider a finite subset of these, for the distances from the north pole given by the values of j. Each of these is characterized by states on a spin-j representation, with a basis $|j,m\rangle$.

This is a neat, complete, finite theory of quantum gravity – but in 3d, where, in fact, not much interesting happens. In the next part we start to address the problem of describing the real world.

6.4 Complements

6.4.1 A few notes on SU(2)$_q$

The quantum group SU(2)$_q$ is not a "group" and there is nothing "quantum" about it. It is *not* a quantum mechanical deformation of the group SU(2); it is a one-parameter deformation of the algebra of the representations of SU(2). The deformation parameter is not to be identified with the Planck constant; if anything, with the cosmological constant.

Consider the standard representations spaces $\mathcal{H}_j$ of SU(2). Each space carries a representation of the su(2) *algebra* given by operators $\vec{L}^{(j)}$; furthermore, these spaces are related by the fact that standard SU(2) representation theory defines a map from $\mathcal{H}_{j_1} \otimes \mathcal{H}_{j_2}$ to $\mathcal{H}_{j_3}$. This map projects on the subspace of $\mathcal{H}_{j_1} \otimes \mathcal{H}_{j_2}$ that transforms in the $\mathcal{H}_{j_3}$ representation. The quantum group SU(2)$_q$ is defined by a deformation of the operators $\vec{L}^{(j)}$ and of the map from $\mathcal{H}_{j_3}$ to $\mathcal{H}_{j_1} \otimes \mathcal{H}_{j_2}$, such that some general algebraic relations between these objects continue to hold. The deformation is characterized by the fact that the map from $\mathcal{H}_{j_1} \otimes \mathcal{H}_{j_2}$ to $\mathcal{H}_{j_3}$ differs from the map from $\mathcal{H}_{j_2} \otimes \mathcal{H}_{j_1}$ to $\mathcal{H}_{j_3}$. In this sense, it is non-commutative.

The first step where this is realized is in the map from $\mathcal{H}_{1/2} \otimes \mathcal{H}_{1/2}$ to $\mathcal{H}_0$. For the standard group SU(2) this map is given by

$$(\phi,\psi) \mapsto \epsilon_{AB}\phi^A\psi^B. \tag{6.30}$$

For SU(2)$_q$, instead,

$$(\phi,\psi) \mapsto \epsilon^q_{AB}\phi^A\psi^B, \tag{6.31}$$

where

$$\epsilon^q_{AB} = \begin{pmatrix} 0 & A \\ -A^{-1} & 0 \end{pmatrix}$$
(6.32)

and

$$A^2 = q$$
(6.33)

is the deformation parameter that characterizes the quantum group. Notice that the map from $\mathcal{H}_{1/2} \otimes \mathcal{H}_{1/2}$ to $\mathcal{H}_0$ is no longer symmetric.

The tensor ϵ^q satisfies the equality

$$\delta^A_C \delta^B_D = A^{-1} \delta^A_B \delta^C_D + A \, \epsilon^{qAC} \epsilon^q_{BD},$$
(6.34)

which can be written graphically in the form

(6.35)

Notice that this is a deformation of the relations (1.47) and (1.50), where we had $A = 1$. See details in the thesis of Haggard (Haggard 2011).

Similarly, for $SU(2)_q$ the map from $2j$ copies of $\mathcal{H}_{\frac{1}{2}}$, that is $\mathcal{H}_{1/2} \otimes \ldots \otimes \mathcal{H}_{\frac{1}{2}}$, $2j$ times to $\mathcal{H}_j$ is simply obtained by projecting in the fully symmetric part of the tensor product. This can be obtained by summing over all permutations σ of the factors. Let P_σ be the operator that realizes the permutation σ of the factors. Then for $SU(2)_q$ the map from $\mathcal{H}_{1/2} \otimes \cdots \otimes \mathcal{H}_{\frac{1}{2}}$, $2j$ times to $\mathcal{H}_j$ is obtained by summing over all permutations σ with a weight. This is given by the projector

$$P_j = \sum_\sigma A^{-3cr(\sigma)} P_\sigma$$
(6.36)

where

$$c^{-1} = \sum_\sigma A^{-4cr(\sigma)},$$
(6.37)

where $cr(\sigma)$ is the (minimal) number of crossings in the permutation. These relations are sufficient to compute the quantum dimension, defined as the trace of P_j. This is

$$\Delta_j = (-1)^{2j} \frac{A^{4j+2} - A^{-4j-2}}{A^2 - A^{-2}} = (-1)^{2j} \frac{\sin((2j+1)h)}{\sin h} \equiv (-1)^{2j} [2j+1],$$
(6.38)

where $q = e^h$. Importantly, the triangular inequalities turn out to be modified and supplemented by the conditions

$$2j_1, 2j_2, 2j_3 \leq j_1 + j_2 + j_3 \leq r - 2,$$
(6.39)

where $q = e^{\frac{i\pi}{r}}$, and these are precisely the triangular inequalities of a triangle on a sphere with a radius determined by r. The geometry of q deformed spin networks is therefore consistent with the geometry of constant curvature space, with the curvature determined by the cosmological constant.

6.4.2 Problem

1. Should we expect an effect of the cosmological constant on the area eigenvalues?

Hint: Recall that these are given by the eigenvalues of the Casimir, which in turn is defined by the SU(2) generators. If SU(2) is deformed to $SU(2)_q$, so are its generators and so is the Casimir.

PART III

THE REAL WORLD

7

The real world: four-dimensional lorentzian theory

"Remember what the dormouse said ..."

This is the main chapter of the book, where the physical theory of quantum gravity is defined. The real world is, as far as all current empirical evidence indicates, four-dimensional and lorentzian. We start from the Holst action of general relativity, given in Section 3.3. Recall that this reads

$$S[e, \omega] = \int e \wedge e \wedge \left(\star + \frac{1}{\gamma} \right) F, \qquad (7.1)$$

where the variables are the tetrad field e and a Lorentz connection ω. On the boundary, the momentum conjugate to ω is the $sl(2, \mathbb{C})$-algebra-valued 2-form

$$\Pi = \frac{1}{8\pi G} B = \frac{1}{8\pi G} \left((e \wedge e)^* + \frac{1}{\gamma}(e \wedge e) \right) \qquad (7.2)$$

where the electric and magnetic parts of B satisfy the linear simplicity constraint (3.46), that is,

$$\vec{K} = \gamma \vec{L}. \qquad (7.3)$$

The theory is invariant under local $SL(2, \mathbb{C})$ transformations. This is a formulation of standard classical general relativity, a theory with very strong empirical support, and we are seeking a quantum theory with this classical limit.

7.1 Classical discretization

We look for the quantum theory following the track of the three-dimensional theory discussed in Section 3.5. For this, we need two ingredients: the generalization to 4d of the discretization on a 2-complex discussed in Section 4.4; and a way to keep track of (7.3) in the quantum theory. The first is straightforward; the second is the technical core of the theory. Let us start with the discretization.

Fix a triangulation Δ of a compact region M of spacetime.[1] That is, chop M into 4-simplices. A 4-simplex can be thought as the convex region of $\mathbb{R}^4$ delimited by 5 points.

---

[1] Later we will generalize the construction to structures more general than a triangulation.

Table 7.1 Bulk terminology and notation	
Bulk triangulation Δ	Two-complex Δ*
4-Simplex (v)	Vertex (v)
Tetrahedron (τ)	Edge (e)
Triangle (t)	Face (f)
Segment (s)	
Point (p)	

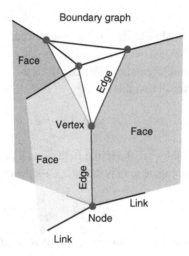

Figure 7.1   Triangulation and 2-complex terminology in 4d.

A 4-simplex is bounded by five tetrahedra. As before, consider the dual Δ* of the triangulation, and focus on its *vertices, edges, and faces*. These objects and their relations, including for the boundary, are listed in Table 7.1 and shown in Figure 7.1.

Notice that the 2-complex objects are exactly the same as those in 3d. They will play similar roles, but they refer to higher-dimensional objects in the triangulation. In particular, 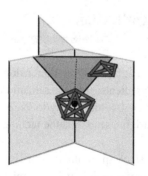 a vertex is still dual to a chunk of spacetime, but now this chunk is a 4-simplex, because we are in four-dimensional. In turn, edges are dual to tetrahedra. See the picture here:

Similarly, on the boundary we still have a graph Γ. Its nodes are still "chunks of space," but now these are tetrahedra, because space is three-dimensional (Table 7.2). The links of the boundary graph connect adjacent tetrahedra and therefore are dual to triangles of the boundary triangulation. A main role is played in the theory by the faces f of the 2-complex (see the picture above). These are dual to triangles of the triangulation (a triangle in the (x, y) plane is dual to a face in the (z, t) plane). A face that touches the boundary is dual to a boundary triangle, and therefore corresponds to a boundary link ℓ. Geometrically, this link is the intersection

Table 7.2 Boundary terminology and notation	
Boundary triangulation $\partial\Delta$	Boundary graph Γ
Tetrahedron (τ)	Node (boundary vertex) (n)
Triangle (t)	Link (boundary edge) (ℓ)

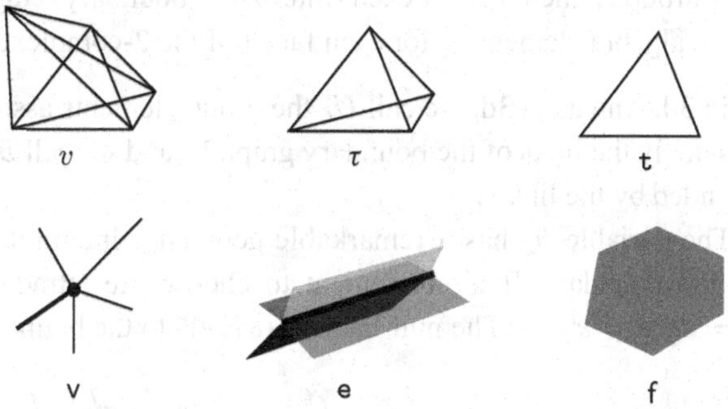

Figure 7.2   A vertex is a point inside the 4-simplex. It has five edges, corresponding to the five tetrahedra bounding the 4-simplex. A triangle is dual to a face that "wraps around it," because in four-dimensional we can go around a triangle in the (x,y) plane, moving in the (t,z) plane.

of the face with the boundary. Therefore a boundary link ℓ is a boundary edge, but is also associated with a face f that touches the boundary. The diligent reader is advised to digest this well, in order not to get lost in the following.

Variables

We discretize the connection, as in three-dimensional, by associating a SL(2, $\mathbb{C}$) group element U_e with each edge e of the 2-complex. We discretize the tetrad by associating an element of the sl(2, $\mathbb{C}$) algebra with each *triangle* of the triangulation (See Figure 7.2). Since triangles are dual to faces, the algebra vectors are associated with faces f, as in three-dimensional. That is

$$\omega \longrightarrow U_e, \tag{7.4}$$

$$e \longrightarrow B_f. \tag{7.5}$$

The formal relation between the continuum and the discrete variables can be taken to be the following:

$$U_e = P\, e^{\int_e \omega} \in \mathrm{SL}(2,\mathbb{C})\,, \tag{7.6}$$

$$B_f = \int_{t_f} B \in sl(2,\mathbb{C})\,. \tag{7.7}$$

The first integral is the holonomy of the connection along the edge, namely, the matrix of the parallel transport generated by the connection along the edge, which we take in the fundamental representation of SL(2, $\mathbb{C}$). The second is the surface integral of the 2-form B across the triangle t_f dual to the face f. Notice that in 3d we had a 1-form e to integrate along a *segment* dual to the faces, while now we have a 2-form B to integrate along a *triangle* dual to the faces. But in either case, what we obtain is an algebra element per *face*. The action can be approximated in terms of these objects.[2]

Thus, the variables of the discretized theory are:

- A group element U_e for each (internal or boundary) edge e of the 2-complex
- An algebra element B_f for each face f of the 2-complex,

as in 3d. And as in 3d, we call U_ℓ the group elements associated with the boundary edges ℓ, namely the links of the boundary graph Γ, and we call B_ℓ the algebra elements of a face bounded by the link ℓ.

The variable B_ℓ has a remarkable geometric interpretation. Consider a triangle lying on the boundary. It is convenient to choose the tetrad field in the time gauge where $e^o = dt$, $e^i = e^i_a dx^a$. The pull-back of $(e \wedge e)^*$ to the boundary vanishes and we have

$$L^i_f = \frac{1}{2\gamma} \epsilon^i{}_{jk} \int_{t_f} e^j \wedge e^k. \tag{7.8}$$

In the approximation in which the metric is constant on the triangle, we see that the norm of this quantity is proportional to the area of the triangle

$$|L_f| = \frac{1}{\gamma} A_{t_f}. \tag{7.9}$$

This can be seen by choosing a gauge and coordinates where $e^i_a = \delta^i_a$. The vector $\vec{L}_f$ itself is normal to the triangle, and has a length proportional to its area. Therefore it is the same quantity (up to the factor $\frac{1}{\gamma}$) as the quantity $\vec{E}_S$ defined in (3.91). We now have all the ingredients to move to the quantum theory.

7.2 Quantum states of gravity

Let us start from the Hilbert space. Since the variables U_e and L_f are associated with edges and faces respectively, the boundary variables are respectively associated with the boundary edges, namely, the links ℓ and the boundary of the faces, which are also links. Therefore, precisely as in the case of the three-dimensional theory, the boundary variables

---

[2] As in three-dimensional, (7.7) can be made more precise to better deal with the gauge. Under a gauge transformation the group elements U defined in (7.6) transform "well," namely as $U_e \mapsto \Lambda_{s_e} U_e \Lambda^{-1}_{t_e}$ where s_e and t_e are the initial ("source") and final ("target") vertices of the (oriented) edge e. Therefore in the discrete theory the continuous local SL(2, $\mathbb{C}$) invariance is reduced to Lorentz transformations at the vertices. Not so for the algebra variables B defined in (7.7), unless a suitable gauge is chosen. To correct this, assume that this definition is taken in a gauge where the connection is constant on the triangle itself, as well as on the adjacent tetrahedron and along the first half of each (oriented) edge. In this way, also the B^i variables are invariant under all gauge transformations except those at the vertices; and they transform covariantly, in fact, in the adjoint representation, under a Lorentz transformation Λ_v at the vertex v.

are $B_\ell \in sl(2,\mathbb{C})$ and $U_\ell \in SL(2,\mathbb{C})$ on each link ℓ of the boundary graph Γ. These become operators in the quantum theory. The natural quantization leads to states $\psi(U_\ell)$, functions on $SL(2,\mathbb{C})^L$, where the operator $B_\ell \in sl(2,\mathbb{C})$ is realized as the generator of $SL(2,\mathbb{C})$ transformations.

However, the components of these generators must satisfy the linear simplicity condition (7.3), namely $\vec{K} = \gamma \vec{L}$, at least in the classical limit. Therefore some limitation of the states is needed for this condition to hold. To realize this, we introduce the Y_γ map.

7.2.1 Y_γ map

Functions on $SL(2,\mathbb{C})$ can be expanded into irreducible *unitary* representations of $SL(2,\mathbb{C})$. The unitary representations of $SL(2,\mathbb{C})$ should not be confused with the common Lorentz representations that are familiar in physics (such as 4-vectors, or spinors), which are not unitary; they are presented in detailed form in Ruhl (1970) and are briefly recalled in the Complements to this chapter. The key facts we need are the following. They are labeled by a positive real number p and non-negative half-integer k. The Hilbert space $V^{(p,k)}$ of the (p,k) representation decomposes into irreducibles of the subgroup $SU(2) \subset SL(2,\mathbb{C})$ as follows:

$$V^{(p,k)} = \bigoplus_{j=k}^{\infty} \mathcal{H}_j, \tag{7.10}$$

where $\mathcal{H}_j$ is a $2j + 1$ dimensional space that carries the spin j irreducible representation of $SU(2)$. In the (p,k) representation we can therefore choose a basis of states $|p,k;j,m\rangle$, with $j = k, k+1, \ldots$ and $m = -j, \ldots, j$. The quantum numbers (p,k) are related to the two Casimir operators of $SL(2,\mathbb{C})$ by

$$|\vec{K}|^2 - |\vec{L}|^2 = p^2 - k^2 + 1, \tag{7.11}$$

$$\vec{K} \cdot \vec{L} = pk, \tag{7.12}$$

while j and m are the quantum numbers of $|\vec{L}|^2$ and L_z respectively.

Now, we want that in the classical limit, namely, in the limit of large quantum numbers,

$$\vec{K} = \gamma \vec{L}. \tag{7.13}$$

If so, the Casimir must satisfy

$$|\vec{K}|^2 - |\vec{L}|^2 = (\gamma^2 - 1)|\vec{L}|^2, \tag{7.14}$$

$$\vec{K} \cdot \vec{L} = \gamma |\vec{L}|^2. \tag{7.15}$$

Passing to quantum numbers, this gives

$$p^2 - k^2 + 1 = (\gamma^2 - 1)j(j+1), \tag{7.16}$$

$$pk = \gamma j(j+1). \tag{7.17}$$

For large quantum numbers this gives

$$p^2 - k^2 = (\gamma^2 - 1)j^2, \tag{7.18}$$

$$pk = \gamma j^2, \tag{7.19}$$

which is solved by the two fundamental equations

$$p = \gamma k, \tag{7.20}$$

$$k = j. \tag{7.21}$$

The first of these two relations is a restriction on the set of the unitary representations. The second picks out a subspace (the lowest spin subspace in the sum (7.10)) within each representation. The states that satisfy these relations have the form

$$|p, k; j, m\rangle = |\gamma j, j; j, m\rangle. \tag{7.22}$$

Notice that these states are in one-to-one correspondence with the states in the representations of SU(2). We can thus introduce a map Y_γ as

$$Y_\gamma : \mathcal{H}_j \rightarrow V^{(p=\gamma j, k=j)}, \tag{7.23}$$

$$|j; m\rangle \mapsto |\gamma j, j; j, m\rangle, \tag{7.24}$$

and all the vectors in the image of this map satisfy the simplicity constraints, in the weak sense. That is,

$$\langle Y_\gamma \psi | \vec{K} - \gamma \vec{L} | Y_\gamma \phi \rangle = 0, \tag{7.25}$$

in the large j limit.[3] Thus we assume that the states of quantum gravity are constructed from the states $|\gamma j, j; j, m\rangle$ alone.

These states are in one-to-one correspondence with the SU(2) states $|j; m\rangle$: the map is given indeed by Y_γ. This map extends immediately to a map from functions over SU(2) to functions over SL(2, $\mathbb{C}$). Explicitly, this is given by

$$Y_\gamma : \qquad L_2[\text{SU}(2)] \rightarrow F[SL(2, \mathbb{C})], \tag{7.26}$$

$$\psi(h) = \sum_{jmn} c_{jmn} D_{mn}^{(j)}(h) \mapsto \psi(g) = \sum_{jmn} c_{jmn} D_{jm\,jn}^{(\gamma j, j)}(g). \tag{7.27}$$

And therefore we have a map from SU(2) spin networks to SL(2, $\mathbb{C}$) spin networks.

We will see soon that the map Y_γ is the core ingredient of the quantum gravity dynamics. It depends on the Einstein–Hilbert action and codes the way SU(2) states transform under SL(2, $\mathbb{C}$) transformations in the theory. This, in turn, codes the dynamical evolution of the quantum states of space.

The physical states of quantum gravity are thus SU(2) spin networks, or, equivalently, their image under Y_γ. Notice that this space carries a scalar product that is well defined:

---

[3] This can be shown as follows. The only non-vanishing matrix element can be $\langle \gamma j, j; j, m | K^i | \gamma j, j; j, m' \rangle$, and these must be proportional to the intertwiner between the representations j, j and 1. Similarly for $\langle \gamma j, j; j, m | L^i | \gamma j, j; j, m' \rangle$. Therefore the two are proportional. The proportionality factor can be determined by reconstructing the Casimirs from the matrix elements.

the one determined by the SU(2) Haar measure. The fact that the scalar product is SU(2) and not SL(2, $\mathbb{C}$) invariant (it is only covariant under SL(2, $\mathbb{C}$)) reflects the fact that the scalar product is associated with a boundary, and this picks up a Lorentz frame.

Notice that the fact that the Hilbert space of quantum gravity is formed by SU(2) spin networks is consistent with the canonical analysis of the theory of Section 3.4.1. There, it was shown that the boundary degrees of freedom are given by the three-dimensional metric. This can be expressed in terms of triads and, in turn, these give rise to the Ashtekar canonical pair (A_a^i, E^{ia}), which form precisely the same kinematical phase space as that of SU(2) Yang–Mills theory. The corresponding quantum states are SU(2) spin networks. All is therefore nicely consistent. Recall also that generators $\vec{L}$ of SU(2) transformations are $1/\gamma$ times E^{ia}, which is the area element. Therefore, restoring physical units from (3.39), we can identify the operator

$$\vec{E}_l = 8\pi\gamma\hbar G \vec{L}_\ell \tag{7.28}$$

associated to each link of the boundary graph as the normal to the corresponding triangle in the boundary, normalized so that the area of the triangle is

$$A_\ell = 8\pi\gamma\hbar G |\vec{L}|^2. \tag{7.29}$$

This gives immediately the eigenvalues of the area of a single triangle as

$$A_j = 8\pi\gamma\hbar G\sqrt{j(j+1)}, \tag{7.30}$$

and the eigenvalues of an arbitrary surface punctured by N punctures $n_1, \ldots, N$ as

$$A_{j_n} = 8\pi\gamma\hbar G\sum_n \sqrt{j_n(j_n+1)}, \tag{7.31}$$

where j and the j_n's are non-negative half-integers.

Notice that this is precisely the result that we obtained in the first chapter, and so we can now fix the loop quantum gravity scale l_o mentioned in the first chapter

$$l_o^2 = 8\pi\gamma\hbar G, \tag{7.32}$$

where γ is now the Barbero–Immirzi constant. The parameter that we left free in Chapter 1 is determined by the action of general relativity and identified with the coupling constant of the Holst term.

7.2.2 Spin networks in the physical theory

In 3d, the area quantum numbers were sufficient to determine a basis of states, without further degeneracy. The same is not true in 4d. This can be seen as follows. Decomposing the Hilbert space as we did in three-dimensional, we have (see Eq. 5.17)

$$L_2[\mathrm{SU}(2)^L] = \otimes_\ell \left[\oplus_j (\mathcal{H}_j \otimes \mathcal{H}_j) \right] = \oplus_{j_\ell} \otimes_\ell (\mathcal{H}_j \otimes \mathcal{H}_j) \tag{7.33}$$

and (see Eq. 5.20)

$$L_2[\mathrm{SU}(2)^L/\mathrm{SU}(2)^N] = \oplus_{j_\ell} \otimes_n \mathrm{Inv}_{\mathrm{SU}(2)} \left(\mathcal{H}_{j_1} \otimes \mathcal{H}_{j_2} \otimes \mathcal{H}_{j_3} \otimes \mathcal{H}_{j_4} \right), \tag{7.34}$$

which differs from (5.20) because now each edge is bounded by four faces, not three. The space $\text{Inv}_{SU(2)}\left(\mathcal{H}_{j_1} \otimes \mathcal{H}_{j_2} \otimes \mathcal{H}_{j_3} \otimes \mathcal{H}_{j_4}\right)$ is not one-dimensional in general. In fact, linearly independent invariant tensors in this space can be constructed as follows

$$i_k^{m_1 m_2 m_3 m_4} = \begin{pmatrix} j_1 & j_2 & k \\ m_1 & m_2 & m \end{pmatrix} g_{mn} \begin{pmatrix} k & j_3 & j_4 \\ n & m_3 & m_4 \end{pmatrix} \tag{7.35}$$

for any k that satisfies the triangular relations both with j_1, j_2 and j_3, j_4, that is, for any k satisfying

$$\max[|j_1 - j_2|, |j_3 - j_4|] \geq k \leq \min[j_1 + j_2, j_3 + j_4]. \tag{7.36}$$

These states are denoted $|k\rangle$. Therefore we have a space

$$\mathcal{K}_{j_1 \ldots j_4} \equiv \text{Inv}_{SU(2)}\left(\mathcal{H}_{j_1} \otimes \mathcal{H}_{j_2} \otimes \mathcal{H}_{j_3} \otimes \mathcal{H}_{j_4}\right) \tag{7.37}$$

with dimension

$$\dim[\mathcal{K}_{j_1 \ldots j_4}] = \min[j_1 + j_2, j_3 + j_4] - \max[|j_1 - j_2|, |j_3 - j_4|] + 1, \tag{7.38}$$

which describes the residual geometrical freedom at each tetrahedron when the area of its faces is sharp.

It follows that a generic gauge-invariant state is a linear combination

$$\psi(U_\ell) = \sum_{j_\ell k_n} C_{j_\ell k_n} \psi_{j_\ell k_n}(U_\ell) \tag{7.39}$$

of the orthogonal states

$$\psi_{j_\ell k_n}(U_\ell) = i_{k_1}^{m_1 m_2 m_3 m_4} \ldots i_{k_N}^{m_{L-3} m_{L-2} m_{L-1} m_L} D_{m_1 n_1}^{j_1}(U_{\ell_1}) \ldots D_{m_L n_L}^{j_L}(U_{\ell_L}), \tag{7.40}$$

which replace (5.33) and (5.34). The difference between the 3d case and the 4d case is that the spin networks of trivalent graphs are labeled only by the spins, while the spin networks of graphs with higher valence are labeled also by an "intertwine" quantum number k associated with each node n. These are the spin-network wave functions. We write them also in the simpler form

$$\psi_{j_\ell k_n}(U_\ell) = \langle U_\ell | j_\ell, k_n \rangle = \bigotimes_n i_{k_n} \cdot \bigotimes_\ell D^{j_\ell}(U_\ell). \tag{7.41}$$

These are the 4d spin-network states in the group representation.

Classically, this residual geometric freedom is described by the phase space of the tetrahedron, namely, the space of the possible shapes of a tetrahedron with fixed areas. This is a two-dimensional space. For instance, one can take two opposite dihedral angles to coordinatize this space. This space can also be seen as the space of the quadruplets of vectors satisfying the closure relation, with given areas, up to global rotations. Once again, the counting of dimensions gives $(4 \times 3 - 4 - 3 - 3 = 2)$. An observable on this space, in particular, is the (oriented) volume V of the tetrahedron, which we have already studied in the first chapter, which is given by

$$V^2 = \frac{2}{9} \epsilon_{ijk} E^i E^j E^k. \tag{7.42}$$

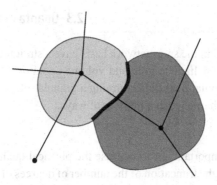

Figure 7.3 Chunks of space with quantized volume are associated to nodes. The area of the surface shared between two cells is quantized as well.

The matrix elements of the volume can be computed explicitly in the $|k\rangle$ basis (see Complements of this chapter), by using standard angular momentum theory. Diagonalizing the resulting matrix, one can then compute the eigenvalues v and the eigenstates $|v\rangle$ of the volume in each Hilbert space $\mathcal{K}_{j_1...j_4}$. In this basis, we write the spin-network states as

$$\psi_{j_\ell v_n}(U_\ell) = \langle U_\ell | j_\ell, v_n \rangle = \bigotimes_n \iota_{v_n} \cdot \bigotimes_\ell D^{j_\ell}(U_\ell), \qquad (7.43)$$

where the ι_{v_n} form a basis that diagonalizes the volume operator in the intertwined space of the node n.

In summary, the truncated state space of quantum gravity is given by the Hilbert space

$$\mathcal{H}_\Gamma = L_2[\mathrm{SU}(2)^L/\mathrm{SU}(2)^N], \qquad (7.44)$$

where the triad operator (or flux operator) associated with each link is

$$\vec{E}_\ell = l_o^2 \vec{L}_\ell \qquad (7.45)$$

and $\vec{L}_\ell$ is the generator of SU(2) transformations on a link, namely, the left invariant vector field. Notice that their algebra is the one that in chapter 1 was simply postulated in (1.12), and the dimensional factor in (1.13) is now determined by the general relativity action, which fixes the scale of the momentum. Furthermore, the closure relation (1.9) is a direct consequence of the SU(2) invariance of the states at each node.

Increasing the complexity of the graph Γ gives a better approximation to the full theory. A basis of states in $\mathcal{H}_\Gamma$ is provided by the spin-network states $|\Gamma, j_\ell, v_n\rangle$, with a spin j_ℓ associated to each link of the graph and a volume eigenvalue v_n associated with each node of the graph. These are eigenstates of the area and the volume operators. This is of course a *separable* Hilbert space,[4] as the explicit discrete basis shows. See Figure 7.3.

7.2.3 Quanta of space

The states in the spin-network basis have a straightforward physical interpretation because they diagonalize the area and volume operators. They are interpreted as quantum states of the gravitational field, forming a granular physical space. This simple intuitive picture must be sharpened in a number of ways.

1. It is important not to confuse the physical quantum discreteness with the discreteness due to the truncation of the number of degrees of freedom. This confusion is so common that it is better to emphasize it again, at the cost of repetition. The fact that we work on a graph is a *classical* truncation, analogous to expanding a field in a box in discrete modes, and truncating the theory considering only a finite numbers of modes. There is nothing related to quantum mechanics in this. In the truncated classical theory, the area of the faces of the triangulation can become arbitrarily small, and the geometry of a discrete triangulation can approximate arbitrarily well a continuum geometry.

   The *quantum* discreteness, on the other hand, is due to the discrete *spectrum* of area and volume. This is analogous to the quantized energy spectrum of each mode of, say, the electromagnetic field. The quanta of energy on a mode are individual photons. This is a genuine quantum phenomenon. In gravity, the area of each triangle and the volume of each tetrahedron are quantized. This is also a genuine quantum phenomenon, with a scale set by $\hbar$.

   Because of this, there is no limit in which the tetrahedra become infinitely small. This is the central point of loop quantum gravity. It is the physical discreteness of space unveiled by loop quantum gravity.

2. The spin-network states do not fully diagonalize the three-dimensional metric. This is because the area and the volume are not sufficient for determining the geometry of a tetrahedron. The geometry of the tetrahedron is determined by six numbers (the six lengths of its sides), while areas and volume are only five numbers. The situation is like that in angular momentum theory: classical angular momentum is determined by three numbers (L_x, L_y, L_x) but only two operators can be diagonalized simultaneously (L^2, L_x), with the consequence that the angular momentum $\vec{L}$ can never be sharp. Similarly, the three-dimensional geometry can never be sharp. Therefore the spatial metric remains always quantum-fuzzy at the Planck scale.

   It follows that the geometrical picture of "tetrahedra," "triangles," etc., must be taken as something meaningful only in some classical approximation and not at the fundamental scale. At the Planck scale there are "tetrahedra" in the same sense in which an electron is a "small rotating pebble." That is, in a very imprecise sense. There are no tetrahedra down at the Planck scale. There are quantum states, formed by quanta of the gravitational field, which have the property of giving rise to something that we describe as a three-dimensional Riemannian geometry in the limit of large quantum numbers.

---

[4] A persistent rumour that the Hilbert of LQG is non-separable is false.

3. The quanta of space described by the spin-network states $|\Gamma, j_\ell, v_n\rangle$ should not be thought of as quanta moving in space. They are not *in space*. They *are* themselves physical space. This can be seen by comparing them to the usual description of the quanta of the electromagnetic field. These are characterized by quantum numbers $|p_1, \ldots, p_n\rangle$ where the p_i's are momenta, namely, Fourier transforms of *position* variables. Position is given with respect to a space in which the photons live. In quantum gravity, instead, the quantum numbers do not include position. They include the intrinsic physical size of the quanta themselves (area and volume) as well as the graph Γ that codes the adjacency relations between these quanta. Therefore the quanta of space are located with respect to one another, and this relation is given by the combinatorial structure of Γ. This is the manner in which loop quantum gravity realizes spatial background independence in the quantum theory. This is the quantum analog of what happens in classical general relativity, where the gravitational field does not live in spacetime: it is itself spacetime.

4. Finally, a warning: the quanta of space of loop quantum gravity should not be taken too naively as actual entities, but rather as modes of interaction. What the theory predicts is that in any interaction where the effect of something depends on its area, this effect will be that of a quantized area. Trying to think too literally in terms of concrete "chunks" forming the quanta of space can be misleading, as always in quantum mechanics. A pendulum is not "made of" its quanta. It is the way it acts in an interaction that is characterized by quantized energy. This is important especially in applications of the theory, where a naively realistic picture, as always in quantum mechanics, can be misleading. A basis in the Hilbert space should not be mistaken for a list of "things," because bases associated with non-commuting observers are equally physically meaningful.

7.3 Transition amplitudes

Finally, we can now complete the construction of the full theory, writing its transition amplitudes.

The experience in three-dimensional gives us the general structure of these amplitudes. Let us recall the three-dimensional theory: this is given in (5.69) and (5.74), which we copy here:

$$W(h_\ell) = \int_{SU(2)} dh_{vf} \prod_f \delta(h_f) \prod_v A_v(h_{vf}), \tag{7.46}$$

where $h_f = h_{vf} h_{v'f} \ldots$ and the vertex amplitude is

$$A_v(h_{vf}) = \sum_{j_f} \int_{SU(2)} dg'_{ve} \prod_f d_{j_f} \mathrm{Tr}_{j_f}[g_{e'v} g_{ve} h_{vf}]. \tag{7.47}$$

Since in 4d the kinematical Hilbert space is essentially the same as in 3d, the first equation can remain unchanged in 4d. In fact, it only reflects the superposition principle of quantum mechanics, for which amplitudes are obtained by summing for amplitudes of individual stories, and a locality principle, for which the amplitude of a story is the product of individual local amplitudes associated with separate regions of spacetime. The actual dynamics is given by the vertex amplitude, which determines the local physics. What is the vertex amplitude in 4d?

It must be a function $A_v(h_{vf})$ of SU(2) group elements living on the graph of the node (namely, on the boundary of a spacetime 4-simplex), but it must be SL(2,$\mathbb{C}$) invariant. The amplitude (7.47) is only invariant under SU(2), and does not know anything about SL(2,$\mathbb{C}$). In order to obtain an amplitude which is SL(2,$\mathbb{C}$) invariant we must replace the SU(2) integrals in (7.47) with SL(2,$\mathbb{C}$) integrals, and somehow map the SU(2) group elements into SL(2,$\mathbb{C}$) ones.

We do have the tool for this, which is the map Y_γ. The map can be used to map a function of SU(2) variables into a function of SL(2,$\mathbb{C}$) variables. These can then be projected on SL(2,$\mathbb{C}$) invariant states and evaluated:

$$A_v(\psi) = \left(P_{SL(2,\mathbb{C})} Y_\gamma \psi \right) (\mathbb{1}). \tag{7.48}$$

Let us write this explicitly. In the group representation:

$$A_v(h_{vf}) = \sum_{j_f} \int_{SL(2,\mathbb{C})} dg'_{ve} \prod_f (2j_f + 1) \, \mathrm{Tr}_{j_f}[Y_\gamma^\dagger g_{e'v} g_{ve} Y_\gamma h_{vf}], \tag{7.49}$$

where the meaning of the Trace notation is

$$\mathrm{Tr}_j[Y_\gamma^\dagger g Y_\gamma h] = \mathrm{Tr}[Y_\gamma^\dagger D^{(\gamma j,j)}(g) \, Y_\gamma \, D^{(j)}(h)] = \sum_{mn} D^{(\gamma j,j)}_{jm,jn}(g) \, D^{(j)}_{nm}(h). \tag{7.50}$$

The vertex amplitude is a function of one SU(2) variable per face around the vertex. Imagine drawing a small sphere surrounding the vertex: the intersection between this sphere and the 2-complex is a graph Γ_v (see Figure 7.8). The vertex amplitude is a function of the states in

$$\mathcal{H}_{\Gamma_v} = L_2[SU(2)^{10}/SU(2)^5]_{\Gamma_v}. \tag{7.51}$$

The graph Γ_v is the complete graph with five nodes (all the nodes connected); see Figure 7.4.

The prime in the SL(2,$\mathbb{C}$) integrations, introduced in (5.73), indicates that we do not integrate in five group variables, but only in four of them. The result of the integration turns out to be independent of the fifth. While this was just cosmetic in 3d, here this procedure is required, because the integration on the last SL(2,$\mathbb{C}$), which is non-compact, would give a divergence if we did not take this precaution. As it is defined, instead, the vertex amplitude can be proven to be finite (Engle and Pereira 2009; Kaminski 2010).

This completes the definition of the covariant formulation of the theory of loop quantum gravity on a given 2-complex.

The dynamics realizes the three criteria that we stated in Section 5.3, which we repeat here, adapted to the physical four-dimensional case:

1. **Superposition principle.** Basic principle of quantum mechanics. Amplitudes are sums of elementary amplitudes (in standard quantum theory, this is Feynman's sum over

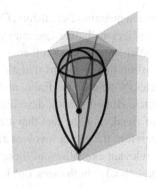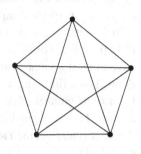

Figure 7.4 The vertex graph in 4d.

paths σ)

$$\langle W|\psi\rangle = \sum_\sigma W(\sigma). \tag{7.52}$$

2. **Locality.** Fundamental discovery of nineteenth- and twentieth-century physics: interactions are local in spacetime. Elementary amplitudes are products of amplitudes associated with spacetime points (in standard quantum field theory, exponential of an integral over spacetime points),

$$W(\sigma) \sim \prod_v W_v. \tag{7.53}$$

3. **Lorentz invariance.**

$$W_v = (P_{SL(2,\mathbb{C})} \circ Y_\gamma \psi_v)(\mathbb{1}), \tag{7.54}$$

where ψ is a boundary state, $P_{SL(2,\mathbb{C})}$ is the projector on its locally $SL(2,\mathbb{C})$-invariant component, and $(\mathbb{1})$ indicates the evaluation of the spin-network state on the identity.

As we shall see later:

1. The theory is related to general relativity in the classical limit.
2. It is ultraviolet finite.
3. With a suitable quantum-deformation which adds the cosmological constant, it is also infrared finite.
4. It leads to the n-point functions of perturbative general relativity, to the Friedmann equation in cosmology, and the (finite) Bekenstein–Hawking black-hole entropy.

7.3.1 Continuum limit

The equations above define a theory with a state space defined on a given graph Γ, and amplitudes determined by a given 2-complex $\mathcal{C}$. This is a theory with a finite number of degrees of freedom. This definition obviously corresponds to a *truncation* of classical general relativity, which is a theory with an infinite number of degrees of freedom. The full theory is approximated by choosing increasingly refined $\mathcal{C}$ and $\Gamma = \partial \mathcal{C}$.

This procedure is a common one in quantum field theory. Concrete physical calculations are mostly performed on *finite* lattices in lattice quantum chomodynamics (QCD), say with N vertices, and at a *finite* order in perturbation theory in quantum electrodynamics (QED). A finite order in perturbation theory means that all Feynman graphs considered have at most a number of particles N, where N is finite, and therefore are de facto fully defined on a finite sum of n-particle Hilbert spaces, describing arbitrarily high momenta, but a finite number of degrees of freedom. (The fact that arbitrarily high momenta enter a single Feynman graph, and therefore the graph involves an infinite number of modes, does not contradict the fact that relevant excitations of these modes are those that can be described by a finite number of particles.) In both cases, one formally considers the $N \to \infty$ limit, but concretely one always computes at finite N, with some arguments that the rest be smaller.

Quantum gravity is defined in the same manner: one considers states and amplitudes of the truncated theory, and chooses a refinement, namely, Γ and $\mathcal{C}$, sufficient for the desired precision.

Let us first consider the state space. The crucial observation in this regard is the following. Let Γ' be a subgraph of Γ, namely, a graph formed by a subset of the nodes and links of Γ. Then it is immediately seen that there is a subspace $\mathcal{H}_{\Gamma'} \subset \mathcal{H}_\Gamma$ which is precisely isomorphic to the loop-gravity Hilbert space of the graph Γ'. Indeed, this is formed by all states $\psi(U_\ell) \in \mathcal{H}_\Gamma$ which are *independent* of the group elements U_ℓ associated with the links ℓ that are in Γ but not in Γ'. Equivalently, $\mathcal{H}_{\Gamma'}$ is the linear span of the spin-network states characterized by $j_\ell = 0$ for any ℓ that is in Γ but not in Γ'.

Exercise 7.1 *Show that the two definitions above are equivalent.*

Therefore if we define the theory on Γ, we have at our disposal a subset of states that captures the theory defined on the smaller graph Γ'. In this sense the step from Γ' to Γ is a refinement of the theory: the refined theory includes the same states, plus others.

One can define a Hilbert space $\mathcal{H}$ that contains all $\mathcal{H}_\Gamma$'s by using projective-limit techniques (Ashtekar and Lewandowski 1995). The family of all spaces $\mathcal{H}_\Gamma$ for all Γ indeed forms a projective family. Since this will not be of much concrete use, we do not go into these mathematical niceties here. We remark, however, that there is a stringent similarity between this structure and Fock space. Fock space can be constructed as follows. First, define the one-particle Hilbert space $\mathcal{H}_1$, the n-particle Hilbert space as the symmetric tensor product $\mathcal{H}_n = \mathcal{H}_1 \otimes_S \cdots \otimes_S \mathcal{H}_1$ of n copies of $\mathcal{H}_1$, and the space with N or fewer particles $\mathcal{H}_N = \oplus_{n=1}^N \mathcal{H}_n$. The space $\mathcal{H}_{N-1}$ is then a subspace of the space $\mathcal{H}_N$ and the Fock space is the formal limit $N \to \infty$. The analogy with loop gravity is between the spaces $\mathcal{H}_\Gamma$, where Γ has N nodes, and $\mathcal{H}_N$. The resulting state space is separable, as the Fock space is separable. Keep in mind that a physicist at CERN who computes scattering amplitudes virtually never cares about Fock space and works entirely within $\mathcal{H}_N$ for some finite N.

Next, consider the transition amplitudes. Again, we can formally define the transition amplitude $W_\Gamma(h_\ell)$ as a limit of $W_\mathcal{C}(h_\ell)$ where $\partial \mathcal{C} = \Gamma$, when $\mathcal{C}$ is refined. The notion of limit is well defined[5] because the set of the two complexes is a partially ordered set with upper limit.[6] Again, it is not clear whether this notion of limit is actually useful or even the relevant one. As far as physics is concerned, the theory is useful to the extent a subclass of graphs and 2-complexes suffice to capture the relevant physics, in the same sense in which low-order Feynman graphs, or QCD lattices with a finite number of sites, are sufficient to compute relevant physics.

This raises the question of the regime of validity of the approximation introduced with a given Γ and $\mathcal{C}$. The answer follows immediately from analyzing the classical theory. The approximation is expected to be good when the discretized theory approximates the continuum theory in the classical context. The regime of validity of a discretization of general relativity is known. The discretization is a good approximation when the curvature on the hinges is small; namely, when the deficit angles, in Regge language, are small. Thus, the expansion in Γ and $\mathcal{C}$ is an expansion which is good around flat space.

The reader might immediately wonder if expanding around flat space was not precisely what went wrong with the conventional *non-renormalizable* quantum-field-theoretical quantization of general relativity. Why are we not finding here the same difficulties? The answer is simple: because in the quantization here the Planck scale discreteness of space, which cuts off the infinities, is taken into account in the formalism. In the conventional Feynman perturbation expansion, it is not.

7.3.2 Relation with QED and QCD

The similarities between the theory defined above and lattice QCD on a given lattice are evident. In both cases, we have a discretization of the classical theory where the connection is replaced by group elements, and a quantum theory defined by an integration over configurations of a configuration amplitude that is a product of local amplitudes. The use of a triangulation instead of a square lattice simply reflects the fact that a square lattice is unnatural in the absence of a flat metric. The more complicated appearance of the gravity amplitude with respect to the Wilson one is simply a reflection of the particular action of gravity.

The substantial difference, as already pointed out, is the fact that the Wilson action depends on an external parameter (as the Yang–Mills theory depends on a fixed external metric), while the gravity action does not (as the Einstein action does not). Accordingly, the continuum limit of lattice Yang–Mills theory involves also the tuning of a parameter (β) to a critical value ($\beta = 0$), while nothing of the sort is needed for gravity.

But there are also striking similarities between the theory introduced here and perturbative QED. The nodes of the graph can be seen as quanta of space and the 2-complex can be read as a history of quanta of space, where these interact, join and split, as in the Feynman graphs. The analogy is strongly reinforced by the fact that the spinfoam amplitude can actually be concretely obtained as a term in a Feynman expansion of an auxiliary field theory, called "group field theory." In this book, we shall not develop the group-field-theory

---

[5] $W = \lim_{\mathcal{C} \to \infty} W_{\mathcal{C}}$ means that there is a $\mathcal{C}_\epsilon$ such that $|W - W_{\mathcal{C}}| < \epsilon$ for any $\mathcal{C} > \mathcal{C}_\epsilon$.

[6] For any $\mathcal{C}$ and $\mathcal{C}'$ there is a $\mathcal{C}''$ such that $\mathcal{C}'' > \mathcal{C}$ and $\mathcal{C}'' > \mathcal{C}'$.

presentation of quantum gravity, for which we refer the reader to for instance Chapter 9 of Rovelli (2004). The specific group field theory that gives the gravitational amplitude has been derived (in the euclidean context) by Krajewski *et al.* (2010). We shall, however, discuss the physical reason for this surprising convergence between the formalisms characteristic of our two most physically successful formulations of fundamental quantum theories: perturbative QED and lattice QCD.

Feynman graphs and the lattice QCD appear to be very different entities. But are they? A Feynman graph is a history of quanta of a field. The lattice of lattice QCD is a collection of chunks of spacetime. But spacetime is a field: it is the gravitational field. And its chunks are indeed quanta of the gravitational field. Therefore when the gravitational field becomes dynamical and quantized, we expect that the lattice of lattice QCD is itself a history of gravitational quanta, namely, a Feynman graph of a quantum field theory. (This, in addition, largely fixes the apparent freedom in the discretization of the theory, which is apparent in lattice QCD.) In both cases, an increased approximation is obtained by increasing the refinement of the graph, or the lattice.

These vague intuitive ideas are realized in concrete in the covariant formulation of loop quantum gravity, where, because of the general covariance of gravity, the QED and the lattice QCD pictures of quantum field theory are beautifully merged (see Figure 7.5).

7.4 Full theory

Summarizing, for any graph Γ we have a Hilbert space

$$\mathcal{H} = L_2[\mathrm{SU}(2)^L / \mathrm{SU}(2)^N]_\Gamma, \tag{7.55}$$

with operators

$$\vec{E}_\ell = l_o^2 \vec{L}_\ell = 8\pi\gamma\hbar G \vec{L}_\ell, \tag{7.56}$$

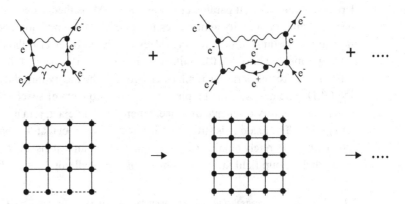

Figure 7.5   QED Feynman graphs and lattice QCD lattices. The two pictures converge in LQG.

interpreted as the flux of the densitized inverse triad on the surfaces bounding the space regions labeled by the nodes. The transition amplitudes between these states can be computed at each order in a truncation determined by a 2-complex $\mathcal{C}$ bounded by Γ. They are given (reinserting a normalization factor dependent on the 2-complex) by

$$W_\mathcal{C}(h_\ell) = N_\mathcal{C} \int_{SU(2)} dh_{vf} \prod_f \delta(h_f) \prod_v A_v(h_{vf}). \tag{7.57}$$

where the vertex amplitude is

$$A_v(h_{vf}) = \sum_{j_f} \int_{SL(2,\mathbb{C})} dg'_{ve} \prod_f (2j_f + 1)\,\mathrm{Tr}_{j_f}[Y_\gamma^\dagger g_{e'v} g_{ve} Y_\gamma h_{vf}]. \tag{7.58}$$

Together with the definition of the Y_γ map, which we repeat here:

$$Y_\gamma : |j,m\rangle \mapsto |\gamma j, j; j, m\rangle. \tag{7.59}$$

Given an observable $\mathcal{O}$ defined on $\mathcal{H}$, we define its expectation value as

$$\langle \mathcal{O} \rangle = \frac{\langle W | \mathcal{O} | \Psi \rangle}{\langle W | \Psi \rangle}. \tag{7.60}$$

This is the full definition of the theory, in the limit in which the cosmological constant vanishes.

For each choice of a graph Γ we have a given truncation of the theory. Refining the graph, we obtain a better approximation. At a given graph, for each choice of 2-complex $\mathcal{C}$ we have a given truncation of the transition amplitudes. Refining the 2-complex, we have a better approximation. The difference with the 3d theory is therefore substantial: in 3d the topological invariance of the theory, namely, the absence of local degrees of freedom, implies that a refinement of the triangulation has no effect on the quantum theory; while in 4d a more refined triangulation can capture more of the infinite number of local degrees of freedom. These refinements determine what is called the "continuum limit" of the theory. A priori, there is no parameter to tune in approaching this limit, for the reasons explained in Chapter 2.

This is the theory that we analyze in detail in the rest of the book. The construction of this theory is the result of a long process, to which a large number of people have contributed. A very condensed historical account with the main references is in the Complements of the last chapter.

7.4.1 Face amplitude, wedge amplitude, and the kernel P

The amplitude above can be equivalently rewritten as a product of face amplitudes instead of a product of vertex amplitudes. For this, it suffices to expand the delta function in representations and perform the dh_{vf} integrals. Then all j_f in the different amplitudes that pertain to the same face are constrained to be equal and we obtain

$$W_C(h_f) = N_C \sum_{j_f} \int_{SL(2,\mathbb{C})} dg'_{ve} \prod_f \mathrm{Tr}_{j_f}[Y^\dagger_\gamma g_{e'v} g_{ve} Y_\gamma \; Y^\dagger_\gamma g_{ev'} g_{v'e''} Y_\gamma \dots], \qquad (7.61)$$

where the trace runs around the face, with the proper orientations. The trace is the face amplitude.

Alternatively, the "wedge amplitude" is defined by

$$W(g,h) = \sum_j (2j+1)\, \mathrm{Tr}_j[Y^\dagger_\gamma g Y_\gamma h]. \qquad (7.62)$$

It is associated with a wedge, as in Figure 7.6. If we Fourier transform this to the basis formed by the Wigner matrix elements,

$$W(g,j,m,m') = \langle j,m'|Y^\dagger g Y|j,m\rangle. \qquad (7.63)$$

and take $g = e^{i\eta K_z}$, this gives the amplitude for a transition between the state $|j,m\rangle$ and the state $|j,m'\rangle$ as observed by an accelerated observer undergoing a boost of lorentzian angle η.

$$W_{m,m'}(\eta) = \langle j,m'|Y^\dagger e^{iK\eta} Y|j,m'\rangle. \qquad (7.64)$$

This matrix element, like several other quantities related to $SL(2,\mathbb{C})$ representations, can be found explicitly in Ruhl (1970), in terms of a hypergeometric function, and reads

$$W_{m,m'}(\eta) = \delta_{mm'} e^{-\eta m} e^{-\eta(1+i\gamma)(j+1)} \,{}_2F_1((1+i\gamma)(j+1), m+j+1, 2j+2; 1-e^{-2\eta}). \quad (7.65)$$

The full amplitude can then be written in terms of the wedge amplitude in the simple wedge form

$$\boxed{W_C(h_\ell) = N_C \int dh_{vf}\, dg'_{ve} \prod_f \delta(h_f) \prod_w W(g_{e'v} g_{ve}, h_{vf}).} \qquad (7.66)$$

Finally, it is often convenient to rewrite the vertex amplitude (7.58) in the form

$$A_v(h_\ell) = \int_{SL(2,\mathbb{C})} dg'_n \prod_\ell P[g_{s(\ell)} g^{-1}_{t(\ell)}, h_\ell], \qquad (7.67)$$

where the ℓ, n notation refers to the nodes and the links of the vertex graph, and the kernel is

$$\boxed{P[g,h] = \sum_j d_j\, \mathrm{Tr}_j[Y^\dagger_\gamma g Y_\gamma h].} \qquad (7.68)$$

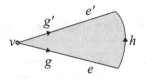

Figure 7.6   The wedge.

Explicitly:

$$P[h,g] = \sum_j (2j+1)\, D^{(\gamma j,j)}_{jm\,jn}(g)\, D^{(j)}_{mn}(h)\,. \tag{7.69}$$

This kernel is a key quantity that appears repeatedly in the calculations. It expresses the link between SU(2) and SL(2, $\mathbb{C}$).

7.4.2 Cosmological constant and IR finiteness

In Section 6.3 we saw that in three dimensions there is a version of the transition amplitudes, given by the Turaev–Viro model, which is finite, and whose classical limit gives general relativity with a positive cosmological constant. The same happens in four dimension. The quantum deformation of the $SL(2,\mathbb{C})$ group needed for defining the theory has been derived by Buffenoir and Roche (Buffenoir and Roche 1999) and Noui and Roche (Noui and Roche 2003), and the theory has been constructed by Fairbairn and Meusburger (Fairbairn and Meusburger 2012) and by Han (Han 2011b), who has also studied the classical limit of the theory (Han 2011b; Ding and Han 2011). In this text we do not enter into the definition of the theory with a cosmological constant, because it would require more detail on quantum groups. We refer to the original papers above.

We only state the two key results:

1. The transition amplitudes of the q-deformed theory are finite.
2. The classical limit of the vertex amplitude is related to the Regge action with cosmological constant.

The theory with cosmological constant depends on *two* dimensionful parameters: the Planck length l_o and the cosmological constant Λ. One of these can be taken as the unit of length, so ultimately quantum gravity depends on a single dimensionless constant, which is a very small number (with respect to our limited human imagination):

$$\Lambda \hbar G \sim 10^{-120}\,. \tag{7.70}$$

Since the cosmological constant is of the order of the cosmological scale, this number is of the order of the ratio between the smallest and the largest observable scales around us.

7.4.3 Variants

There are a number of possible alternatives in the definition of the theory, and aspects that we have left open, which we mention here because they might play a role in the future.

- We have not specified how the normalization factor N_C depends on C. We discuss this below, when dealing with the continuous limit. Recall that a non-trivial N_C was needed to define the finite Turaev–Viro theory in three-dimensional.
- We have defined the theory with $p = \gamma j$. But the alternative

$$p = \gamma(j+1), \tag{7.71}$$

first suggested by Alexandrov (Alexandrov 2010), is of substantial interest and may be preferable; with this choice the linear simplicity condition holds weakly exactly, and not just for large j (Ding and Rovelli 2010b). See the last section in the Complements of this chapter.

- We have defined the theory on 2-complexes and graphs dual to a triangulation. However, more general complexes can be used, and these might be relevant. First, we can take a generalized triangulation, obtained by gluing 4-simplices, but dropping the requirement that two 4-simplices be connected by at most one tetrahedron. This is a generalization that we definitely make in the following.

  More generally, we can take an arbitrary 2-complex, not necessarily dual to a triangulation – in particular, where vertices and edges have arbitrary valence. It is easy to see that the formulas above extend immediately to this case. This generalization of the theory has been pointed out and developed by the Polish school (Kaminski *et al.* 2010a). The theory is then defined on generic 2-complexes. In particular, the nodes of the graphs need not be four-valent. The geometrical interpretation of such nodes continues to hold, thanks to a theorem proved by Minkowski which states that a set of n vectors satisfying $\sum_{\ell=1}^{n} \vec{E}_\ell = 0$ *uniquely* defines a polyhedron with n faces, where the vectors are normal to these faces and have a length proportional to the area of the faces. The corresponding geometry, and in particular the volume, has been explored in particular in Bianchi *et al.* (2011b) and Doná and Speziale (2013). One of the advantages of this version of the theory is that it becomes closer to the canonical formulation of loop quantum gravity, where nodes have arbitrary valence.

- In the euclidean version of the theory, the Polish school has also explored the possibility of modifying the face amplitude term given by the delta function in (7.57); it is not yet clear if a corresponding lorentzian version of this exists (Bahr *et al.* 2011).

- The amplitude associated with a given 2-complex can be obtained as a Feynman amplitude in the perturbative expansion of an associated quantum field theory with fields defined on a group (Krajewski *et al.* 2010). For a general introduction to group field theories (GFT), see Krajewski (2011) and references therein. Full GFT amplitudes are given by *sums* over 2-complexes, and not by *refining* the 2-complex, as here. In spite of the apparent difference, the two definitions agree under some conditions (in particular, on the normalization constants N_C), as shown in Rovelli and Smerlak (2012). It is not known, at present, if these conditions are satisfied for the physical quantum gravity theory.

- An alternative definition of the Y_γ map has been defined in Baratin *et al.* (2010), starting from group field theory. In the same paper a derivation of the formalism much closer to a simple path integral formulation of the discretized theory was developed.

- Although it does not appear to be a viable physical theory, we mention here also the Barrett–Crane theory (Barrett and Crane 2000), because it has played a major role in the development of the current theory. The current theory was in fact found as a modification of the Barrett–Crane theory, in order to address some shortcomings it had. The Barrett–Crane model can be seen as an appropriate $\gamma \to \infty$ limit of the theory defined in this chapter. We refer to Perez (2012) for a simple introduction to this model.

- In the next chapter we discuss coherent-states techniques. These techniques can also be used to directly define the theory directly in coherent-state basis. One of the (independent) paths in which the theory presented here was first found, in the euclidean context, is using these techniques (Freidel and Krasnov 2008). In the euclidean context, this approach gives a slightly different theory if $\gamma > 1$; not so in the lorentzian context, where the direct use of these techniques is not entirely clear yet. Again, see Perez (2012) for a detailed discussion.

The theory defined by the equations above is of course not written in stone. Perhaps it needs to be adjusted or modified, in order to describe Nature – that is, in order to be fully consistent and to agree with experiments. At present, however, it seems to yield a possibly consistent and complete theory of non-perturbative quantum gravity, finite and with a classical limit, consistent with what we know about the world.[7]

7.5 Complements

7.5.1 Summary of the theory

Kinematics

Hilbert space: $\mathcal{H}_\Gamma = L_2 \left[\mathrm{SU}(2)^L / \mathrm{SU}(2)^N \right]$, where the action of the quotient is given by the combinatoric structure of the graph Γ.

Operators: Triad operator $\vec{E}_\ell = 8\pi G \hbar \gamma \vec{L}_\ell$ where $\vec{L}_\ell$ is the left-invariant vector field (the derivative in the algebra), acting on the group elements on the links ℓ.

　　On each node we have the operators $G_{\ell\ell'} = \vec{E}_\ell \cdot \vec{E}_{\ell'}$. The norm $A_\ell = \sqrt{G_{\ell\ell}} = |\vec{E}_\ell|$ is the area of the faces of the tetrahedron punctured by the link ℓ.

　　Volume associated to a node: $V_\mathsf{n}^2 = \frac{2}{9} \vec{E}_1 \cdot (\vec{E}_2 \times \vec{E}_3)$; does not depend on the triplet used because of the gauge invariance at the node.

A_ℓ and V_n form a complete set of commuting observables.

States: $|\Gamma, j_\ell, v_\mathsf{n}\rangle$ spin-network basis. The *quantum geometry* is discrete and fuzzy in the small scale and gives a Riemannian geometry in the large.

Dynamics

Transition amplitude: Depends on a 2-complex $\mathcal{C}$

$$W_\mathcal{C}(h_\ell) = N_\mathcal{C} \int_{\mathrm{SU}(2)} dh_\mathsf{f} \prod_\mathsf{f} \delta\left(\prod_{\mathsf{v} \in \mathsf{f}} h_{\mathsf{fv}} \right) \prod_\mathsf{v} A_\mathsf{v}(h_{\mathsf{fv}}).$$

---

[7] Supergravity might still yield a finite theory, but at present we do not know how to extract from it the non-perturbative information needed to describe Planck-scale physics in the four-dimensional world without exact supersymmetry that we experience. String theory is likely to be finite, but its connection with the world we experience, and therefore its predictive power, is even frailer. Theoreticians working on these theories have long predicted low-energy supersymmetry, which so far has not shown up.

Vertex amplitude: Calling $h_\ell = h_{vf}$ the variables on the links of the vertex graph, and n the nodes of the vertex graph,

$$A_v(h_\ell) = \int_{SL(2,\mathbb{C})} dg_{n'} \prod_\ell \sum_j d_j\, D_{mn}^{(j)}(h_\ell) D_{jmjn}^{(\gamma j,j)}(g_n g_{n'}^{-1}).$$

The integration is over one g_n for each node (edge of v), except one. The 4-product is over 10 faces f, and $D^{(j)}$ and $D^{(p,k)}$ are matrix elements of the SU(2) and SL(2, $\mathbb{C}$) representations. These are connected by the simplicity map.

Simplicity map:

$$Y_\gamma: \quad \mathcal{H}_j \to \mathcal{H}_{p,k} \qquad \text{with} \quad p = \gamma j, \quad k = j,$$

$$|j; m\rangle \mapsto |\gamma j, j; j, m\rangle. \tag{7.72}$$

7.5.2 Computing with spin networks

A good and detailed introduction to spin-network theory and techniques is that of Haggard (2011), to which we refer the reader. Here we give a few basics. Recall that the dimension of the invariant part of the tensor product of three SU(2) irreducible representations is

$$\dim\left[Inv\left(\mathcal{H}_{j_1} \otimes \mathcal{H}_{j_2} \otimes \mathcal{H}_{j_3}\right)\right] \begin{cases} 1 & \text{if the triangular inequalities are satisfied} \\ 0 & \text{otherwise.} \end{cases} \tag{7.73}$$

And recall that if the triangular inequalities are satisfied, there exists one invariant state

$$|\iota\rangle = \sum_{m_1, m_2, m_3} \iota^{m_1\, m_2\, m_3} |j_1, m_1\rangle \otimes |j_2, m_2\rangle \otimes |j_3, m_3\rangle. \tag{7.74}$$

Its components are proportional to the Wigner 3j-symbol:

$$\iota^{m_1\, m_2\, m_3} = \begin{pmatrix} j_1 & j_2 & j_3 \\ m_1 & m_2 & m_3 \end{pmatrix}. \tag{7.75}$$

To get invariant objects, we can contract magnetic indices by means of the Wigner matrix, which we use to raise and lower magnetic indices:

$$g_{mm'} = (-1)^{j-m} \delta_{m,-m'}. \tag{7.76}$$

The Wigner 3j-symbol is normalized, that is:

$$\iota^{m_1\, m_2\, m_3}\, \iota_{m_1\, m_2\, m_3} = 1. \tag{7.77}$$

A diagrammatic notation is very convenient to keep track of the pattern of index contraction. For this, we write

$$\iota^{m_1\, m_2\, m_3} = \quad \vcenter{\hbox{}} . \tag{7.78}$$

Then equation (7.77) reads

$$\vcenter{\hbox{}} = 1 . \tag{7.79}$$

And in this notation, the Wigner 6j symbol reads

$$
\begin{array}{c}
\text{(diagram)}
\end{array}
=
\left\{
\begin{array}{ccc}
j_1 & j_2 & j_3 \\
j_4 & j_5 & j_6
\end{array}
\right\}.
$$

With these tools, we can find the matrix elements of the generator of the rotation group L^i in any representation j. These are tensors with two indices in the j representation and one index in the adjoint. Therefore they must be proportional to the Wigner 3j-symbols

$$
\langle j,m'|L^i|j,m\rangle = \left(L^i\right)_{m'}{}^m = c\, \iota^i{}_{m'}{}^m,
\tag{7.80}
$$

or

$$
L^{i\,m'm} = c\, \iota^{i\,m'\,m},
\tag{7.81}
$$

where we have changed basis from the magnetic $m = -1,0,-1$ basis to the real $i = 1,2,3 = x,y,z$ basis in the adjoint representation of SU(2). The proportionality constant c can be found by recalling that the Casimir is

$$
\vec{L}^2 = j(j+1)\mathbb{1};
\tag{7.82}
$$

therefore, the trace of the Casimir is $j(j+1)$ times the dimension of the representation, which is $2j+1$. Thus

$$
j(j+1)(2j+1) = \mathrm{Tr}\vec{L}^2 = \delta_{ik}L^i_m{}^{m'}L^k_{m'}{}^m = c^2\, \iota^{mm'i}\iota_{mm'i} = c^2.
\tag{7.83}
$$

This fixes the normalization. Thus we conclude

$$
L^{imm'} = \sqrt{j(j+1)(2j+1)}
\quad
\begin{array}{c}
\text{(diagram)}
\end{array}
.
\tag{7.84}
$$

Intertwiners

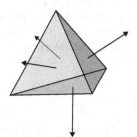

Let us now study the Hilbert space $\mathcal{H}_n$ associated with a tetrahedron. This is the invariant part of the tensor product of *four* SU(2) irreducibles

$$
\mathcal{H}_n = \mathcal{K}_{j_1...j_4} = Inv\left(\mathcal{H}_{j_1}\otimes\mathcal{H}_{j_2}\otimes\mathcal{H}_{j_3}\otimes\mathcal{H}_{j_4}\right).
\tag{7.85}
$$

Tensors in this space are called intertwiners. They have the form

$$
|\iota\rangle = \sum_{m_1...m_4}\iota^{m_1 m_2 m_3 m_4}|j_1,m_1\rangle\otimes\ldots\otimes|j_4,m_4\rangle
\tag{7.86}
$$

and are invariant under the diagonal action of SU(2). That is,

$$
D^{(j_1)m_1}{}_{n_1}(U)...D^{(j)m_1}{}_{n_1}(U)\,\iota^{n_1 n_2 n_3 n_4} = \iota^{m_1 m_2 m_3 m_4}.
\tag{7.87}
$$

If $U = e^{i\vec{\alpha}\cdot\vec{\tau}}$ and we take $|\alpha| << 1$ we can expand

$$
D(U)|\iota\rangle = (1 + i\vec{\alpha}\cdot(L_{j_1}+L_{j_2}+L_{j_3}+L_{j_4}))|\iota\rangle = |\iota\rangle,
\tag{7.88}
$$

where each L_j acts on a different factor space. That is, intertwiners satisfy the closure constraint:

$$\sum_{\ell=1}^{4} \vec{L}_\ell |\iota\rangle = 0. \tag{7.89}$$

This equation says that the sum of the normals to the triangles of a tetrahedron, normalized to their areas, vanishes.

It is not difficult to find a basis of invariant tensors, starting from the trivalent intertwiners. Indeed,

$$\iota_k^{m_1 m_2 m_3 m_4} = \sqrt{2k+1} \begin{pmatrix} m_1 & m_2 & m \\ j_1 & j_2 & k \end{pmatrix} g_{mm'} \begin{pmatrix} m' & m_3 & m_4 \\ k & j_3 & j_4 \end{pmatrix} \tag{7.90}$$

is clearly invariant. Equation (7.90) can be expressed diagrammatically as

$$\tag{7.91}$$

with a "virtual link" with spin k in the node.

Exercise 7.2 *Given the graph Γ_1 with two nodes connected by three links oriented from the first to the second node, and the graph Γ_2 formed by four nodes all connected among themselves, write the corresponding spin-network states $\psi_{j_1 j_2 j_3}(U_l)$ and $\psi_{j_1 j_2 j_3 j_4 j_5 j_6}(U_l)$.*

Exercise 7.3 *Consider now the first graph and the state $\psi_1 = Tr(UVWV)$, where $U_l = (U, V, W)$. Is it a physical state? What about $\psi_2 = Tr(UV^{-1}WV^{-1})$? If so, expand it in the spin-network basis.*

Exercise 7.4 *Consider the second graph and the following states:*

$$\psi_1 = Tr(UWV^{-1}), \quad \psi_2 = Tr(VYZW), \quad \psi_3 = Tr(X^{-1}UWY), \quad \psi_3 = Tr\left[(X^{-1}VY)^7\right],$$

$$\tag{7.92}$$

Which of them belong to physical Hilbert space, with what identification between the group elements and the oriented links?

Exercise 7.5 *Consider the state $|\Gamma_1, \frac{1}{2}, 1, \frac{1}{2}\rangle$. Write this state as a sum of products of "loops," where a loop is the trace of a product of a sequence of U's along a closed cycle on the graph. Any spin-network state can be written in this way. This is the historical origin of the denomination "loop quantum gravity" for the theory. [Hint: You may want to read a discussion on p. 237 of Rovelli (2004).]*

Observables

Here we introduce the observables in $\mathcal{H}_n$. An observable O commutes with the constraint (7.89), $\mathcal{C} = \sum_{\ell=1}^{4} \vec{L}_\ell |\iota\rangle = 0$, for the generator of the rotations. That is, $[O, \mathcal{C}] = 0$. Two of these observables are $(\vec{L}_1 + \vec{L}_2)^2$ and $\vec{L}_\ell^2$.

$$[\vec{L}_\ell^2, \mathcal{C}] = 0, \tag{7.93}$$

$$[(\vec{L}_1 + \vec{L}_2)^2, \mathcal{C}] = 0, \tag{7.94}$$

$$[(\vec{L}_1 + \vec{L}_2)^2, \vec{L}_\ell^2] = 0. \tag{7.95}$$

They are invariant under rotations. They can be diagonalized simultaneously. They form a maximal set of commuting observables

$$\vec{L}_\ell^2 |\iota\rangle = j_\ell(j_\ell + 1)|\iota\rangle. \tag{7.96}$$

$(\vec{L}_1 + \vec{L}_2)^2$ measures k,

$$(\vec{L}_1 + \vec{L}_2)^2|\iota_k\rangle = k(k+1)|\iota_k\rangle \qquad \text{where} \qquad |j_1 - j_2| \le k \le j_1 + j_2 \tag{7.97}$$

$$(\vec{L}_3 + \vec{L}_4)^2|\iota_k\rangle = k(k+1)|\iota_k\rangle \qquad \text{where} \qquad |j_3 - j_4| \le k \le j_3 + j_4. \tag{7.98}$$

The range of k is dictated by the triangular inequalities. Thus $dim\ \mathcal{H}_n = k_{max} - k_{min} + 1$.

The states $|\iota_k\rangle$ form an orthogonal basis because they are eigenstates of a hermitian operator.

Problem

Verify explicitly that the states $|i_k\rangle$ are orthogonal and normalized.

Solution

The proportionality constant in

$$\text{(diagram)} = \alpha \tag{7.99}$$

can be determined by closing both sides of the equation (Wigner–Eckart theorem). The left-hand side gives unit, so that $\alpha = 1/(2k+1)$. Then

$$\langle \iota_k | \iota_{k'} \rangle = \sqrt{2k+1}\sqrt{2k'+1}\ \text{(diagram)}\ k' = \sqrt{2k+1}\sqrt{2k'+1}\ \frac{1}{2k+1}\ \delta_{kk'}. \tag{7.100}$$

7.5.3 Spectrum of the volume

Recall that the volume operator reads

$$\hat{V}_n = \frac{\sqrt{2}}{3}(8\pi G\hbar\gamma)^{\frac{3}{2}}\sqrt{|\vec{L}_1 \cdot (\vec{L}_2 \times \vec{L}_3)|}. \tag{7.101}$$

We have four vectors $\vec{L}_i$ but any (oriented) triplet gives the same result because of the constraint

$$\vec{L}_1 + \vec{L}_2 + \vec{L}_3 + \vec{L}_4 = 0. \tag{7.102}$$

The volume operator has eigenstates $|\iota_v\rangle$,

$$\hat{V}_n|\iota_v\rangle = v|\iota_v\rangle. \tag{7.103}$$

In order to compute $|\iota_v\rangle$ we start by computing

$$\vec{L}_1 \cdot (\vec{L}_2 \times \vec{L}_3)|\iota_q\rangle = q|\iota_q\rangle. \tag{7.104}$$

Then we will have

$$\hat{V}_n|\iota_q\rangle = \frac{\sqrt{2}}{3}(8\pi G\hbar\gamma)^{\frac{3}{2}}\sqrt{|q|}|\iota_q\rangle. \tag{7.105}$$

We need the matrix elements of this operator,

$$Q_k^{k'} = \langle\iota_k|\vec{L}_1\cdot(\vec{L}_2\times\vec{L}_3)|\iota_{k'}\rangle; \tag{7.106}$$

these form a $d\times d$ matrix with eigenvectors $\langle\iota_k|\iota_q\rangle$. We can compute the matrix elements and diagonalize.

Let us start by observing that

$$\frac{1}{2}\left[(\vec{L}_1+\vec{L}_2)^2,\vec{L}_1\cdot\vec{L}_3\right] = [\vec{L}_1\cdot\vec{L}_2,\vec{L}_1\cdot\vec{L}_3] = [L_{1i},L_{1j}]\,L_2^i L_3^j = i\epsilon_{ijk}L_1^i L_2^j L_3^k = i\vec{L}_1\cdot(\vec{L}_2\times\vec{L}_3). \tag{7.107}$$

Using this,

$$Q_k^{k'} = \langle\iota_k|\vec{L}_1\cdot(\vec{L}_2\times\vec{L}_3)|\iota_{k'}\rangle \tag{7.108}$$

$$= -\frac{i}{2}\langle\iota_k|[(\vec{L}_1+\vec{L}_2)^2,\vec{L}_1\cdot\vec{L}_3]|\iota_{k'}\rangle \tag{7.109}$$

$$= -\frac{i}{2}\langle\iota_k|(\vec{L}_1+\vec{L}_2)^2\vec{L}_1\cdot\vec{L}_3|\iota_{k'}\rangle - \langle\iota_k|\vec{L}_1\cdot\vec{L}_3(\vec{L}_1+\vec{L}_3)|\iota_{k'}\rangle \tag{7.110}$$

$$= -\frac{i}{2}(k(k+1)-k'(k'+1))\langle\iota_k|\vec{L}_1\cdot\vec{L}_3|\iota_{k'}\rangle. \tag{7.111}$$

The last term can be computed diagrammatically (Figure 7.7) using

$$\tag{7.112}$$

That is, cutting the graph by inserting a resolution of the identity in terms of trivalent intertwiners. The eigenvalues are complex conjugate (see $k+1\to k-1$ above). We need to compute just one so that we automatically get the other. This gives

$$\langle\iota_k|\vec{L}_1\cdot\vec{L}_3|\iota_{k'}\rangle = \sqrt{2k+1}\sqrt{2k'+1}\sqrt{j_1(j_1+1)(2j_1+1)}\sqrt{j_3(j_3+1)(2j_3+1)} \tag{7.113}$$

times the two tetrahedral graphs in Figure 7.7.

Notice that the triangular inequalities in these graphs force k and k' to differ precisely by one unit. A little algebra gives finally that the the only non-vanishing matrix elements are

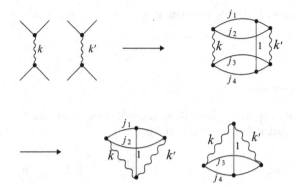

Figure 7.7 A grasping is a connection made by a line carrying spin 1, using (7.84).

$$a_k = Q_k{}^{k-1} = -\frac{1}{2}\left(k(k+1)-(k-1)k\right)\sqrt{2k+1}\sqrt{2k-1} \tag{7.114}$$

$$\times \sqrt{j_1(j_1+1)(2j_1+1)}\sqrt{j_3(j_3+1)(2j_3+1)}$$

$$\times \left\{ \begin{array}{ccc} j_1 & 1 & j_1 \\ k & j_2 & k-1 \end{array} \right\}\left\{ \begin{array}{ccc} j_3 & 1 & j_3 \\ k & j_4 & k-1 \end{array} \right\}$$

and their transposes, and the matrix has the form

$$Q_k{}^{k'} = \begin{pmatrix} 0 & +a_1 & 0 & \cdots \\ -a_1 & 0 & +a_2 & \\ 0 & -a_2 & 0 & \\ \vdots & & & \ddots \end{pmatrix}. \tag{7.115}$$

Matrices of this form have several peculiar properties: in particular they have a non-degenerate spectrum. The diagonalization of these matrices, however, is not straightforward, and must be done numerically.

Since the spectrum is not degenerate, we can use the eigenbasis of the volume and use the eigenvalues of $\vec{L}_1 \cdot (\vec{L}_2 \times \vec{L}_3)$ as good labels for the basis. Notice that in the matrix the eigenvalues have opposite sign, but when we take the modulus in the volume we are going to miss this sign: at each node we need to specify the corresponding volume and orientation.

There are special values of j for which the $\{6j\}$ have a simple analytic form. They can be found in Landau's book or with Mathematica.

$$\left\{ \begin{array}{ccc} a & 1 & a \\ c & b & c-1 \end{array} \right\} = (-1)^{-(a+b+c)}\frac{2\Delta(a+\frac{1}{2},b+\frac{1}{2},c)}{\sqrt{a(a+1)(2a+1)}\sqrt{2c+1}\sqrt{2c-1}\sqrt{c}} \tag{7.116}$$

where $\Delta(a,b,c) = \frac{1}{4}\sqrt{(a+b+c)(a+b-c)(a-b+c)(-a+b+c)}$ is *Heron's formula* for the area of a triangle of sides (a,b,c). Using this,

$$a_k = -\frac{1}{2}2k\frac{2\Delta(j_1+\frac{1}{2},j_2+\frac{1}{2},k)}{\sqrt{2k+1}\sqrt{2k-1}\sqrt{k}}\frac{2\Delta(j_3+\frac{1}{2},j_4+\frac{1}{2},k)}{\sqrt{2k+1}\sqrt{2k-1}\sqrt{k}} \tag{7.117}$$

$$= \frac{-4}{\sqrt{4k^2-1}}\Delta(j_1+\frac{1}{2},j_2+\frac{1}{2},k)\Delta(j_3+\frac{1}{2},j_4+\frac{1}{2},k) \tag{7.118}$$

Example 7.1 Take the case with all spin equal $j_1 = j_2 = j_3 = j_4 = j$ Then the above formulas simplify to

$$a_k = \frac{1}{4} \frac{k^2(k^2 - (2j-1)^2)}{\sqrt{4k^2 - 1}}. \tag{7.119}$$

In particular:

- If $j = \frac{1}{2}$, then $a_1 = -\frac{\sqrt{3}}{4}$, the possible virtual spins are $k = 0, 1$, the dimension of the node Hilbert space is $d = 2$, and the matrix is

$$Q_k{}^{k'} = -\frac{\sqrt{3}}{4} \begin{pmatrix} 0 & +i \\ -i & 0 \end{pmatrix}.$$

This has eigenvalues $q_- = -\frac{\sqrt{3}}{4}$ and $q_+ = +\frac{\sqrt{3}}{4}$ and eigenvectors

$$|q_\pm\rangle = \frac{1}{\sqrt{2}} (|k=0\rangle \pm i|k=1\rangle). \tag{7.120}$$

Therefore the full (oriented) volume operator has these eigenvectors with eigenvalues (the sign codes the orientation)

$$V_\pm = \pm \frac{\sqrt{2}}{3} (8\pi G\hbar\gamma)^{\frac{3}{2}} \sqrt{\frac{\sqrt{3}}{4}}. \tag{7.121}$$

- If $j = 1$, then $a_1 = -\frac{2}{\sqrt{3}}$, $a_2 = -\frac{\sqrt{5}}{\sqrt{3}}$. The possible virtual spins are $k = 0, 1, 2$, the dimension of the node Hilbert space is $d = 3$, and the matrix is

$$Q_k{}^{k'} = -\frac{1}{\sqrt{3}} \begin{pmatrix} 0 & 2i & 0 \\ -2i & 0 & \sqrt{5}i \\ 0 & -\sqrt{5}i & 0 \end{pmatrix}. \tag{7.122}$$

The eigenvalues are $q_o = 0$ and $q_\pm = \pm\sqrt{3}$ and the eigenvectors $|q_o\rangle = \frac{\sqrt{5}}{2} + |2\rangle$ and $|q_\pm\rangle = -\frac{2}{\sqrt{5}}|0\rangle \pm \frac{3i}{\sqrt{5}}|1\rangle + |2\rangle$. The eigenvalues of the volume are

$$v_0 = 0, \qquad v_I = \pm\frac{\sqrt{2}}{3}(8\pi G\hbar\gamma)^{\frac{3}{2}}3^{\frac{1}{4}}. \tag{7.123}$$

Problem

Recall the volume formula

$$V^2 = c\epsilon_{ijk}L_{j_1}^i L_{j_2}^j L_{j_3}^k \mathbf{1}_{j_4}, \tag{7.124}$$

where L_{j_i} are in j_i representation of SU(2) and similarly $\mathbf{1}_{j_4}$ is the identity matrix in j_4 representation, and c is some constant.

1. Derive the value of c. [*Hint:* Take a tetrahedron having three sides equal to the three orthogonal basis vectors in $\mathbb{R}^3$.]
2. Prove that the volume does not depend on which three (out of four) links we choose. In other words, if we denote the volume operator in (7.124) as V_{123}^2, this gives the same result as if we had chosen V_{124}^2 or V_{234}^2. [*Hint.* Recall that on physical states we have $(L_1^{j_1} + L_2^{j_2} + L_3^{j_3} + L_4^{j_4})\psi = 0$.]
3. Calculate the volume eigenvalues for $K_{11\frac{1}{2}\frac{1}{2}}$ and K_{1111}.

7.5.4 Unitary representation of the Lorentz group and the Y_γ map

We review here the unitary representation of the Lorentz group; more precisely, the representations of $SL(2,\mathbb{C})$, the double cover of $SO(3,1)$, the component of the Lorentz group connected to the identity. The Lie algebra of $SL(2,\mathbb{C})$ is

$$[J^{IJ}, J^{KL}] = -\eta^{IK}J^{JL} + \eta^{IL}J^{JK} + \eta^{JK}J^{IL} - \eta^{JL}J^{IK}. \tag{7.125}$$

We fix an $SU(2)$ subgroup of $SL(2,\mathbb{C})$ (as groups of matrices, $SU(2)$ is simply the subgroup of the unitary matrices, of course; in the abstract groups, this depends on the choice of a Lorentz frame, namely, a timelike vector $t^I = (1,0,0,0)$. $SU(2)$ is the *little group* that preserves t^I). The generators J^{IJ} of $SL(2,\mathbb{C})$ then split into rotations and boosts

$$L^I = \frac{1}{2}\epsilon^I{}_{JKL}J^{KL}t^L = (0, L^i), \tag{7.126}$$

$$K^I = J^{IJ}t_J = (0, K^i), \tag{7.127}$$

with the Lie algebra

$$[L^i, L^j] = \epsilon^{ij}{}_k L^k, \quad [L^i, K^j] = \epsilon^{ij}{}_k K^k, \quad [K^i, K^j] = \epsilon^{ij}{}_k L^k. \tag{7.128}$$

Finite-dimensional representations

The unitary representations of $SL(2,\mathbb{C})$, should not be confused with the well-known finite-dimensional representations of the Lorentz group, which are ubiquitous in fundamental theoretical physics. These are for instance:

1. The fundamental, of Weyl, or spinor representation. This is defined on $\mathbb{C}^2$. Recall that we write spinors in $\mathbb{C}^2$ as $\mathbf{z} = z^A = (z^0, z^1)$. The generators of the rotations are the $L^i = \tau_i = -i\frac{\sigma_i}{2}$ matrices, precisely as the fundamental $SU(2)$ representation. The generators of boosts are the non-hermitian operators $K^i = -\frac{\sigma_i}{2}$.
2. The vector representation. This is the fundamental of $SO(3,1)$, formed by the Minkowski vectors: $X'^I = \Lambda^I{}_J X^J$, with $I, J = 0, 1, 2, 3$.
3. The adjoint representation. This is given by tensors with two antisymmetric Minkowski indices, like the electromagnetic field F^{IJ}. If a specific frame is fixed, or a specific time direction n_I, then these tensors split into a magnetic and an electric part, as we have seen for B. In particular, the generators of the algebra, which are of course in the adjoint representation, split into boosts and rotations.
4. Et cetera.

These representations are all non-unitary. There is no scalar product defined on them. (The Minkowski "scalar product" $X^I X_I = (X^0)^2 - |\vec{X}|^2$ is not positive definite.) Let us therefore move to the unitary representations.

Infinite-dimensional representations

The representations that are of interest for us are the unitary irreducible representations in the principal series. These are described in detail in the book by Ruhl (1970). $SL(2,\mathbb{C})$ has rank 2 and therefore two independent Casimirs. These can be taken to be

$$C_1 = \frac{1}{2}J_{IJ}J^{IJ} = \vec{K}^2 - \vec{L}^2, \tag{7.129}$$

$$C_2 = \frac{1}{8}\epsilon_{IJKL}J^{IJ}J^{KL} = \vec{K}\cdot\vec{L}. \tag{7.130}$$

Accordingly, the unitary representations are labeled by two quantum numbers: $p \in \mathbb{R}$ and $k \in \frac{1}{2}\mathbb{N}$. The representation space is $V^{(p,k)}$ and is infinite dimensional: $dim\, V^{(p,k)} = \infty$. On this space act the hermitian generators $J^{IJ} = -J^{JI}$. The value of the Casimirs in these representations is given above in (7.12). $V^{(p,k)}$ is a *reducible* representation of SU(2). It admits the decomposition (7.10) into an infinite number of (different) SU(2)-invariant blocks For instance, if $U \in$ SU(2),

$$
D^{(p,\frac{1}{2})}(U) = \begin{pmatrix} \boxed{\begin{matrix} \cdot & \cdot \\ \cdot & \cdot \end{matrix}} & & & \\ & \boxed{\begin{matrix} \cdot & \cdot & \cdot \\ & & \end{matrix}} & & \\ & & \ddots & \end{pmatrix}, \tag{7.131}
$$

where there first 2×2 matrix is $D^{(\frac{1}{2})}$ and so on. The generators of rotations preserve this decomposition (the blocks are invariant under SU(2) transformation) while the boost generator K^i "takes out" of these blocks. It sends $\mathcal{H}_i$ into $\mathcal{H}_{j-1} \oplus \mathcal{H}_j \oplus \mathcal{H}_{j+1}$.

The orthonormal basis $|p,k;j,m\rangle$ of $V^{(p,k)}$, called the canonical basis, is obtained by diagonalizing the operators C_1, C_2, $\vec{L}^2$ and L_z. The generators in this basis are (Gel'fand *et al.* (1963), pages 187–189):

$$
L^3|j,m\rangle = m|j,m\rangle,
$$

$$
L^+|j,m\rangle = \sqrt{(j+m+1)(j-m)}|j,m+1\rangle,
$$

$$
L^-|j,m\rangle = \sqrt{(j+m)(j-m+1)}|j,m-1\rangle,
$$

$$
K^3|j,m\rangle = -\alpha_{(j)}\sqrt{j^2-m^2}|j-1,m\rangle - \beta_{(j)}m|j,m\rangle + \alpha_{(j+1)}\sqrt{(j+1)^2-m^2}|j+1,m\rangle,
$$

$$
K^+|j,m\rangle = -\alpha_{(j)}\sqrt{(j-m)(j-m-1)}|j-1,m+1\rangle - \beta_{(j)}\sqrt{(j-m)(j+m+1)}|j,m+1\rangle \tag{7.132}
$$
$$
\qquad - \alpha_{(j+1)}\sqrt{(j+m+1)(j+m+2)}|j+1,m+1\rangle,
$$

$$
K^-|j,m\rangle = \alpha_{(j)}\sqrt{(j+m)(j+m-1)}|j-1,m-1\rangle - \beta_{(j)}\sqrt{(j+m)(j-m+1)}|j,m-1\rangle
$$
$$
\qquad + \alpha_{(j+1)}\sqrt{(j-m+1)(j-m+2)}|j+1,m-1\rangle,
$$

where

$$
L^{\pm} = L^1 \pm iL^2, \qquad K^{\pm} = K^1 \pm iK^2
$$

and

$$
\alpha_{(j)} = \frac{i}{j}\sqrt{\frac{(j^2-k^2)(j^2+p^2)}{4j^2-1}}, \qquad \beta_{(j)} = \frac{kp}{j(j+1)}. \tag{7.133}
$$

Problem

Show that if we define the map Y_γ by

$$
p = \gamma(j+1), \qquad k = j \tag{7.134}
$$

(instead of $p = \gamma j$, $k = j$), then the simplicity conditions are weakly realized *exactly* and not just in the large j limit.

Solution

Recall that $\langle jm | L^I | jm \rangle = c\iota^{im}{}_{m'}$ where $c = \sqrt{j(j+1)(2j+1)}$. Let us compute these matrix elements on the image of Y: $\langle p,j;jm' | L^i | p,j;j,m \rangle$ is invariant in the space $Inv(j \otimes j \otimes 1)$. There is only one such object – that is the $\{3j\}$ – and the same computation that fixes the coefficient in the SU(2) representations holds here. Thus $\langle p,j;j,m' | L^i | p,j;j,m \rangle = c\iota^{im}{}_{m'}$.

To do the same for K^i, we need to take care because K^i "goes out" of the SU(2) representation spaces. But what we actually need is to compute this on the image of the map Y_γ

$$\langle p,j;j,m' | K^i | p,j;j,m \rangle = \alpha\iota^{im'}{}_m. \tag{7.135}$$

On the left and on the right we have the same spin, therefore we are *still* inside a block of (7.131), in fact the one with the lowest weight. We have

$$L^{im}{}_{m'} K^i_m{}^{m'} = \vec{L} \cdot \vec{K} \, \mathrm{Tr}\, \mathbf{1}_j = (2j+1)pj = \alpha c \iota^{imm'} \iota_{imm'} = \alpha c \tag{7.136}$$

because the intertwiner is normalized. The ratio of the two proportionality factors gives

$$\frac{\alpha}{c} = \frac{(2j+1)pj}{j(j+1)(2j+1)} = \frac{p}{j+1}. \tag{7.137}$$

If $p = \gamma(j+1)$, this $\frac{\alpha}{c} = \gamma$ and the ratio does not depend on j. Thus, if we consider the map Y_γ defined by $p = \gamma(j+1)$ and we consider the states $|\phi\rangle, |\psi\rangle \in Imm\, Y_\gamma$ then $\langle \phi | \vec{K} | \psi \rangle = \frac{\alpha}{c} \langle \phi | \vec{L} | \psi \rangle$, namely,

$$\langle \phi | \vec{K} | \psi \rangle = \gamma \langle \phi | \vec{L} | \psi \rangle . \tag{7.138}$$

On the image of Y_γ we have $k = j$ and $p = \gamma(j+1)$, therefore the matrix elements of $\vec{K}$ and $\vec{L}$ satisfy the relation

$$\boxed{\vec{K} = \gamma\vec{L}}. \tag{7.139}$$

Because of this intriguing result, the theory may be better defined with $p = \gamma(j+1)$ instead of $p = \gamma j$ (Alexandrov 2010; Ding and Rovelli 2010a). The question is open.

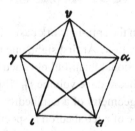

The graph Γ_v, the boundary graph of the vertex, or pentagram, was called ὑγιεία (Hugieia) by the Pythagoreans. They saw mathematical perfection in it. Hugieia was a goddess of health and the expression "Hugieia" was used as a greeting.

8

Classical limit

The theory defined in the previous chapter does not much resemble general relativity. Where are the Einstein equations? Where is Riemannian geometry? Where is curvature? And so on. In this chapter, we show how the classical limit emerges from the quantum theory defined in Chapter 7.

If we start from the Schrödinger equation of a non-relativistic particle, we can recover the dynamics of a classical particle by studying the evolution of wave packets. A wave packet is a state in which both position and momentum are not too quantum-spread. In the limit where the actions in play are much larger than the Planck constant, wave packets behave like classical particles. A useful class of "wave packet" states is given by the coherent states, which are a family of states labeled by classical variables (position and momenta) that minimize the spread of both.

Coherent states are the basic tool for studying the classical limit also in quantum gravity. They connect the quantum theory defined in the previous chapter with classical general relativity. In this chapter we study the coherent states in the Hilbert space of the theory, and their use in proving the large-distance behavior of the vertex amplitude and connecting it to the Einstein equations.

8.1 Intrinsic coherent states

Consider a state in the spin-network basis $|\Gamma, j_\ell, v_n\rangle$. Can we associate a three-dimensional classical geometry to it? At first sight, one is tempted to answer as follows: take a triangulated three-dimensional manifold obtained by gluing flat tetrahedra where triangles have areas j_ℓ and tetrahedra have volume v_n. But a moment of reflection shows that this strategy is weak: the geometry of a tetrahedron is only partially specified by giving areas and volume. The shape of a tetrahedron is specified by six numbers, not five. Therefore a state $|\Gamma, j_\ell, v_n\rangle$ is in fact highly non-classical, in the sense that some variables of the intrinsic geometry it describes are maximally quantum-spread.

The same problem can be seen from a different perspective. Fix the four areas of a tetrahedron. The resulting quantum state space is the Hilbert space $\mathcal{H}_n$ associated with a single node of the graph. A basis in this space is given by the states $|i_k\rangle$, defined in (7.35). We have seen that these are eigenstates of $\vec{L}_1 \cdot \vec{L}_2$. They diagonalize the dihedral angle θ_{12} between the faces 1 and 2. On the other hand, the angle, say θ_{13} between the faces 1 and 3, namely the observable $\vec{L}_1 \cdot \vec{L}_3$, is spread.

Given a classical tetrahedron, can we find a quantum state in $\mathcal{H}_\Gamma$ such that all the dihedral angles are minimally spread around the classical values? The solution to this problem is represented by the intrinsic coherent states, which we describe below.

8.1.1 Tetrahedron geometry and SU(2) coherent states

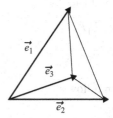

Let us reconsider the geometry of a classical tetrahedron sketched in the first chapter. A tetrahedron in flat space can be determined by giving three vectors, $\vec{e}_a$ with $a = 1, 2, 3$, representing three of its sides emanating from a vertex P. If we choose a non-orthogonal coordinate system where the axes are along these vectors and the vectors determine the unit of coordinate length, then e_a^i is the triad and

$$h_{ab} = \vec{e}_a \cdot \vec{e}_b \tag{8.1}$$

is the metric in these coordinates. The three vectors

$$\vec{E}^a = \frac{1}{2} \epsilon^{abc} \, \vec{e}_b \times \vec{e}_c \tag{8.2}$$

are normal to the three triangles adjacent to P and their length is the area of these faces. The products

$$\vec{E}^a \cdot \vec{E}^b = (\det h) \, h^{ab} \tag{8.3}$$

define the matrix h^{ab} which is the inverse of the metric $h_{ab} = \vec{e}_a \cdot \vec{e}_b$. The volume of the tetrahedron is

$$V = \frac{1}{3} \vec{e}_1 \cdot (\vec{e}_2 \times \vec{e}_3) = \frac{1}{3!} \sqrt{2^3 |\vec{E}_1 \cdot (\vec{E}_2 \times \vec{E}_3)|}. \tag{8.4}$$

We extend the range of the index a to $1, 2, 3, 4$, and denote all the four normals, normalized to the area, as $\vec{E}^a$. It is easy to show that these satisfy the *closure* condition

$$\sum_{a=1}^{4} \vec{E}_a = \sum_{f=1}^{4} A_a \vec{n}_a = 0. \tag{8.5}$$

(Physical intuition: a vector quantity normal to a face and proportional to the area is the force due to the *pressure*. If we put the tetrahedron in a gas and increase the pressure, the tetrahedron does not move, that is, the sum of all the forces gives zero. This is what is indicated by the closure equation (8.5).)

The dihedral angle between two triangles is given by

$$\vec{E}_1 \cdot \vec{E}_2 = A_1 A_2 \vec{n}_1 \cdot \vec{n}_2 = A_1 A_2 \cos\theta_{12}. \tag{8.6}$$

Now we move to the quantum theory. Here, the quantities $\vec{E}^a$ are quantized as

$$\vec{E}_a = 8\pi G \hbar \gamma \vec{L}_a \tag{8.7}$$

in terms of the four operators $\vec{L}_a$, which are the (hermitian) generators of the rotation group (recall (1.12)!):

$$[L_a^i, L_b^j] = i\delta_{ab}\,\epsilon^{ij}{}_k L_a^k. \tag{8.8}$$

The commutator of two angles is

$$[\vec{E}_1 \cdot \vec{E}_2, \vec{E}_1 \cdot \vec{E}_3] = (8\pi G\hbar\gamma)^4 [L_{1i}, L_{1j}] L_2^i L_3^j \tag{8.9}$$

$$= (8\pi G\hbar\gamma)^4 i\epsilon_{ijk} L_1^i L_2^j L_3^k \tag{8.10}$$

$$= (8\pi G\hbar\gamma)\vec{E}_1 \cdot (\vec{E}_2 \times \vec{E}_3). \tag{8.11}$$

From this commutation relation, the Heisenberg relation follows:

$$\Delta(\vec{E}_1 \cdot \vec{E}_2) \cdot \Delta(\vec{E}_1 \cdot \vec{E}_3) \geq \frac{1}{2} 8\pi G\hbar\gamma \, |\langle\vec{E}_1 \cdot (\vec{E}_2 \times \vec{E}_3)\rangle|, \tag{8.12}$$

where $\langle A \rangle = \langle \iota | A | \iota \rangle$ and $\Delta A = \sqrt{\langle \iota | A^2 | \iota \rangle - (\langle \iota | A | \iota \rangle)^2}$. Now we want to look for states whose dispersion is small compared with their expectation value: semiclassical states where

$$\frac{\Delta(\vec{E}_a \cdot \vec{E}_b)}{|\vec{E}_a||\vec{E}_b|} \ll 1 \qquad \forall a, b. \tag{8.13}$$

SU(2) coherent states

Consider a single rotating particle. How to write a state for which the dispersion of its angular momentum is minimized? If j is the quantum number of its total angular momentum, a basis of states is

$$|j, m\rangle \in \mathcal{H}_j. \tag{8.14}$$

Since

$$[L_x, L_y] = iL_z, \tag{8.15}$$

we have the Heisenberg relations

$$\Delta L_x \, \Delta L_y \geq \frac{1}{2} |\langle L_z \rangle|. \tag{8.16}$$

Every state satisfies this inequality. Can we saturate it? A state that saturates (8.15), namely, for which $\Delta L_x \, \Delta L_y = \frac{1}{2} |\langle L_z \rangle|$, is given by $|j, j\rangle$. This can be shown as follows. $|j, j\rangle$ is an

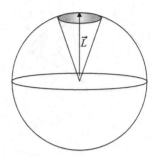

Figure 8.1 The angular size of the cone that gives the spread goes as $\sim 1/\sqrt{j}$.

eigenstate of L_z

$$L_z|j,j\rangle = j|j,j\rangle, \tag{8.17}$$

so that $\delta L_z = 0$. Then

$$\langle L_x\rangle = \langle L_y\rangle = 0, \tag{8.18}$$

$$\langle L_x\rangle^2 = \frac{1}{2}\langle L_x^2 + L_y^2\rangle = \frac{1}{2}\langle L^2 - L_z^2\rangle = \frac{1}{2}(j(j+1) - j^2) = \frac{j}{2}. \tag{8.19}$$

Therefore we conclude that

$$\Delta L_x = \Delta L_y = \sqrt{\frac{j}{2}}. \tag{8.20}$$

Therefore the state $|j,j\rangle$ saturates the uncertainty relations

$$\Delta L_x\,\Delta L_y = \frac{j}{2}, \qquad \frac{1}{2}|\langle L_z\rangle| = \frac{j}{2}. \tag{8.21}$$

In the large-j limit we have

$$\frac{\Delta L_x}{\sqrt{\langle \vec{L}^2\rangle}} = \frac{\sqrt{\frac{j}{2}}}{\sqrt{j(j+1)}} = \frac{1}{\sqrt{2(j+1)}} \to_{j\to\infty} 0. \tag{8.22}$$

Therefore this state becomes sharp for large j.

The geometrical picture corresponding to this calculation is transparent: the state $|j,j\rangle$ represents a spherical harmonic maximally concentrated on the north pole of the sphere, and the ratio between the spread and the radius decreases with the spin (Figure 8.1).

Thus we have found a good candidate for a coherent state: $|j,j\rangle$. What about the other states of the family? Well, these are easily obtained by rotating the state $|j,j\rangle$ into an arbitrary direction $\vec{n}$. Let us introduce Euler angles θ, ϕ to label rotations, as in Figure 8.2. Then let $z^i = (0,0,1)$ and define the matrix $R \in SO(3)$ of the form $R = e^{-i\phi L_z}\,e^{-i\theta L_y}$, by $R^i_{\ j}z^j = n^i$. With this, define

$$|j,\vec{n}\rangle = D_{\vec{n}}(R)|j,j\rangle. \tag{8.23}$$

The states $|j,\vec{n}\rangle$ form a family of states, labeled by the continuous parameter $\vec{n}$, which saturate the uncertainty relations for the angles.

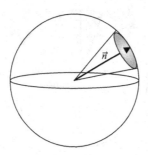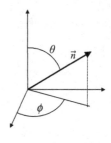

Figure 8.2   For a generic direction $\vec{n} = (n_x, n_y, n_z)$.

Some of their properties are the following.

$$\vec{L} \cdot \vec{n} |j, \vec{n}\rangle = (D(R) L_z D^{-1}(R)) |j, \vec{n}\rangle \tag{8.24}$$

$$= (D(R) L_z D^{-1}(R)) D(R) |j, j\rangle = D(R) L_z |j, j\rangle \tag{8.25}$$

$$= j D(R) |j, j\rangle = j |j, \vec{n}\rangle. \tag{8.26}$$

Therefore

$$\langle j, \vec{n} | \vec{L} | j, \vec{n}\rangle = j \, \vec{n} \tag{8.27}$$

and

$$\Delta(\vec{L} \cdot \vec{m}) = \sqrt{1 - (\vec{n} \cdot \vec{m})^2} \sqrt{\frac{j}{2}}. \tag{8.28}$$

The expansion of these states in terms of L_z eigenstates is

$$|j, \vec{n}\rangle = \sum_m \phi_m(\vec{n}) |j, m\rangle \qquad \text{where } \phi_m(\vec{n}) = \langle j, m | D(R) | j, j\rangle = D^{(j)}(R)^j_m. \tag{8.29}$$

The most important property of the coherent states is that they provide a resolution of the identity. That is,

$$\mathbb{1}_j = \frac{2j+1}{4\pi} \int_{S^2} d^2\vec{n} \, |j, \vec{n}\rangle \langle j, \vec{n}|. \tag{8.30}$$

The left-hand side is the identity in $\mathcal{H}_j$. The integral is over all normalized vectors, therefore over a two sphere, with the standard $\mathbb{R}^3$ measure restricted to the unit sphere.

Exercise 8.1 *Prove this. (Hint: use the definition of the coherent states in terms of Wigner matrices, note the invariance under* U(1) *and promote the* S^2 *integral to an* SU(2) *integral, use the Peter–Weyl theorem.)*

Finally, observe that by taking tensor products of coherent states, we obtain coherent states. This follows from the properties of mean values and variance under tensor product.

8.1.2 Livine–Speziale coherent intertwiners

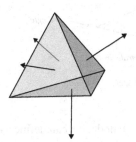

Armed with these tools, it is now easy to introduce "coherent tetrahedra" states. A classical tetrahedron is defined by the four areas A_a and the four (normalized) normals $\vec{n}_a$, up to rotations. These satisfy $\sum_{a=1}^{4} j_a \vec{n}_a = 0$. Therefore let us consider the coherent state

$$|j_1, \vec{n}_1\rangle \otimes |j_2, \vec{n}_2\rangle \otimes |j_3, \vec{n}_3\rangle \otimes |j_4, \vec{n}_4\rangle \tag{8.31}$$

in $\mathcal{H}_1 \otimes \ldots \otimes \mathcal{H}_4$ and project it down to its invariant part in the projection

$$P: \mathcal{H}_1 \otimes \cdots \otimes \mathcal{H}_4 \to Inv(\mathcal{H}_1 \otimes \cdots \otimes \mathcal{H}_4). \tag{8.32}$$

The resulting state

$$||j_a, \vec{n}_a\rangle \equiv P(|j_1, \vec{n}_1\rangle \otimes |j_2, \vec{n}_2\rangle \otimes |j_3, \vec{n}_3\rangle \otimes |j_4, \vec{n}_4\rangle) \tag{8.33}$$

is the element of $\mathcal{H}_\gamma$ that describes the semiclassical tetrahedron. The projection can be explicitly implemented by integrating over SO(3),

$$||j_a, \vec{n}_a\rangle = \int_{SO(3)} dR(|j_1, R\vec{n}_1\rangle \otimes |j_2, R\vec{n}_2\rangle \otimes |j_3, R\vec{n}_3\rangle \otimes |j_4, R\vec{n}_4\rangle). \tag{8.34}$$

Or, more properly, over the SU(2) action on the states,

$$||j_a, \vec{n}_a\rangle = \int_{SU(2)} dh \; (D^{j_1}(h)|j_1, \vec{n}_1\rangle \otimes D^{j_2}(h)|j_2, \vec{n}_2\rangle \otimes D^{j_3}(h)|j_3, \vec{n}_3\rangle \otimes D^{j_4}(h)|j_4, \vec{n}_4\rangle). \tag{8.35}$$

These are called the Livine–Speziale coherent intertwiners, and are essential tools in analyzing the theory.

It can be shown that if we expand this state in any intertwiner basis

$$||j_a, \vec{n}_a\rangle = \sum_k \Phi_k(\vec{n}_a)|\iota_k\rangle, \tag{8.36}$$

we find that for large j the coefficients $\Phi_k(\vec{n}_a) = \iota^{m_1 m_2 m_3 m_4} \psi_{m_1}(\vec{n}_1) \ldots \psi_{m_4}(\vec{n}_4)$ have the form $\Phi_k(\vec{n}_a) \sim e^{-\frac{1}{2}\frac{(k-k_0)^2}{\sigma^2}} e^{ik\psi}$: they are concentrated around a single value k which determines the value of the corresponding dihedral angle, and have a phase such that when changing basis to a different intertwined basis, we still obtain a state concentrated around a value.

Some properties of these states are the following. For large j,

$$\langle \iota(n_\ell) | E_a \cdot E_b | \iota(n_\ell) \rangle \approx j_a j_b \vec{n}_a \vec{n}_b \quad \text{and} \quad \frac{\Delta(\vec{E}_A \cdot \vec{E}_b)}{|E_a||E_b|} \ll 1. \tag{8.37}$$

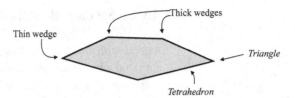

Figure 8.3 Thin and thick wedges between spacelike tetrahedra in Minkowski space.

Finally, by combining coherent intertwiners at each node, we can define a coherent state in $\mathcal{H}_\Gamma$ which is a wave packet peaked on a classical triangulated geometry:

$$\psi_{j_\ell, \vec{n}_{s_\ell}, \vec{n}_{t_\ell}}(\vec{U}_\ell) = \otimes_\ell D^{(j_\ell)}(U_\ell) \cdot \otimes_n \iota_n(\vec{n}_\ell). \tag{8.38}$$

8.1.3 Thin and thick wedges and time-oriented tetrahedra

If we consider a compact region in a pseudo-Riemannian manifold bounded by a spacelike surface Σ, the outgoing normal to Σ will be partially future oriented and partially past oriented. In a triangulation of Σ, some of the tetrahedra will be future oriented and some past oriented. In such a lorentzian context, a coherent intertwiner can be obtained from the euclidean one (8.35) by time-reversing the normals of past-pointing tetrahedra (Bianchi and Ding 2012). Time reversal is a well-defined operator in the representations we use and its effect on a coherent state is given by

$$T|j, \vec{n}\rangle = (-1)^j |j, -\vec{n}\rangle, \tag{8.39}$$

and thus on the (past-pointing) coherent intertwiner by

$$T||j_a, \vec{n}_a\rangle = (-1)^{\sum_{b>a} j_{ab}} ||j_a, -\vec{n}_a\rangle. \tag{8.40}$$

A link of the graph between two nodes represents a triangle joining two tetrahedra. In spacetime, this triangle is the edge of a wedge, which can then be of two types: a *thick wedge* if the incident tetrahedra have the same time orientation, a *thin wedge* otherwise (Barrett *et al.* 2010) (Figure 8.3).

When dealing with three-dimensional geometries embedded in spacetime, it is convenient to introduce a quantity Π_ℓ to denote the lorentzian geometry of the link ℓ in the following way:

$$\Pi_\ell = \begin{cases} 0, & \text{thick wedge} \\ \pi, & \text{thin wedge} \end{cases}. \tag{8.41}$$

We shall use this in the last chapter.

Before using these states to discuss the classical limit of quantum gravity, let us pause to introduce some facts about spinors, since these coherent states simplify considerably when written in spinor language.

8.2 Spinors and their magic

The fundamental representation of SL(2, $\mathbb{C}$) is defined on the same vector space as the fundamental representation of SU(2), namely, $\mathcal{H}_{\frac{1}{2}} = \mathbb{C}^2$, the space of couples of complex numbers. The elements of this space are called *spinors* and are fundamental and ubiquitous objects. Something magical about spinors is how they encode three-dimensional and four-dimensional physical objects, rotations, and Lorentz transformations. Michael Atiyah, one of the great living mathematicians has written about spinors:

> *Nobody fully understands spinors. Their algebra is clear, but their geometrical significance remains mysterious. In a sense, they describe the "square root" of geometry. To fully understand the square root of -1 has taken centuries, the same could be true for spinors.*[1]

We have already mentioned spinors repeatedly in the text. Here we summarize and expand on their properties and illustrate their use to describe coherent quantum states of physical space.

Recall that a spinor $\mathbf{z} \in \mathbb{C}^2$ is a two-dimensional complex vector

$$\mathbf{z} = \begin{pmatrix} z^0 \\ z^1 \end{pmatrix}, \qquad z^0, z^1 \in \mathbb{C}. \tag{8.42}$$

There is a large and confusing variety of notations used to denote spinors. Depending on the context, we use the notations

$$\mathbf{z} = \begin{pmatrix} z^0 \\ z^1 \end{pmatrix} = z^A = |z\rangle. \tag{8.43}$$

The space of these objects carries the fundamental representation of SL(2, $\mathbb{C}$)

$$g\mathbf{z} = \begin{pmatrix} a & c \\ b & d \end{pmatrix} \begin{pmatrix} z^0 \\ z^1 \end{pmatrix}, \qquad g \in \text{SL}(2, \mathbb{C}), \tag{8.44}$$

where $ac - bd = 1$, and of course a representation of the subgroup SU(2)

$$h\mathbf{z} = \begin{pmatrix} a & -\bar{b} \\ b & \bar{a} \end{pmatrix} \begin{pmatrix} z^0 \\ z^1 \end{pmatrix}, \qquad h \in \text{SU}(2). \tag{8.45}$$

From now on g will always indicate an SL(2, $\mathbb{C}$) element and h an SU(2) element. There are two bilinear forms which are defined on this space. The first is

$$(\mathbf{w}, \mathbf{z}) = n^A z^B \epsilon_{AB} = w^0 z^1 - w^1 z^0. \tag{8.46}$$

---

[1] Quoted in Farmelo (2009), Chapter 31.

It is antisymmetric and invariant under SL(2, $\mathbb{C}$),

$$(g\mathbf{w}, g\mathbf{z}) = (\mathbf{w}, \mathbf{z}). \tag{8.47}$$

The second is the scalar product of $\mathcal{H}_{\frac{1}{2}}$,

$$\langle \mathbf{w} \,|\, \mathbf{z} \rangle = \overline{w^A} z^B \delta_{AB} = \overline{w^0} z^0 + \overline{w^1} z^1. \tag{8.48}$$

This is invariant under SU(2),

$$\langle h\mathbf{w} \,|\, h\mathbf{z} \rangle = \langle \mathbf{w} \,|\, \mathbf{z} \rangle, \tag{8.49}$$

but *not* under SL(2, $\mathbb{C}$). (This is the reason why the fundamental representation of SL(2, $\mathbb{C}$) is not unitary.) Indeed,

$$\langle g\mathbf{w} \,|\, g\mathbf{z} \rangle = \langle \mathbf{w} \,|\, (g^\dagger g)\mathbf{z} \rangle. \tag{8.50}$$

If we write g as a rotation times a boost, the rotation cancels in $(g^\dagger g)$ but not the boost. Therefore the scalar product depends on the choice of a given frame, namely, an SU(2) subgroup of SL(2, $\mathbb{C}$). It has information about the choice of a Lorentz frame. The same information can be coded in the map $J : \mathbb{C}^2 \to \mathbb{C}^2$ defined by

$$J\begin{pmatrix} z^0 \\ z^1 \end{pmatrix} = \begin{pmatrix} \overline{z^1} \\ -\overline{z^0} \end{pmatrix}. \tag{8.51}$$

In fact, it is easy to see that

$$\langle \mathbf{w} \,|\, \mathbf{z} \rangle = (J\mathbf{w}, \mathbf{z}). \tag{8.52}$$

In words: we can define the (frame-dependent) scalar product in terms of the (frame-independent) bilinear form and the (frame-dependent) map J. These are the basic structures of spinor space.[2]

---

[2] We write $\overline{z^A}$ instead of the lighter notation $\bar{z}^A$ to avoid confusion with the notation used for instance by Wolfgang Wieland, where $\bar{z}^0$ and $\bar{z}^1$ do not indicate the complex conjugate of z^0 and z^1 respectively, but rather the complex conjugate of z^1 and $-z^0$ respectively. Wieland notation is

$$z^A = z^A, \quad z_A = \epsilon_{AB} z^B, \quad \bar{z}^A = (Jz)^A, \quad \bar{z}_A = (Jz)_A, \tag{8.53}$$

which leads to

$$\langle z \,|\, w \rangle = \bar{z}_A w^A, \quad (z, w) = z_A w^A. \tag{8.54}$$

Another notation used in the field is that used by Freidel and Speziale, who write

$$|z\rangle = z, \quad |z] = Jz, \tag{8.55}$$

so that

$$[z|w\rangle = (z, w). \tag{8.56}$$

One advantage of this notation is that it renders explicit the fact that (,) is not symmetric.

8.2.1 Spinors, vectors, and bivectors

Spinors and 3-vectors

Let us now see why spinor space is "the square root of geometry," in the words of Atiyah. The point is that with each spinor $\mathbf{n} \in \mathbb{C}^2$ we can associate a three-dimensional real vector.

$$\vec{n} = \langle \mathbf{n} | \vec{\sigma} | \mathbf{n} \rangle. \tag{8.57}$$

The Pauli matrices are hermitian and this vector is real. Explicitly, it is given by

$$\vec{n} = \begin{pmatrix} 2Re(n^0 \overline{n}^1) \\ 2Im(n^0 \overline{n}^1) \\ |n^0|^2 - |n^1|^2 \end{pmatrix}, \tag{8.58}$$

with a norm

$$|\vec{n}|^2 = \langle \mathbf{n} | \mathbf{n} \rangle, \tag{8.59}$$

and vice versa,

$$|\mathbf{n}\rangle\langle\mathbf{n}| = |\vec{n}| \mathbb{1} + \vec{n} \cdot \vec{\sigma} \tag{8.60}$$

where the left-hand side is the matrix $n^B \overline{n}_A$. As you can see, the map from spinors to vectors is quadratic; therefore a vector is the "square" of a spinor, and a spinor is the "square root" of a vector. Now a real number x has two square roots: $\pm\sqrt{x}$. Similarly, a vector has more than one spinor "square root." This is simple to see since $\mathbf{n}$ and $e^{i\phi}\mathbf{n}$ clearly give the same vector. Therefore a spinor has more information than a vector: in addition there is the phase ϕ. Roger Penrose pictures spinors as arrows with a little flag: the arrow indicates the vector, the flag (which can rotate around the arrow) indicates the phase (Figure 8.4).

The important point to absorb here is that we can associate a normalized spinor $\mathbf{n}$ (with a free phase) to a normalized vector $\vec{n}$.

Spinors and 4-vectors

Spinors also have a four-dimensional interpretation in Minkowsky space. For this, it is sufficient to construct the quadruplet of matrices $\sigma^I = (\mathbb{1}, \vec{\sigma})$ and extend the definition (8.57) to

$$n^I = \langle \mathbf{n} | \sigma^I | \mathbf{n} \rangle. \tag{8.61}$$

Figure 8.4   Penrose representation of a spinor: a vector with a flag.

Remarkably, this transforms as a vector under the action of SL(2, $\mathbb{C}$) on the spinor (try). The vector (8.61) is *null*, because it follows from (8.59) that

$$n^0 = |\vec{n}|, \tag{8.62}$$

and therefore $n_I n^I = (n^0)^2 - |\vec{n}|^2 = 0$. Therefore a spinor determines a vector in space and a null vector in spacetime.

Spinors and bivectors

But suppose now that we have now a spinor in $H_{\frac{1}{2}}$ (as opposed to a spinor in $\mathbb{C}^2$). That is, we have a spinor, but also a scalar product defined in $\mathbb{C}^2$, or, equivalently, the map J. Then a spinor defines a spacelike *bivector* in Minkowski space. This is quite obvious, since a 3-vector and a Lorentz frame define a timelike two-plane in Minkowski space, and its dual defines a spacelike plane. Equivalently, in three-dimensional a vector determines a plane, and this plane can be seen as a plane in Minkowski space if space is a slice of Minkoski space. For instance, the vector $(0, 0, 1)$ in the frame where the time direction is $(0, 0, 0, 1)$ determines the spacelike plane $(0, x, y, 0)$. Let us see how this works explicitly with spinors. The bivector is defined by

$$n^{IJ} = \epsilon^{IJ}{}_{KL} \langle \mathbf{n} | \sigma^{[K} | \mathbf{n} \rangle \langle J\mathbf{n} | \sigma^{L]} | J\mathbf{n} \rangle. \tag{8.63}$$

Indeed, recall that J acts as parity. Then, for instance, the spinor $n = (1, 0)$ determines the 3-vector $\vec{n} = (0, 0, 1)$, the null 4-vector $\langle \mathbf{n} | \sigma^K | \mathbf{n} \rangle = (1, 0, 0, 1)$, the parity reversed null 4-vector $\langle J\mathbf{n} | \sigma^L | J\mathbf{n} \rangle = (1, 0, 0, -1)$, and therefore the bivector $n^{IJ} = \delta_2^{[I} \delta_3^{J]}$ that determines the plane 23.

Therefore a spinor determines a spacelike bivector. This relation plays an important role in the theory. On the boundary, spinors are associated with (oriented) links, and represent the vector normal to a triangle. In the bulk, they are associated with faces, and represent the bivector parallel to the triangles of the bulk triangulation. All this is realized by the coherent states, where spinors play a crucial role.

8.2.2 Coherent states and spinors

The spinor $\mathbf{n} = (1, 0)$ is the eigenvector of L^z with eigenvalue $\frac{1}{2}$ and unit norm. Therefore it is the state $|j = \frac{1}{2}, m = \frac{1}{2}\rangle$, which is a coherent state. Since all other coherent states in the $j = \frac{1}{2}$ representation are obtained by rotating $|\frac{1}{2}, \frac{1}{2}\rangle$ and since rotation preserves the norm of the spinor, it follows that all normalized spinors $\mathbf{n}$ describe coherent states in the fundamental representation, that is,

$$|\mathbf{n}\rangle = |1/2, \vec{n}\rangle, \tag{8.64}$$

where $\mathbf{n}$ and $\vec{n}$ are related by (8.57). Indeed, from (8.57) it follows that

$$\langle \mathbf{n} | \vec{L} | \mathbf{n} \rangle = \langle \mathbf{n} | \frac{\vec{\sigma}}{2} | \mathbf{n} \rangle = \frac{1}{2} \vec{n} = j\vec{n}. \tag{8.65}$$

Normalized spinors are coherent states for the normalized 3-vectors they define. This is quite remarkable, and even more remarkable is that it extends to any representation.

Indeed, the tensor product of coherent states is a coherent state, as can be checked from the definitions. Consider the state

$$|j, \mathbf{n}\rangle = n \otimes \cdots \otimes n \qquad (8.66)$$

$2j$ times. This is in the spin-j representation, and is precisely the coherent state $|j, \vec{n}\rangle$ that satisfies

$$\langle j, \mathbf{n} | \vec{L} | j, \mathbf{n} \rangle = j \vec{n} \qquad (8.67)$$

and is minimally spread.

8.2.3 Representations of SU(2) and SL(2,$\mathbb{C}$) on functions of spinors and Y_γ map

In the Complements of the first chapter we constructed the irreducible representations of SU(2) by tensoring spinor space. The spin-j representation space $\mathcal{H}_j$ can be realized by symmetric tensors $y^{A_1 \cdots A_{2j}}$ with $2j$ spinor indices. There is another realization of the spin spaces $\mathcal{H}_j$ that is particularly useful. It is defined on a space of *functions* $f(z)$ of spinors. More precisely, the finite-dimensional vector space $\mathcal{H}_j$ can be realized as the space of the totally symmetric polynomial functions $f(z)$ of degree $2j$. This is indeed quite immediate, starting from the $y^{A_1 \cdots A_{2j}}$ states: the corresponding polynomial function of z is simply

$$f(\mathbf{z}) = y^{A_1 \cdots A_{2j}} z_{A_1} \cdots z_{A_{2j}} \qquad (8.68)$$

and satisfies the homogeneity condition

$$f(\lambda \mathbf{z}) = \lambda^{2j} f(\mathbf{z}). \qquad (8.69)$$

Clearly SU(2) acts on these functions via its transpose action

$$(Uf)(\mathbf{z}) = f(U^T \mathbf{z}). \qquad (8.70)$$

Let us see how coherent states look in this representation. For spin 1/2, clearly a coherent state is represented very simply by the linear function

$$f_{\mathbf{n}}(\mathbf{z}) \sim n^A z_A \sim \langle z | \mathbf{n} \rangle \qquad (8.71)$$

up to a normalization. If we take the symmetrized tensor product of this state with itself $2j$ times, we obtain the coherent state in the j representation in the surprisingly simple form

$$f_{\mathbf{n}}^{(j)}(\mathbf{z}) \sim \langle \mathbf{z} | \mathbf{n} \rangle^{2j}. \qquad (8.72)$$

The normalization can be computed, giving the normalized states

$$f_{\mathbf{n}}^{(j)}(\mathbf{z}) = \sqrt{\frac{2j+1}{\pi}} \langle \mathbf{z} | \mathbf{n} \rangle^{2j}. \qquad (8.73)$$

This is a very straightforward representation of the coherent state with angular momentum $j\vec{n}$.

This concrete realization of SU(2) representation spaces turns out to be very useful to relate SU(2) representations with SL(2,$\mathbb{C}$) unitary representations, because a similar representation exists for SL(2,$\mathbb{C}$). In the previous chapter, indeed, we constructed the SL(2,$\mathbb{C}$) representation space $V^{(p,k)}$ in the canonical basis $|p, j; jm'\rangle$. This is conceptually simple, but not easy to work with. A realization of the representation space of $V^{(p,k)}$ which is more convenient for computing is in terms of functions of spinors $f(\mathbf{z})$, where $\mathbf{z} \in \mathbb{C}^2$. The representation (p,k) is defined on the space of the homogeneous functions of spinors that have the property

$$f(\lambda \mathbf{z}) = \lambda^{-1+ip+k} \, \bar{\lambda}^{-1+ip-k} \, f(\mathbf{z}). \tag{8.74}$$

The representation is given by the transpose action

$$gf(\mathbf{z}) = f(g^T \mathbf{z}). \tag{8.75}$$

The translation between the canonical basis and the spinor basis is computed explicitly in Barrett *et al.* (2010), where it is found to be

$$f^j_m(\mathbf{z}) = \langle \mathbf{z} | p,k;j,m \rangle = \sqrt{\frac{2j+1}{\pi}} \, \langle \mathbf{z} | \mathbf{z} \rangle^{\,ip-1-j} \, D^j_{mk}(g(\mathbf{z})) \tag{8.76}$$

with

$$g(\mathbf{z}) = \begin{pmatrix} z_0 & \bar{z}_1 \\ z_1 & \bar{z}_0 \end{pmatrix}. \tag{8.77}$$

In these representations (both for SU(2) and SL(2, $\mathbb{C}$)) the scalar product between two functions is given by an integral in spinor space. The invariant integral in spinor space is defined by

$$\langle f | g \rangle = \int \bar{f} g \, \Omega \tag{8.78}$$

with

$$\Omega = \frac{i}{2}(z^0 \mathrm{d}z^1 - z^1 \mathrm{d}z^0) \wedge (\bar{z}^0 \mathrm{d}\bar{z}^1 - \bar{z}^1 \mathrm{d}\bar{z}^0). \tag{8.79}$$

This can be expressed more easily by exploiting the homogeneity properties of the functions. Indeed, by homogeneity these are uniquely determined by their value on $z^1 = 1$. Thus we can define

$$F(z) = f\begin{pmatrix} z \\ 1 \end{pmatrix}, \tag{8.80}$$

which has the same information as $f(\mathbf{z})$. Then the scalar product has the simple form

$$\langle f | g \rangle = \int_{\mathbb{C}} d^2 z \, \overline{G(z)} F(z), \tag{8.81}$$

where $d^2 z = dz\,d\bar{z} = dx\,dy$ if ($z = x + iy$) is the standard integration measure on the complex plane. In exchange, the group action becomes a bit more complicated (see Gel'fand *et al.* (1963), p. 247):

$$gf(z) = (az+b)^{k-1+ip}(\bar{a}\bar{z}+\bar{b})^{-k-1+ip} f\left(\frac{az+b}{cz+d}\right), \tag{8.82}$$

where

$$g = \begin{pmatrix} a & b \\ c & d \end{pmatrix}. \tag{8.83}$$

These spinor representations are particularly convenient because the Y_γ map takes a particularly simple form in this language. Since the embedding of H_j in $V^{(p,k)}$ is given by

$$f(\mathbf{z}) \rightarrow \langle \mathbf{z} | \mathbf{z} \rangle^{-1+ip-k} f(\mathbf{z}), \qquad (8.84)$$

we have simply

$$Y_\gamma f(\mathbf{z}) = \langle \mathbf{z} | \mathbf{z} \rangle^{-1+(i\gamma-1)j} f(\mathbf{z}). \qquad (8.85)$$

This is the map at the basis of loop gravity in this representation!

This allows us to write immediately the action of the Y_γ map on coherent states: from (8.85) we obtain

$$\langle \mathbf{z} | Y_\gamma | j, \vec{n} \rangle = \sqrt{\frac{2j+1}{\pi}} \langle \mathbf{z} | \mathbf{z} \rangle^{-1+(i\gamma-1)j} \langle \mathbf{z} | \mathbf{n} \rangle^{2j}. \qquad (8.86)$$

Or, for later convenience,

$$\langle \mathbf{z} | Y_\gamma | j, \vec{n} \rangle = \frac{\sqrt{2j+1}}{\sqrt{\pi} \langle \mathbf{z} | \mathbf{z} \rangle} e^{j[(i\gamma-1)\ln\langle \mathbf{z} | \mathbf{z} \rangle + 2\ln\langle \mathbf{z} | \mathbf{n} \rangle]}. \qquad (8.87)$$

Notice how straightforward is the form of the Y_γ map when acting on spinor states. This expression gives the form of the coherent state $|j, \vec{n}\rangle$ after it is mapped by Y_γ into the appropriate SL(2, $\mathbb{C}$) representation.

8.3 Classical limit of the vertex amplitude

We now use the tools developed above to rewrite the vertex amplitude using spinors. Recall the vertex amplitude is given by (7.58), which we repeat here:

$$A_v(h_{vf}) = \sum_{j_f} \int_{\mathrm{SL}(2,\mathbb{C})} dg'_{ve} \prod_f (2j_f + 1) \mathrm{Tr}_{j_f}[Y_\gamma^\dagger g_{e'v} g_{ve} Y_\gamma h_{vf}]. \qquad (8.88)$$

It is convenient to drop the subscript v, label the edges emerging from the vertex with labels $a, b = 1, \ldots, 5$ and the faces adjacent to the vertices as (ab), where a and b are the two edges bounding the face at the vertex. Then the amplitude reads

$$A_v(h_{ab}) = \sum_{j_{ab}} \int_{\mathrm{SL}(2,\mathbb{C})} dg'_a \prod_{ab} (2j_{ab} + 1) \mathrm{Tr}_{j_{ab}}[Y_\gamma^\dagger g_a^{-1} g_b Y_\gamma h_{ab}]. \qquad (8.89)$$

This is the quantity we want to study.

8.3.1 Transition amplitude in terms of coherent states

The trace in the last equation can be written inserting two resolutions of the identity in terms of coherent states

$$\mathrm{Tr}_j[Y_\gamma^\dagger g g' Y_\gamma h] = \int_{S^2} d\vec{n}, d\vec{m} \, \langle j, \vec{m} | Y_\gamma^\dagger g g' Y_\gamma | j, \vec{n} \rangle \, \langle j, \vec{n} | h | j, \vec{m} \rangle. \qquad (8.90)$$

Let us focus on the first of these matrix elements, which is the crucial one. Writing it in the spinor basis gives

$$\langle Y_\gamma j, \vec{m} | gg' | Y_\gamma j, \vec{n} \rangle = \int_{\mathbb{C}^2} d\Omega \, \langle Y_\gamma j, \vec{m} | g\mathbf{z} \rangle \, \langle g'^\dagger \mathbf{z} | Y_\gamma j, \vec{n} \rangle.$$

Using the representation (8.86) of coherent states in the spinor basis, and introducing the (common but a bit confusing) notation

$$\mathbf{Z} = g\mathbf{z}, \quad \mathbf{Z}' = g'^\dagger \mathbf{z}, \tag{8.91}$$

this gives

$$\langle Y_\gamma j, \vec{m} | gg' | Y_\gamma j, \vec{n} \rangle = \frac{(2j+1)}{\pi} \int_{\mathbb{C}^2} \frac{d\Omega}{\langle \mathbf{Z} | \mathbf{Z} \rangle \langle \mathbf{Z}' | \mathbf{Z}' \rangle} \, e^{j \, S(\mathbf{n}, \mathbf{m}, \mathbf{Z}, \mathbf{Z}')}, \tag{8.92}$$

where

$$S(\mathbf{n}, \mathbf{m}, \mathbf{Z}, \mathbf{Z}') = \ln \frac{\langle \mathbf{Z} | \mathbf{m} \rangle^2 \langle \mathbf{Z}' | \mathbf{n} \rangle^2}{\langle \mathbf{Z} | \mathbf{Z} \rangle \langle \mathbf{Z}' | \mathbf{Z}' \rangle} + i\gamma \ln \frac{\langle \mathbf{Z} | \mathbf{Z} \rangle}{\langle \mathbf{Z}' | \mathbf{Z}' \rangle}. \tag{8.93}$$

Let us now insert this into the vertex amplitude (8.89). We need unit vectors $\vec{n}_{ab}$ (where $\vec{n}_{ab}$ and $\vec{n}_{ba}$ indicate different objects). The amplitude is a functional on the space $\mathcal{H}_{\Gamma_V}$ of the states living on the boundary of the vertex graph Γ_V, namely, on the boundary of the 4-simplex dual to the vertex. Let us choose a *coherent* state in $\mathcal{H}_{\Gamma_V}$. In particular, pick a quadruplet of normalized vectors $\vec{n}_{ab}$ for each node a of Γ_V. These define a state $|j_{ab}, \vec{n}_{ab}\rangle$. A moment of reflection shows that the amplitude of this state is

$$A_V(j_{ab}, \vec{n}_{ab}) \equiv \langle A_V | j_{ab}, \vec{n}_{ab} \rangle = \int_{\mathrm{SL}(2,\mathbb{C})} dg'_a \prod_f (2j_{ab} + 1) \, \langle j_{ab}, \mathbf{n}_{ab} | Y_\gamma^\dagger g_a^{-1} g_b Y_\gamma | j_{ba}, \mathbf{n}_{ba} \rangle. \tag{8.94}$$

To obtain a coherent tetrahedron state there is no need to integrate over SU(2), since the integration over $\mathrm{SL}(2,\mathbb{C})$ already does the job. Using the result above, this gives

$$A_V(j_{ab}, \vec{n}_{ab}) = \mu(j_{ab}) \int_{\mathrm{SL}(2,\mathbb{C})} dg'_a \int_{\mathbb{C}^2} \frac{d\Omega_{ab}}{|\mathbf{Z}_{ab}| |\mathbf{Z}_{ba}|} \, e^{\sum_{ab} j_{ab} \, S(\mathbf{n}_{ab}, \mathbf{n}_{ba}, \mathbf{Z}_{ab}, \mathbf{Z}_{ba})}, \tag{8.95}$$

with $\mu(j_{ab}) = \prod_{ab} \frac{(2j_{ab}+1)^2}{\pi}$ and $\mathbf{Z}_{ab} = g_a \mathbf{z}_{ab}$ and $\mathbf{Z}_{ba} = g_b \mathbf{z}_{ab}$. Notice that $j_{ab} = j_{ba}$ and $z_{ab} = z_{ba}$ but $\vec{n}_{ab}$ and $\mathbf{Z}_{ab}$ are in general different from $\vec{n}_{ba}$ and $\mathbf{Z}_{ba}$.

The last equation gives the form of the vertex amplitude in terms of coherent states. This is the starting point for studying the classical limit of the dynamics, which is what we do in the next section.

Saddle point

In any quantum theory, the classical limit is the limit of large quantum numbers. In general, this is the regime where actions are large with respect to $\hbar$ and therefore quantum effects can be disregarded. To study the classical limit of the dynamics, we have therefore to study what happens to the transition amplitude in the limit of large quantum numbers. Let us see in particular what happens when j is large.

If the spins j_{ab} are large, the integral (8.95) can be evaluated using the saddle-point approximation (Barrett *et al.* 2010). The saddle-point approximation for a one-dimensional integral is

$$\int_R dx\, g(x)\, e^{jf(x)} = \sqrt{\frac{2\pi}{jf(x_o)}}\, g(x_o)\, e^{\lambda f(x_o)}\left[1 + o\left(\frac{1}{j}\right)\right], \qquad (8.96)$$

where x_o is the saddle point, namely a point where the first derivative of $f(x)$ vanishes (here assumed to be unique for simplicity). The generalization of this formula to integrals in d dimension is

$$\int_{R^d} dx^d\, g(x)\, e^{jf(x)} = \left(\frac{2\pi}{j}\right)^{\frac{d}{2}} (\det H_2 f)^{-\frac{1}{2}}\, g(x_o)\, e^{\lambda f(x_o)}\left[1 + o\left(\frac{1}{j}\right)\right], \qquad (8.97)$$

where $H_2 f$ is the hessian (the matrix formed by the second derivatives) of f at the saddle point, which is now a point where the gradient of f vanishes. For this to hold, the real part of f must be negative. Intuitively, what happens is rather simple: if f is real, a large j gives a narrow gaussian around the maximum of f. If f is imaginary, when j is large, the exponential oscillates very fast, and the integral is canceled out by the fast oscillations. The points where this does not happen are the ones where the derivative of f vanishes, because in these points the fast variation of the exponential in x is suppressed. The two cases are illustrated in Figure 8.5.

The imaginary case is of particular interest because it is what gives the standard classical limit of quantum mechanics from the Feynman path-integral formulation: the classical trajectories are given by an extremum of the action because in the limit where $\hbar$ is small, quantum interference cancels out the amplitude except for the saddle point of the path integral. It is good to keep this general fact in mind to follow what happens here.

We do not give here the full details of the calculation to find the saddle point of the integral (8.95). These can be found in the classic paper of Barrett *et al.* (2010). See also Magliaro and Perini (2011b); Han and Zhang (2013); Ding and Han (2011) for recent analyses. We only sketch the main steps. We have to find the stationary points of the action (8.93). Let us start from the real part. This is given by

$$Re[S] = \sum_{ab} \log \frac{|\langle \mathbf{Z}_{ab} \mid \mathbf{n}_{ab} \rangle|^2\, |\langle \mathbf{Z}_{ba} \mid \mathbf{n}_{ba} \rangle|^2}{\langle \mathbf{Z}_{ab} \mid \mathbf{Z}_{ab} \rangle\, \langle \mathbf{Z}_{ba} \mid \mathbf{Z}_{ba} \rangle}. \qquad (8.98)$$

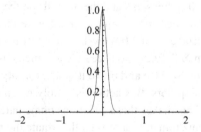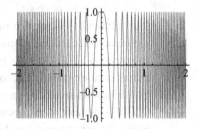

Figure 8.5 $e^{jf(x)}$ for large j in the real and the imaginary case.

This is always smaller than unit, because the scalar product $|\langle \mathbf{Z}_{ab} | \mathbf{n}_{ab} \rangle|^2$ is necessarily smaller than the norm $\langle \mathbf{Z}_{ab} | \mathbf{Z}_{ab} \rangle$ as coherent states are normalized. The maximum is therefore when the log vanishes. An obvious solution is when

$$\mathbf{n}_{ab} = \frac{\mathbf{Z}_{ab}}{|\mathbf{Z}_{ab}|}, \qquad \mathbf{n}_{ba} = \frac{\mathbf{Z}_{ba}}{|\mathbf{Z}_{ba}|}, \tag{8.99}$$

but this is not the only solution because there might be a phase that cancels in the products. The general solution is instead

$$\mathbf{n}_{ab} = e^{i\phi_{ab}} \frac{\mathbf{Z}_{ab}}{|\mathbf{Z}_{ab}|}, \qquad \mathbf{n}_{ba} = e^{i\phi_{ab}} \frac{\mathbf{Z}_{ba}}{|\mathbf{Z}_{ba}|}. \tag{8.100}$$

Recalling the meaning of $\mathbf{Z}$, this gives

$$g_a^{-1} \mathbf{n}_{ab} = \frac{\mathbf{Z}_{ba}}{\mathbf{Z}_a b} e^{i\theta_{ab}} g_a^{-1} \mathbf{n}_{ba}. \tag{8.101}$$

Let us now look at the extrema of the action under variations of the spinor variables z_{ab}. The explicit calculation gives an equation similar, but not identical, to the last one:

$$g_a \mathbf{n}_{ab} = \frac{\mathbf{Z}_{ba}}{\mathbf{Z}_a b} e^{i\theta_{ab}} g_a \mathbf{n}_{ba}. \tag{8.102}$$

Finally, we now look at the extrema of the action under variations of the group elements g_a and the integration variables z_{ab}. The first is relatively easy. The group variables enter $\mathbf{Z}_{ab}$. A first-order variation of the group element gives the action of the algebra element. Therefore the saddle-point equations for the group elements give the vanishing of the action of an infinitesimal $SL(2, \mathbb{C})$ transformation on the action. Such an action can be decomposed into boosts and rotations, but by construction the action of the boosts is proportional to the action of the rotations in the relevant representations. Therefore the needed invariance is only for rotations. This can be moved from the $\mathbf{Z}$ to the normals n by change of variables, and the result is

$$\sum_b j_{ab} | \mathbf{n}_{ab} \rangle = 0 \tag{8.103}$$

This is beautiful and surprising: the saddle-point equations for the group integral are precisely the closure conditions for the normal at each of the boundary nodes of the vertex graph. A priori, we have chosen an arbitrary set of normals, but the dynamics suppresses all possible sets of $\mathbf{n}_{ab}$ unless these satisfy the closure constraint at each node – that is, unless they define a proper tetrahedron τ_a at each node a of the vertex graph.

Now, because of equation (8.103), the spins and the normals on the vertex graph define four tetrahedra. These are three-dimensional objects, and can be thought of as lying in a common three-dimensional surface Σ of Minkowski space left invariant by the $SU(2)$ subgroup of $SL(2, \mathbb{C})$. A vector in Σ defines a surface in Σ to which it is normal. A Lorentz transformation can then act on this surface and move it to an arbitrary (spacelike) surface in Minkowski space. In terms of spinors, this action is simply given by the action of an $SL(2, \mathbb{C})$ element on the spinor. This makes these equations immediately transparent: they imply the existence of five Lorentz transformations g_a that rotate the five tetrahedra τ_a in Minkowski space, in such a way that the b face of the tetrahedron τ_a is parallel to the a

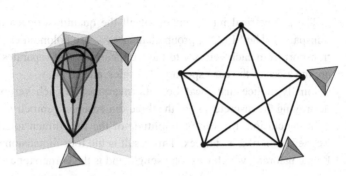

Figure 8.6   Two tetrahedra in the boundary of a vertex are connected by a face.

face of the tetrahedron τ_b. That is, the group elements g_a rotate the five tetrahedra from a common spacelike surface into place to form the boundary of a 5-simplex (Figure 8.6).

Therefore at the critical point the variables describe precisely a 4-simplex in Minkowski space. The value of the action at the saddle point is then

$$S = i\gamma \sum_{ab} j_{ab} \Theta_{ab} \qquad (8.104)$$

plus a term that determines only a sign in the amplitude. What is Θ_{ab}? It is the difference between the Lorentz transformations to the opposite side of each side of a triangle in the 4-simplex. Therefore it is the dihedral angle between two tetrahedra. But γj_{ab} (in units $8\pi G\hbar = 1$) is the area of the boundary faces of the 4-simplex. It follows that S on the critical point is the Regge action of the 4-simplex having the boundary geometry determined by the 10 areas j_{ab}.

More precisely, since it is possible to construct two 4-simplices with opposite orientation from the same boundary data, the vertex amplitude, in the large j limit, is proportional to

$$A(j_{ab}, n_{ab}) \sim c e^{iS_{Regge}(j_{ab})} + c' e^{-iS_{Regge}(j_{ab})}. \qquad (8.105)$$

The constants c and c' are not equal, due to the presence of the same Maslov index that appears in three-dimensional. Here we have assumed that the boundary tetrahedra define a non-degenerate lorentzian 4-simplex. For the other possible case and a detailed derivation and discussion, including the details neglected here, see Barrett *et al.* (2010).

Summarizing, if the state on the boundary of a vertex represents a geometry that can be the boundary geometry of a 4-simplex, the amplitude of the vertex is the Regge amplitude associated with the corresponding 4-simplex. Importantly, if the normals are not the ones determined by the geometry of the 4-simplex, the amplitude is suppressed.

The Regge action of a 4-simplex is the Hamilton function of the 4-simplex. The full amplitude over a 2-complex is then obtaining by summing over spins. The spin configurations such that no normals exist for which the tetrahedra match are suppressed. The remaining ones are those for which there are lengths at the boundaries of the 4-simplices, that determine the corresponding areas at the faces. The sum over these is equivalent to a sum over truncated geometries, where the weight is determined by the Regge action. The result gives, in the classical limit, the Regge Hamilton function of the boundary data.

The geometrical interpretation of all the quantities appearing in the integral is then transparent. The SL$(2, \mathbb{C})$ group elements g_{ve} are holonomies of the spin-connection that transport from each vertex v to the tetrahedron e that separates two vertices. The spins j_f are the areas of the corresponding triangles, and so on.

Finally, notice that local Lorentz invariance at each vertex is implemented without destroying the discreteness of the three-dimensional geometry.

In conclusion, the vertex amplitude of the four-dimensional theory approximates the Regge action of a 4-simplex. This result is the four-dimensional version of the Ponzano–Regge theorem, which was long sought and is the cornerstone of the theory.

8.3.2 Classical limit versus continuum limit

In the previous section we discussed the large-j limit, which is a *classical* limit. In the previous chapter we discussed the limit in which the 2-complex $\mathcal{C}$ is refined, which is a *continuous* limit. The two limits are conceptually and practically very different and should not be confused. Unfortunately, confusion between the two limits abounds in the literature. So, let us try to get some clarity here. The relation between the two limits is summarized in Table 8.1.

The transition amplitudes on a fixed 2-complex that we have defined sit in the top left corner of Table 8.1. They yield an increasingly better approximation by refining the 2-complex. As we refine the 2-complex, we move toward the bottom of the table, namely, toward the (ideal) exact transition amplitudes of quantum gravity.

If we look at increasingly large spin values of the boundary state at fixed triangulation, we move toward the right in the table. At each fixed 2-complex $\mathcal{C}$, dual to a triangulation Δ, the classical (large j) limit of the transition amplitudes is related to the Hamilton function of the Regge truncation of general relativity defined on that triangulation Δ.

If we refine the triangulation, the Regge discretization is known to converge to classical general relativity. Therefore as we (correspondingly) refine the 2-complex, the transition amplitudes define a quantum theory with more degrees of freedom, whose classical limit is closer and closer to continuum general relativity. This is the way amplitudes are defined in covariant loop gravity.

Recall again that also in QED and QCD, concrete calculations are generally performed on the truncated theory. This theory provides an approximation that is good in certain regimes.

The regimes where the *classical limit* is good in quantum gravity are those involving scales L that are much larger than the Planck scale.

$$L \gg L_{\text{Planck}}. \tag{8.106}$$

The regimes where the *truncation* is good are suggested by the Regge approximation. This is good when the deficit angles are small. This happens when the scale of the discretization is small with respect to curvature scale $L_{\text{curvature}}$,

$$L \ll L_{\text{curvature}}. \tag{8.107}$$

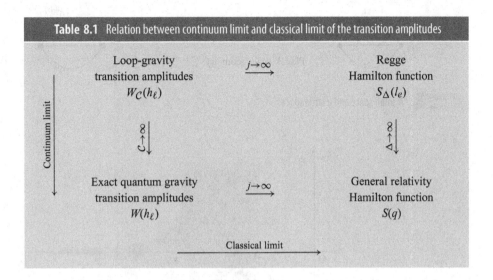

Table 8.1 Relation between continuum limit and classical limit of the transition amplitudes

Therefore a triangulation with few cells, and, correspondingly, a 2-complex with few vertices, provide an approximation in the regimes (determined by the boundary data) where the size of the cells considered is small with respect to the curvature scale (of the classical solution of the Einstein equation determined by the given boundary data).

Refining the triangulation leads to including shorter length-scale degrees of freedom. But the physical scale of a spinfoam configuration is not given by the graph or the 2-complex. It is given by the size of its geometrical quantities, which is determined by the spins (and intertwiners). The same triangulation can represent both a small and a large size of spacetime. A large chunk of nearly flat spacetime can be well approximated by a coarse triangulation, while a small chunk of spacetime where the curvature is very high requires a finer triangulation. In other words, triangulations are like all engineers' discretizations: they do not need to be uselessly fine: they need to be just so fine as to capture the relevant curvature.[3]

On the same Hilbert space $\mathcal{H}_\Gamma$, determined by the same graph, there are states representing small and large chunks of space (Figure 8.7).

A detailed analysis of the continuous limit for large trianguations, taking into account the need for working at a scale intermediate between the Planck scale and the curvature scale for recovering the classical theory, has been developed in particular by Muxin Han (Han and Zhang 2013; Han 2013). Han's analysis, which we do not report here, studies the expansion that leads from the full loop quantum gravity amplitudes to the effective action as a two-parameter expansion: $1/\lambda \sim L_{\text{Planck}}^2/L^2$ measures quantum corrections, while $\Theta \sim L^2/L_{\text{curvature}}^2$ measures the high-energy corrections. Interestingly, Han uncovers a relation between γ and these two parameters, of the form

$$|\Theta| \le \gamma^{-1}\lambda^{\frac{1}{2}}, \tag{8.108}$$

illustrated in Figure 8.8.

---

[3] Misunderstanding of this point has generated substantial confusion in the literature.

Figure 8.7   A small space and a large space

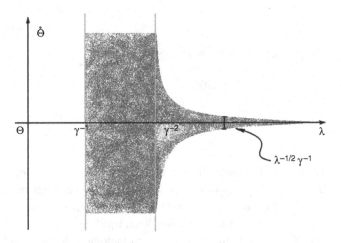

Figure 8.8   The Einstein sector of loop quantum gravity.

The shaded region in Figure 8.8 is where the expansion is good.[4] It is called the *Einstein sector* of loop gravity. It defines a subspace of spinfoam configurations, where configurations are interpreted as lorentzian geometries satisfying

$$L_{\text{Planck}} \ll L \ll L_{\text{curvature}}. \tag{8.109}$$

When the fluctuations of the spinfoam variables are in this sector, the perturbative expression of the spinfoam amplitude is then given by

$$W \sim \sum_{(j_{\text{f}}, g_{\text{ve}}, z_{\text{vf}})} e^{\frac{i}{L_{\text{Plank}}^2} \int_M d^4x \sqrt{-g} R[g_{\mu\nu}] \left(1 + o(L^2/L_{\text{curvature}}^2) + o(L_{\text{Planck}}/L)\right)} \tag{8.110}$$

where $g_{\mu\nu}$ is the lorentzian metric approximated by the spinfoam data $(j_{\text{f}}, g_{\text{ve}}, z_{\text{vf}})$. The leading contributions come from the configurations $(j_{\text{f}}, g_{\text{ve}}, z_{\text{vf}})$ which give $g_{\mu\nu}$ satisfying the Einstein equations.

---

[4] These results clarify the source of the apparent "flatness" notices by same authors (Bonzom 2009; Hellmann and Kaminski 2013) in the spinfoam amplitudes: a naive large-scale limit cannot be taken, because it leads outside the correct regime of validity of the low-energy approximation: the naive $j \to \infty$ limit on a fixed triangulation yields a flat geometry.

8.4 Extrinsic coherent states

There is one additional topic we have to cover before concluding this chapter. The astute reader may wonder why the intrinsic coherent states introduced in Section 8.23 were called "intrinsic." The reason is that they are coherent states for the operators $\vec{L}_\ell$ that represent the intrinsic geometry of space, but they are not coherent states for the operators representing the extrinsic geometry. The quantum spread of the intrinsic geometry of space is minimized by these states, but not its conjugate momentum, namely, its time variation. Therefore these are not truly semiclassical states in the sense in which a common wave packet is. In fact, these states are eigenstates of the area variables, and therefore the variable conjugate to the area is completely quantum-spread. The geometrical interpretation of the variable conjugate to the area of a triangle between two tetrahedra is the four-dimensional dihedral angle between the four-dimensional normals of the two tetrahedra. This is clearly the discrete version of the extrinsic curvature, which is the variable conjugate to the 3-metric in the ADM variables. The states that are both coherent in the intrinsic and extrinsic geometry are the *extrinsic* coherent states.

There are different manners in which to construct the extrinsic coherent states (Bahr and Thiemann 2009; Freidel *et al.* 2009; Freidel and Speziale 2010b; Bianchi *et al.* 2010a). Here we present a simple one, which makes their meaning transparent (Bianchi *et al.* 2010a,c). To see what goes on, let us start from a simple example. Consider a one-dimensional quantum system with position q and momentum p. A wave packet peaked on the phase space point (q,p) is (in the Schrödinger representation and disregarding normalizations)

$$\langle x \,|\, q,p \rangle \equiv \psi_{q,p}(x) = e^{-\frac{(x-q)^2}{2\sigma^2} + \frac{i}{\hbar} px}. \tag{8.111}$$

This state has spread $\Delta x \sim \sigma$ in position and $\Delta p \sim \hbar/\sigma$ in momentum, since its Fourier transform is (up to a normalization)

$$\langle k \,|\, q,p \rangle \sim e^{-\frac{(k - p/\hbar)^2}{2/\sigma^2} + iqk}. \tag{8.112}$$

For later convenience, observe that this state can be written in the form (always disregarding normalization)

$$\psi_{q,p}(x) = e^{-\frac{(x-z)^2}{2\sigma^2}}, \tag{8.113}$$

where z is the complex variable

$$z = q - i\frac{\sigma^2}{\hbar}p. \tag{8.114}$$

That is, a wave packet peaked on the phase space point (q,p) can be written as a gaussian function with a *complex* position.

We need an analog of this state in $\mathcal{H}_\Gamma$, peaked both on group variables and on their conjugate. Let us first consider $L_2[\mathrm{SU}(2)]$. A state in this space which is completely peaked on group variables is clearly given by a delta function on the group. The (generalized) state

$$\psi(U) = \delta(U) \tag{8.115}$$

is sharp on the origin $U = 1$, while the state

$$\psi(U) = \delta(Uh^{-1}) \tag{8.116}$$

is sharp on the group element $h \in SU(2)$. These states are of course completely spread in the conjugate variable, as is clear by "Fourier transforming" on the group:

$$\delta(U) = \sum_j d_j Tr_j[U]. \tag{8.117}$$

But we can "spread this out" around the origin by adding a gaussian factor in momentum space. To see how this can be done, recall that a gaussian can be obtained by acting on the delta function with the operator $exp\{-t\nabla^2\}$, where ∇^2 is the Laplacian and $t = \sigma^2$. Doing the same on the group and using the Laplacian $\vec{L}^2$ gives the state

$$\psi_{1,0}(U) = \sum_j d_j \, e^{-tj(j+1)} \, Tr_j[U], \tag{8.118}$$

which is peaked on $U = 1$ with spread $\sqrt{t}$ as well as on $j = 0$ with spread $1/\sqrt{t}$. Similarly,

$$\psi_{h,0}(U) = \sum_j d_j \, e^{-tj(j+1)} \, Tr_j[Uh^{-1}], \tag{8.119}$$

is peaked on $U = h$ and $j = 0$ with the spread $\sqrt{t}$ and $1/\sqrt{t}$, respectively. How then do we obtain a state peaked on $j \neq 0$? In the case of the particle, this is done by adding the oscillating factor $exp\{ipx/\hbar\}$. But we have also seen that this can be obtained by complexifying the argument. Let us do the same on the group. A complexification of $SU(2)$ is given by $SL(2,\mathbb{C})$. Let us therefore consider the state

$$\psi_H(U) = \sum_j d_j \, e^{-tj(j+1)} \, TrD^{(j)}(UH^{-1}). \tag{8.120}$$

labeled by a variable $H \in SL(2,\mathbb{C})$ such that

$$H = e^{it\frac{E}{l_o^2}} h \tag{8.121}$$

where $h \in SU(2)$ and $E = \vec{E} \cdot i\frac{\vec{\sigma}}{2}$ is in the algebra su(2). Notice that the Wigner matrix is here evaluated on the $SL(2,\mathbb{C})$ element H: by this we mean its analytic continuation on the complex plane. This is a good guess for a wave packet peaked both in the group variable and on its conjugate variable. In fact, an explicit calculation gives

$$\frac{\langle \psi_H | U | \psi_H \rangle}{\langle \psi_H | \psi_H \rangle} = h \,, \qquad \frac{\langle \psi_H | \vec{E} | \psi_H \rangle}{\langle \psi_H | \psi_H \rangle} = \vec{E}, \tag{8.122}$$

with the same spread as before. The spread is $\Delta U_\ell \sim \sqrt{t}$ and $\Delta E_\ell \sim l_o^2 \sqrt{1/t}$.

If we study a phenomenon at a scale $E \sim L^2 \sim l_o^2 j \ll l_o^2$, then

$$\frac{\Delta E}{E} \sim \frac{1}{\sqrt{t}\, j}. \tag{8.123}$$

For this to be small, we need

$$\sqrt{t} \gg \frac{1}{j}. \tag{8.124}$$

Therefore we must have at the same time

$$\frac{1}{j} \ll \sqrt{t} \ll 1. \tag{8.125}$$

Notice then that $\sqrt{t}$ determines an intermediate scale $L_{wp}^2 = l_o^2 \sqrt{t}$ such that

$$l_o \ll L_{wp} \ll L. \tag{8.126}$$

The scale L_{wp} is a scale intermediate between the scale of the phenomenon L and the Planck scale l_o. It gives the size of the wave packet. In order to be in a semiclassical regime, this must be much smaller that the scale considered, but still much larger than the Planck scale.

It is then easy to generalize these states to spin-network states, by making them invariant under SU(2) at the nodes. An extrinsic coherent state on a graph Γ is then labeled by an SL(2, $\mathbb{C}$) variable H_ℓ associated with each link ℓ of the graph, and is defined by

$$\psi_{H_\ell}(U_\ell) = \int_{\text{SU}(2)} dh_{\text{n}} \prod_\ell \sum_{j_\ell} d_{j_\ell} \, e^{-t j_\ell(j_\ell+1)} \, \text{Tr} D^{(j_\ell)}(U_\ell h_{s_\ell} H_\ell^{-1} h_{t_\ell}^{-1}), \tag{8.127}$$

where, for each ℓ, the nodes s_ℓ and t_ℓ are the source and the target. It is convenient to introduce the integral kernel

$$K(U, H) = \sum_j d_j \, e^{-t j(j+1)} \, \text{Tr} D^{(l)}(UH^{-1}), \tag{8.128}$$

called the "heat kernel", on the group, because it is the analytic continuation on SL(2, $\mathbb{C}$) of the solution of the diffusion equation for the Laplacian on the group, and write

$$\psi_{H_\ell}(U_\ell) = \int_{\text{SU}(2)} dh_{\text{n}} \prod_\ell K(U_\ell, h_{t_\ell} H_\ell h_{s_\ell}^{-1}). \tag{8.129}$$

The kernel intertwines between SU(2) and SL(2, $\mathbb{C}$), but should not be confused with the Y_γ map, to which it is unrelated.

The states ψ_{H_ℓ} are the extrinsic coherent states of loop quantum gravity. They are peaked both on the intrinsic and on the extrinsic curvature of a three-dimensional triangulation. Because of the SU(2) integrations in their definition, what matters is not the specific direction of each normal, but the SU(2) invariant quantities, which are the lengths of the normals, namely the areas of the faces, and the angles between any two normals of a node. The relation between the data H_ℓ and the geometry can be clarified as follows.

A Lorentz transformation H can always be written as the product of a rotation R, a boost $e^{p\frac{\sigma_3}{2}}$ in the z direction and another rotation R'. In turn, any rotation R can be written as the product of $R_{\vec{n}}$ and a rotation $e^{iq\frac{\sigma_3}{2}}$ around the z axis, where $R_{\vec{n}} \in$ SU(2) is the rotation matrix that rotates a unit vector pointing in the $(0,0,1)$ direction into the unit vector $\vec{n}$, around the axis normal to both. Using this we can always write

$$H_\ell = R_{\vec{n}_{s(\ell)}} \, e^{-iz_\ell \frac{\sigma_3}{2}} \, R_{\vec{n}_{t(\ell)}}^{-1} \tag{8.130}$$

for a *complex* variable

$$z_\ell = k_\ell + i\frac{2t}{l_o^2}A_\ell. \tag{8.131}$$

The geometrical interpretation of H_ℓ is then given by the four quantities $(A_\ell, k_\ell, \vec{n}_s, \vec{n}_t)$. In fact, it is not difficult to show (Bianchi *et al.* 2010a) that the state ψ_{H_ℓ} can be rewritten in the form

$$\psi_{H_\ell} = \sum_{j_\ell}\prod_\ell d_{j_\ell}e^{-t(j_\ell - j_\ell^0)^2}e^{ik_\ell}\psi_{j_\ell, \vec{n}_{s_\ell}, \vec{n}_{t_\ell}}, \tag{8.132}$$

where $j_\ell^0 = A_\ell/l_o^2$ and $\psi_{j_\ell, \vec{n}_{s_\ell}, \vec{n}_{t_\ell}}$ are the intrinsic coherent states (8.23) defined in Section 8.1.2. The two vectors $\vec{n}_s$ and $\vec{n}_t$ represent the normals to the triangle ℓ, in the two tetrahedra bounded by this triangle, A_ℓ is the area of this triangle, because the spin, which determines the area, is peaked on $j_\ell^0 = A_\ell/l_o^2$, and k_ℓ determines the variable conjugate to the area, which is related to the extrinsic geometry at the triangle, namely, the angle between the four-dimensional normals to the two tetrahedra.[5] This geometrical interpretation allows us to build discretized spaces with intrinsic and extrinsic geometry, from the H_ℓ.

If we assume the metric to be flat within each tetrahedron, then the metric of the triangulation can be discontinuous at each triangle, because the area and the normal of the triangle match from one side to the other, but not necessarily to the length of the triangle sides. These discretized geometries are called "twisted geometries" and have been studied in depth by Speziale and Freidel (Freidel and Speziale 2010b). See also Freidel and Speziale (2010a); Rovelli and Speziale (2010); Magliaro and Perini (2012) and Haggard *et al.* (2013).

These states are the starting point of a full semiclassical analysis of the dynamics, analogous to the analysis of standard quantum mechanics in terms of wave packets. We will use them repeatedly in the physical applications in the last chapters.

---

[5] More precisely, k_ℓ is determined by the holonomy of the Ashtekar connection along the link, which depends both on the dihedral angle and the component of the three-dimensional connection normal to the triangle (Rovelli and Speziale 2010).

9

Matter

The universe is not made by the gravitational field alone. There are also fermions and Yang–Mills fields, and, so it seems, scalar fields. In this chapter we illustrate how fermions and Yang–Mills fields can be coupled to the gravitational field, following Bianchi *et al.* (2010b) (see also Han and Rovelli (2013)). The resulting theory is still ultraviolet-finite, because the quantization of the geometry acts as a physical cut-off on the high-momentum Feynman integrals, also for the matter fields. In a sense, fermions and Yang–Mills behave as if they were on a Planck-sized lattice.

9.1 Fermions

The dynamical coupling of a fermion to quantum gravity is simple in the spinfoam formalism. Consider first four-dimensional euclidean space, for simplicity. The Dirac action reads

$$S_D = i \int d^4x \, \overline{\psi}_D \, \slashed{\partial} \psi_D \tag{9.1}$$

plus complex conjugate, always understood here and in what follows. Let us focus on a chiral spinor $\psi \in \mathbb{C}^2$ with 2-complex components $\psi^A, A = 0, 1$. Its action is the projection of S_D on one of its two helicity components. This reads

$$S = i \int d^4x \, \overline{\psi} \sigma^I \partial_I \psi, \tag{9.2}$$

where $\sigma^I = (\mathbb{1}, \vec{\sigma})$, $\vec{\sigma}$ are the Pauli matrices and $\overline{\psi}$ denotes the hermitian conjugate of ψ. This can be rewritten in form notation as

$$S = i \int \overline{\psi} \sigma^I d\psi \wedge e^J \wedge e^K \wedge e^L \, \epsilon_{IJKL}, \tag{9.3}$$

where e^I is a tetrad field, for the moment just euclidean.

Fix a coordinate system $x = (x^I)$ and discretize spacetime by chopping it into the union of 4-cells v. Consider the dual complex, with vertices v (with coordinates x_v) connected by (oriented) edges e. Call $|v|$ the number of edges bounded by v. Approximate the field $\psi(x)$ by its values $\psi_v = \psi(x_v)$ at each vertex. Discretize the derivative on each edge as

$$\psi_{t_e} - \psi_{s_e} \sim (x^I_{t_e} - x^I_{s_e}) \partial_I \psi, \tag{9.4}$$

where s_e and t_e are the source and the target of the edge e. This gives

$$u_e^I \partial_I \psi \sim \frac{\psi_{t_e} - \psi_{s_e}}{\ell_e}, \tag{9.5}$$

where $\ell_e = |x_{t_e}^I - x_{s_e}^I|$ is the length of the edge e and $u_e^I = (x_{t_e}^I - x_{s_e}^I)/\ell_e$ is the unit vector parallel to e. The action (9.2) can then be discretized as a sum over 4-cells: $S \to \sum_v S_v$.

$$S_v \sim \frac{4iV_v}{|v|} \sum_{e \in v} \overline{\psi}_e u_e^I \sigma_I \frac{\psi_{v+e} - \psi_v}{\ell_e}, \tag{9.6}$$

where the sum is over the edges bounded by v, V_v is the volume of the 4-cell v, and the factor $\frac{4}{|v|}$ is included to take into account the fact that multiple edges over-count the derivative. The second term above cancels in subtracting the complex conjugate. Now consider the 3-cell τ_e dual to the edge e. Assume this is orthogonal to the edge. Each 4-cell v can be partitioned into the union of $|v|$ pyramids with base τ_e and height $h_e = \ell_e/2$. The 4-volume of these is $\frac{1}{4} h_e v_e$, where v_e is the 3-volume of τ_e. Using this,

$$S_v \sim \frac{i}{2} \sum_{e \in v} \overline{\psi}_e v_e \sigma_e \psi_{v+e}, \tag{9.7}$$

where

$$\sigma_e \equiv \sigma^I u_{eI} \tag{9.8}$$

is the σ-matrix "in the direction of the edge e." This can be written as

$$v_e \sigma_e \equiv \sigma^I \int_{\tau_e} \epsilon_{IJKL} \, e^J \wedge e^K \wedge e^L, \tag{9.9}$$

where $e^I = dx^I$ is the tetrad 1-form. Notice now that each term of the sum depends only on edge quantities. This suggests considering writing the full discretized action as a sum of edge terms

$$S = i \sum_e \overline{\psi}_{s_e} v_e \sigma_e \psi_{t_e}. \tag{9.10}$$

More generally, say that the collection of edges e, together with the data (v_e, u_{eI}), approximate a flat metric at a scale a if (Ashtekar *et al.* 1992):

$$\int d^4x \, \delta_I^\mu \omega_\mu^I(x) = \sum_e v_e \int_e \omega^I u_{eI} \tag{9.11}$$

for any quadruplet of 1-forms $\omega^I = \omega_\mu^I dx^\mu$ that varies slowly at the scale a. Then (9.10) is a discretization of the Weyl action if the 2-complex approximates a flat metric. Equation (9.10) is a very simple expression that discretizes the fermion action.[1] Its simplicity recalls the simplicity of the free particle hamiltonian on a graph (Rovelli and Vidotto 2010). It could have also been directly guessed from (9.9) and the form (9.3) of the action.

---

[1] Fermions on a lattice suffer from the fermion-doubling problem and the related chiral anomaly. Because of the absence of a regular triangulation and the integration on the gravitational variables that we introduce below, however, here fermions are essentially on a random lattice, where the (obvious) species doubling problem does not arise.

In view of the coupling with gravity, and in order to better understand boundary states, it is convenient to move from vertex variables to edge variables. Let x_e be the intersection between e and τ_e and introduce edge variables $\psi_e = \psi(x_e)$. In the approximation in which we are working, where the second derivative of the field can be neglected, $\psi_e \sim \frac{1}{2}(\psi_{s_e} + \psi_{t_e})$. Using this and the fact that the sum of quadratic terms averages to zero, we can write

$$S = \frac{i}{4} \sum_{ve} \overline{\psi}_v v_e \sigma_e \psi_e, \tag{9.12}$$

where the sum is over all couples of adjacent vertices-edges. Then observe that (always in this approximation) we can express the vertex fermion as an average over the corresponding boundary edge fermions: $\psi_v = \frac{1}{|v|} \sum_{e' \in v} \psi_{e'}$. This gives

$$S \sim \sum_v \frac{i}{4|v|} \sum_{e,e' \in v} \overline{\psi}_{e'} v_e \sigma_e \psi_e. \tag{9.13}$$

Thus, we consider the discretization of the fermion action defined by associating with each 4-cell v with boundary fields ψ_e the action

$$S_v = i \sum_{ee'} \overline{\psi}_{e'} v_e \sigma_e \psi_e \tag{9.14}$$

where we have assumed here for simplicity that all vertices have the same valence and we have absorbed a constant in a redefinition of the field. This is an expression that can be used to couple the fermion to quantum gravity.

Quantum fermion field on a 2-complex

Consider the fermion partition function of a 2-complex characterized by the quantities (v_e, σ_e). This will be given by the Berezin integral

$$Z = \int D\psi_e \, e^{iS}. \tag{9.15}$$

Choose the integration measure to be

$$D\psi_e = \frac{1}{v_e^2} \, d\psi_e d\overline{\psi}_e \, e^{-v_e \overline{\psi}_e \psi_e}, \tag{9.16}$$

which realizes the scalar product at each edge, seen as a boundary between two 4-cells, and where we interpret the field ψ_e as an anticommuting variable, in order to take the Pauli principle into account. The volume v_e in the exponent is needed for dimensional reasons and to keep in mind the fact that in classical theory

$$\langle \psi | \psi' \rangle = \int d^3x \sqrt{q} \, \overline{\psi(x)} \psi'(x) \tag{9.17}$$

contains the 3-volume factor $\sqrt{q}$ (Rovelli and Vidotto 2010). The definition of the Berezin integral is that the only non-vanishing integral is

$$\int d\psi d\overline{\psi} \, \psi^a \overline{\psi}^b \psi^c \overline{\psi}^d = \epsilon^{ac} \epsilon^{bd} = (\delta^{ab} \delta^{cd} - \delta^{ad} \delta^{bc}). \tag{9.18}$$

Expand the vertex action in Taylor series,

$$e^{iS_V} = 1 + \sum_{e_1 e_2} \overline{\psi}_{e_1} v_{e_2} \sigma_{e_2} \psi_{e_2}$$

$$+ \sum_{e_1 e_2 e_3 e_4} (\overline{\psi}_{e_1} v_{e_2} \sigma_{e_2} \psi_{e_2})(\overline{\psi}_{e_3} v_{e_4} \sigma_{e_4} \psi_{e_4}) + \cdots \tag{9.19}$$

The series stops because there can be at most four fermions (two ψ and two $\overline{\psi}$) per edge. Each edge integration in (9.15) gives zero unless on the edge there is either no fermion, or a term $\overline{\psi}_e \psi_e$, or a term $\overline{\psi}_e \psi_e \overline{\psi}_e \psi_e$. The volume factors in the measure cancel those in the vertex action. The result is that (9.15) becomes a sum of terms, each being the product of traces of the form

$$Z = \sum_{\{c\}} \prod_c A_c, \tag{9.20}$$

where $\{c\}$ is a collection of oriented cycles c, each formed by a closed sequence of oriented edges on the 2-complex: $c = (e_1, \ldots, e_N)$, and each trace is

$$A_c = (-1)^{|c|} \, Tr[\sigma_{e_1} \ldots \sigma_{e_N}], \tag{9.21}$$

where $|c|$ is the number of negative signs from (9.18). For every edge, there cannot be more than two fermionic lines (Pauli principle) and if there are two lines, these are anti-symmetrized (by (9.18)). Explicitly, each term reads

$$A_c = (-1)^{|c|} \, Tr[\sigma_{I_1} \ldots \sigma_{I_n}] \, u_{e_1}^{I_1} \ldots u_{e_n}^{I_n} \tag{9.22}$$

and is therefore a Lorentz-invariant contraction of the normals to the 3-cells.[2]

Fermions in interaction with gravity

Let us now return to the lorentzian framework, and come to our first key technical observation. Assume that all edges are timelike. Consider an $SL(2,\mathbb{C})$ matrix g_e that rotates the unit vector $u^I = (1,0,0,0)$ into the vector u_e^I (the phase is irrelevant, and can be fixed by requiring that g_e is a pure boost)

$$u_J \Lambda_{g_e}^J{}_I = u_{eI}, \tag{9.24}$$

where Λ_g is the vector representation of $SL(2,\mathbb{C})$. Recall the transformation properties of the σ^I matrices,

$$\Lambda_g^I{}_J \sigma^J = g^\dagger \sigma^I g, \tag{9.25}$$

and observe that

$$u_e^I \sigma_I = g_e^\dagger \sigma^0 g_e = g_e^\dagger g_e, \tag{9.26}$$

---

[2] If we do not include the volume factors in (9.16), then (9.21) is replaced by

$$A_c = (-1)^{|c|} \, v_{e_1} \ldots v_{e_N} \, Tr[\sigma_{e_1} \ldots \sigma_{e_N}]. \tag{9.23}$$

This alternative will be studied elswhere.

because $\sigma^0 = 1$. Therefore we can write the discretized action (9.10) in the form

$$S = i \sum_e v_e \, \overline{\psi}_{s_e} g_e^\dagger g_e \psi_{t_e}, \tag{9.27}$$

and the vertex action as

$$S_v = i \sum_{ee'} v_e \, \overline{\psi}_{e'} g_e^\dagger g_e \psi_e. \tag{9.28}$$

What is the geometrical interpretation of these $SL(2, \mathbb{C})$ group elements?

Consider a single chiral fermion. This can be thought as a quantum excitation of a single mode of a fermion field, or as a fermionic particle with spin $\frac{1}{2}$ at one space point.[3] The fermion ψ transforms in the fundamental representation $H^{\frac{1}{2}} \sim \mathbb{C}^2$ of SU(2). The vector space $\mathbb{C}^2$ is also the carrier space $H^{(\frac{1}{2}, 0)}$ of the fundamental representation of $SL(2, \mathbb{C})$, and this determines the Lorentz transformation properties of the fermion. But $H^{\frac{1}{2}}$ is a unitary representation, while $H^{(\frac{1}{2}, 0)}$ is not; in other words, $\mathbb{C}^2$ can be equipped with a scalar product which is SU(2) invariant, but this scalar product is not $SL(2, \mathbb{C})$ invariant.[4] This becomes particularly transparent if we write $\langle \psi | \phi \rangle = \overline{\psi} \sigma^0 \phi$, which shows that the scalar product is the time component of a 4-vector $\overline{\psi} \sigma^I \phi$. The Dirac action exploits this dependence by constructing a Lorentz scalar contracting the Lorentz vector $s^I = \overline{\psi} \sigma^I d\psi$ with the covector u_I giving the direction of the derivative, where $\partial_I = d(u_I)$.

In the light of this discussion, the geometrical interpretation of the matrices g_e is clear. In the discretized theory, a Lorentz frame is determined at each 3-cell by the 4-normal to the 3-cell, which is the direction along which the derivative is computed. The matrices g_e parallel-transport the fermion from the fixed reference frame at the center of the 4-cell to a frame at the center of a 3-cell, namely, at the boundary of the 4-cell, where the normal to the 3-cell is oriented in the time-direction $(1, 0, 0, 0)$. The scalar product that defines the action is taken in that frame. The construction is Lorentz invariant, since the preferred frame of the scalar product is determined by the normals along which the variation of the fermion is computed.

Now, (9.28) is of particular interest for generalizing the discretization to a curved spacetime. In fact, we can discretize a curved geometry in terms of flat 4-cells glued along flat 3-cells, as in Regge calculus. Curvature is then confined on 2-cells. This implies that the holonomy of the spin connection around a 2-cell can deviate from unity. Therefore in general on a curved space there is no way of choosing a reference frame in each 4-cell which is parallel-transported to itself across all 3-cells. In other words, the parallel transport from s_e to e may be different from the parallel transport from t_e to e. Therefore the generalization of the fermion action to a curved spacetime, discretized à la Regge, can be obtained by replacing (9.27) with

$$S = i \sum_e v_e \, \overline{\psi}_{s_e} g_{es_e}^\dagger g_{et_e} \psi_{t_e}, \tag{9.29}$$

---

[3] On the notion of particle in the absence of Poincaré invariance, see Colosi and Rovelli (2009).

[4] If we fix a basis in $\mathbb{C}^2$, the scalar product is given by $\langle \psi | \phi \rangle = \overline{\psi}^a \phi^b \delta_{ab}$ and the tensor δ_{ab} is invariant under SU(2) but not under $SL(2, \mathbb{C})$.

where g_{ev} is the holonomy of the spin connection from a coordinate patch covering the (flat) 4-cell v to one covering the (flat) 3-cell e; or equivalently, replacing the vertex amplitude (9.28) by

$$S_v = i \sum_{e'e} \mathsf{v}_e \, \overline{\psi}_{e'} \, g^{\dagger}_{es_e} g_{et_e} \psi_e. \tag{9.30}$$

The fermion partition function on the 2-complex representing a curved spacetime is therefore as before, with the only difference being that the amplitude (9.22) of each cycle is replaced by

$$A_c = (-1)^{|c|} \, Tr[g^{\dagger}_{e_1 v_1} g_{e_1 v_2} \cdots g^{\dagger}_{e_n v_n} g_{e_n v_1}], \tag{9.31}$$

where $(v_1, e_1, v_2, e_2, \ldots, v_n, e_n)$ is the sequence of vertices and oriented edges crossed by the cycle c. This can be written in a form more easy to read by defining

$$g^* = (g^{-1})^{\dagger} = -\epsilon g^{\dagger} \epsilon. \tag{9.32}$$

Notice that for a rotation $g^* = g$, while for a boost $g^* = g^{-1}$. Using this,

$$A_c = (-1)^{|c|} \, Tr[g_{v_1} e_1^* g_{e_1 v_2} \cdots g_{v_n e_n}^* g_{e_n v_1}], \tag{9.33}$$

where the sequence of vertices and edges is in the cyclic order. The full partition function (9.20) becomes

$$Z = \sum_{\{c\}} \prod_c (-1)^{|c|} \, \mathrm{Tr} \frac{1}{2} \left(\prod_{e \in c} (g^*_{s_e e} g_{et_e})^{\epsilon_{ec}} \right), \tag{9.34}$$

where $\epsilon_{ec} = \pm 1$ according to whether the orientations of the edge and the cycle match.

The action (9.27) and the amplitude (9.33) are of particular value for coupling the fermion field to quantum gravity because they depend on the geometry only via the two quantities v_e and g_{ve}, which are precisely the quantities that appear in the gravitational spinfoam amplitude. This we do in the next section.

Coupling to quantum gravity

Recall that the gravity transition amplitudes can be written as

$$W_C(h_\ell) = N_C \sum_{j_f} \int_{SL(2,C)} dg'_{ve} \prod_f \mathrm{Tr}_{j_f}[Y^{\dagger}_{\gamma} g_{e'v} g_{ve} Y_{\gamma} \ Y^{\dagger}_{\gamma} g_{ev'} g_{v'e''} Y_{\gamma} \ldots]. \tag{9.35}$$

For simplicity, consider here the partition function (no boundaries)

$$Z_C = N_C \sum_{j_f} \int_{SL(2,C)} dg'_{ve} \prod_f \mathrm{Tr}_{j_f}[Y^{\dagger}_{\gamma} g_{e'v} g_{ve} Y_{\gamma} \ Y^{\dagger}_{\gamma} g_{ev'} g_{v'e''} Y_{\gamma} \ldots]. \tag{9.36}$$

It is pretty obvious now how to couple fermions to quantum gravity: it suffices to write the fermion partition function above on the quantum sum over states. The $SL(2, \mathbb{C})$ matrices g_{ev} in this expressions can be identified with the $SL(2, \mathbb{C})$ matrices g_{ev} in (9.27) and (9.33). They have the same geometrical interpretation as parallel-transport operators from the edge to the vertex; and second, more importantly, the asymptotic analysis of the vertex amplitude in Barrett *et al.* (2010) shows that the saddle-point approximation of the integral is on the value of g_{ev} that rotates the (arbitrary) Lorentz frame of the 4-cell into a Lorentz frame at the 3-cell where the time direction is aligned with the normal of the 3-cell, which is precisely the geometry described in the previous paragraph. Therefore in the limit in which we move away from the Planck scale, these group elements take precisely the value needed to yield the fermion action.

The obvious ansatz for the dynamics in the presence of fermions is therefore

$$
Z = N_{\mathcal{C}} \sum_{\{c\}} \sum_{j_{\mathsf{f}}} \int_{SL(2,\mathbb{C})} dg'_{\mathsf{ve}} \prod_{\mathsf{f}} \mathrm{Tr}_{j_{\mathsf{f}}} \left(\prod_{\mathsf{e}\in\mathsf{f}} Y_{\gamma}^{\dagger} g_{\mathsf{e}'\mathsf{v}} g_{\mathsf{ve}} Y_{\gamma} \right)
$$

$$
\times \prod_{c} (-1)^{|c|} \, \mathrm{Tr} \left(\prod_{\mathsf{e}\in c_n} (g_{\mathsf{es}_\mathsf{e}} g_{\mathsf{et}_\mathsf{e}}^{\dagger})^{\epsilon_{\mathsf{e}c}} \right); \tag{9.37}
$$

where $\{c\}$ labels families of worldlines running along the edges of the foam. The sum is over all families that do not overlap more than once. Notice that the definition of the $*$ in this expression depends on the choice of a specific SU(2) subgroup at each edge, but this dependence drops from the total expression, because of the $SL(2,\mathbb{C})$ integrations, precisely as discussed in Rovelli and Speziale (2011). Therefore Lorentz invariance is implemented in the bulk.

This expression defines a quantum theory of gravity interacting with fermions.[5]

9.2 Yang–Mills fields

Suppose now that the fermion lives in the fundamental representation of a compact group G. Then the above theory is invariant under global G trasformations. To make it invariant under local gauge transformations we can introduce a group element $U_{\mathsf{ve}} \in G$ associated to each wedge (ve), and replace (9.10) by

$$
S = i \sum_{\mathsf{e}} \overline{\psi}_{s_\mathsf{e}} U_{s_\mathsf{e}\mathsf{e}}^{\dagger} \mathsf{v}_{\mathsf{e}} \sigma_{\mathsf{e}} U_{t_\mathsf{e}\mathsf{e}} \psi_{t_\mathsf{e}}. \tag{9.38}
$$

The quantum kinematics on the boundary is then evident: spinfoams carry representations of $SL(2,\mathbb{C})$ and intertwiners at the nodes have a possible extra leg representing fermions in (antisymmetric products of) the fundamental representation of $SL(2,\mathbb{C}) \times G$.

---

[5] This naturalness of the fermion dynamics in gravity was already observed early in the loop quantum gravity literature (Morales-Tecotl and Rovelli 1994, 1995) and is surprising. A fermion is essentially an extra "face" of spin $\frac{1}{2}$ of a quantum of space, which is non-local over the 2-complex. At fixed time, it can be seen as a "non-local" loop that disappears outside spacetime, to reappear far away, like a Wheeler–Smolin "Kerr–Newman fermion": the picture of fermions as wormholes suggested by John Wheeler long ago (Wheeler 1962; Sorkin 1977), and considered by Lee Smolin (Smolin 1994) in the context of loop gravity.

What is the dynamics? One possibility of obtaining it is simply to keep only the gravity and fermion terms in the action. The Yang–Mills action is then generated by the one-loop radiative corrections to the fermion action in the Yang–Mills field, as suggested by Zel'dovich Adler (1982),

$$
Z = \sum_{\{c\}} \sum_{j_f} \int_{\mathrm{SL}(2,\mathbb{C})} dg'_{ve} \int_G dU_{ve} \prod_f \mathrm{Tr}_{j_f} \Big(\prod_{e \in f} Y_\gamma^\dagger g_{e'v} g_{ve} Y_\gamma \Big)
$$

$$
\times \prod_c (-1)^{|c|} \, \mathrm{Tr} \left(\prod_{e \in c_n} (g_{es_e} U_{es_e} U_{et_e}^\dagger g_{et_e}^\dagger)^{\epsilon_{ec}} \right). \tag{9.39}
$$

This expression defines a minimally-coupled spinfoam formulation of the Einstein–Weyl–Yang–Mills system. In principle, it can be used to compute all quantum gravity amplitudes order by order. The analysis of the the theory of loop gravity with matter is in a primitive stage and little has been done so far (Han and Rovelli 2013).

PART IV

PHYSICAL APPLICATIONS

10

Black holes

This chapter and the next two introduce the main current applications of the theory to physical situations. Results have been obtained mainly in three directions:

- Thermal properties of black holes
- Early cosmology
- Scattering theory.

The theory is evolving rapidly in these directions and therefore these are only briefly described in the following, covering only basic facts and pointing to some relevant literature. We begin with black holes, objects that have now been found to be numerous in the sky, but that remain among the most fascinating and mysterious in the universe.

10.1 Bekenstein–Hawking entropy

Stephen Hawking made the remarkable theoretical discovery that quantum field theory predicts that a black hole is hot (Hawking 1974, 1975). More precisely, a quantum field on a spacetime where a black hole forms evolves from its vacuum state to a stationary state with a thermal distribution of outgoing radiation at temperature

$$T_H = \frac{\hbar c^3}{k 8\pi G M}.$$

(10.1)

The surprising aspect of this result is not that particles are created: there is particle creation any time a quantum field interacts with a non-stationary potential; what is surprising is the *thermal* nature of the outgoing radiation.

This result adds credibility to a series of previous physical intuitions (Bekenstein 1973) and precise mathematical results (Bardeen *et al.* 1973) indicating that classical black holes behave in a way that is akin to thermal systems. They rapidly evolve to an equilibrium state characterized by a quantity (the horizon area) that cannot decrease in (classical) physical processes, just like the entropy of a statistical system. This is not completely surprising either: entropy measures the amount of information about the microstates that is not captured by the macroscopic variables describing a system with many degrees of freedom. A black hole is a system with a horizon that screens information, precisely as the description of a system by macroscopic parameters does. Therefore, it is reasonable to expect similarities between the physics of the observables accessible from outside the horizon and statistical mechanics. In a stationary context, it is then reasonable to expect thermal properties for black holes.

The thermal properties of a macroscopic state are captured by giving its entropy as a function of macroscopic variables. In the microcanonical setting, entropy measures the volume of the region of the microscopic phase space determined by these macroscopic variables. That is, entropy measures the information lost in the coarse graining. The entropy S that yields the temperature computed by Hawking's quantum field theory calculation is given by a celebrated formula: the Bekenstein–Hawking entropy:

$$S_{BH} = \frac{kc^3}{4\hbar G} A, \tag{10.2}$$

where A is the area of the black horizon and the subscript BH stands either for "black hole," or for "Bekenstein–Hawking." This is a beautiful formula. It contains the Boltzmann constant k, the speed of light c, the Planck constant $\hbar$, and the Newton constant G; therefore it pertains to a regime where all fundamental theories are relevant: statistical mechanics, special relativity, quantum mechanics, and general relativity. The curious factor "4" appears a bit strange at first, but it is not so if we remember that the coupling constant of general relativity is not G but rather $8\pi G$, which is in the denominator of the Einstein–Hilbert action and on the right-hand side of the Einstein equations. The origin of the curious "4" is therefore $4 = 8\pi/2\pi$. That is, in natural units $k = c = \hbar = 8\pi G = 1$, the Bekenstein–Hawking formula reads

$$S_{BH} = 2\pi A; \tag{10.3}$$

the "4" is now a more presentable 2π.

In standard thermodynamics, the expression of the entropy as a function of the macroscopic variables can be computed using statistical mechanics. That is, it is possible to *compute* the value of the entropy. For instance, we can compute the volume of the phase space corresponding to given macroscopic variables. Can the same be done for the Bekenstein–Hawking entropy? This is an open challenge for any theory of quantum gravity.

Much fuss has sometimes been made about the fact that $\hbar$ enters (10.2) and the fact that it enters in the denominator. But this is wrong. The entropy of an electromagnetic field in equilibrium in a cavity of volume $V = L^3$, expressed as a function of the relevant mechanical macroscopic quantity, which is the energy E, is

$$S = k \frac{4}{3} \sqrt[4]{\frac{\pi^2 L^3 E^3}{15c^3 \hbar^3}}, \tag{10.4}$$

which contains $\hbar$ "downstairs" as well. The presence of $\hbar$ depends on the fact that $\hbar$ determines the size of the physical quantum granularity that underlies the statistical behavior, and this is true both for electromagnetism and for gravity. If we take $\hbar$ to zero, both the entropy of a black hole and the entropy of a black body diverge. The reason is that they are no longer bound by quantum discreteness. This yields the ultraviolet catastrophe contemplated by the theoreticians who tried to apply statistical mechanics to classical fields, at the end of the nineteenth century.

In fact, the fact that there is no ultraviolet catastrophe was precisely the evidence that let Planck and Einstein realize the discreteness of the photons (Einstein 1905b). In the same manner, the fact that the entropy of a black hole is finite and $\hbar$ appears in the denominator of (10.2) is evidence that there is a similar granularity in quantum spacetime, preventing a gravitational ultraviolet catastrophe.

Let us make the analogy sharper. From (10.4) we can see what is the size of a single cell in phase space: this is given when S/k is of order unity, that is, by $LE/c \sim \hbar$, but L is also the wavelength, therefore a single cell is characterized by

$$E \sim \hbar \nu, \tag{10.5}$$

where ν is the frequency. In other words, we can read out (or actually, Einstein did) the size of the scale from the expression of the entropy. In the same manner, the size of the quanta of gravity can be read out from the Bekestein-Hawking entropy by equating it to a number of order unity, which gives a quantization of the area:

$$A \sim \hbar G. \tag{10.6}$$

In agreement with the results of loop quantum gravity.

In the case of electromagnetism, we can compute the entropy directly from the quantum theory of the electromagnetic field. Can we do the same for gravity?

10.2 Local thermodynamics and Frodden–Ghosh–Perez energy

An important step in the right direction has been taken by Frodden, Ghosh, and Perez (Ghosh and Perez 2011). The mass M of a Schwartzshild black hole can be identified with energy measured at infinity. The same energy measured at a Schwartzshild radius R is scaled by the redshift factor

$$E(R) = \sqrt{g^{oo}(R)}M = \frac{M}{\sqrt{1 - \frac{2GM}{R}}}. \tag{10.7}$$

Consider an observer at coordinate distance $r \ll 2GM$ from the horizon. The energy of the black hole for this observer is

$$E(r) = \frac{M}{\sqrt{1 - \frac{2GM}{2GM+r}}} \sim M\sqrt{\frac{2GM}{r}}. \tag{10.8}$$

Near the horizon the coordinate distance r is related to the physical distance d by

$$d = \sqrt{g_{rr}}r = \frac{1}{\sqrt{1 - \frac{2GM}{2GM+r}}} \sim r\sqrt{\frac{2GM}{r}} = \sqrt{2GMr}. \tag{10.9}$$

Therefore the energy measured at a physical distance d from the horizon is

$$E(d) = \frac{2GM^2}{d}. \tag{10.10}$$

Now recall that the area of the black hole horizon is $A = 4\pi(2GM)^2$. Also, a standard calculation in classical general relativity shows that an observer staying at a fixed distance d from the horizon maintains an acceleration $a = 1/d$ (in its own frame). Thus, for a static observer near the horizon, with acceleration a, the black-hole energy is

$$E = \frac{aA}{8\pi G}. \tag{10.11}$$

This beautiful formula has been confirmed by Frodden, Ghosh, and Perez in a number of ways. For instance, let dE be the energy of a particle of small mass m falling from infinity into the black hole, as measured by the observer close to the horizon. When the mass enters the black hole, the area of the hole increases by an amount dA where

$$dE = \frac{a\,dA}{8\pi G}. \tag{10.12}$$

The interest of this formula is that it is local on the horizon. An observer near the horizon can interact at most with an infinitesimal patch of the horizon. If this patch has area dA, the observer can associate an energy dE to this patch, given by the last equation. A variation of energy dE entering or leaving the hole near the observer is reflected in a local variation dA of the area, and dE and dA are related by (10.12).

Several alternative derivations of the Frodden–Gosh–Perez equations can be given. For instance, for a uniformly accelerated observer in a flat region, the proper time is just the dimensionless boost parameter η for a boost centred in the horizon, scaled by the acceleration a. (See Exercise 10.1, in Section 10.5.1.) The energy E is the generator of evolution in proper time, and for this observer $K = E/a$ is a boost generator. Inserting this in (3.95) we get immediately the Frodden–Gosh–Perez relation (10.11). This shows that the relation is not constrained to black holes, but in fact is valid for any event horizon: an accelerated observer sees an event horizon, and a local change of area dA of this horizon is related to an energy density dE crossing it by (10.11), as a consequence of the Einstein equations. Still another derivation of this relation can be obtained by studying the corner term associated with an action of a finite region of spacetime (Carlip and Teitelboim 1995; Bianchi and Wieland 2012; Smolin 2012).

Bill Unruh has shown (Unruh 1976) that an observer moving with acceleration a in the vacuum state of a quantum field observes a temperature

$$T = \frac{\hbar a}{2\pi k}. \tag{10.13}$$

Combining this with the above result, we can use the Clausius definition of entropy to obtain

$$dS = \frac{dE}{T} = \frac{2\pi k}{\hbar a}\frac{a\,dA}{8\pi G} = k\frac{dA}{4\hbar G}. \tag{10.14}$$

The r.h.s. is precisely the Bekenstein–Hawking entropy S_{BH}. Therefore the Unruh temperature T and the Frodden–Ghosh–Perez energy E are related by the thermodynamical relation

$$dE = T dS_{BH}. \tag{10.15}$$

If we red-shift the energy and the temperature from the distance d of the horizon all the way to infinity, this relation becomes

$$dM = T_H \, dS_{BH}. \tag{10.16}$$

where M is the Arnowitt–Deser–Misner (ADM) mass of the black hole and T_H is the Hawking temperature. Therefore the Hawking temperature is nothing else than the Unruh temperature which an accelerated observer measures in the vicinity of the horizon, red-shifted to infinity.

This derivation of the Bekenstein–Hawking entropy relies on the classical Einstein equations and the properties of the vacuum of a quantum field on flat spacetime; namely, on the separate use of classical general relativity on the one hand and quantum field theory on the other. Can it be repeated in the context of the full quantum theory of gravity, without relying on semiclassical approximations?

Two avenues have been studied in order to address this question. The first is based on using statistical arguments: counting the number of states *on* the horizon. The second is based on the identification of the relevant entropy with the entanglement entropy *across* the horizon. As we shall see later, the two are not alternatives, as they may seem at first sight; in fact, they are two sides of the same coin. Let us first briefly sketch them separately.

10.3 Kinematical derivation of the entropy

The first possibility is to realize that an observer external to the black hole does not have access to the degrees of freedom within the interior of the hole. Therefore she sees the hole simply as a two-dimensional surface evolving in time. The interior of the hole does not affect the external evolution. Therefore the horizon can be treated as an independent dynamical system interacting with the exterior.

Consider for simplicity a non-rotating and non-charged black hole. At equilibrium, this system is formed by a spherical surface with area A. In a thermodynamical context, the system will undergo statistical fluctuation. In other words, the horizon can "shake" in a thermal context, since Heisenberg uncertainty allows it to do so, and this is the origin of its thermal behavior.

We can address the problem in the micro-canonical framework, and compute the entropy by counting the states with given energy. The relevant energy is not the energy measured at infinity, which includes whatever is happening in the exterior, but rather the energy measured from the vicinity of the horizon. As shown above, this is determined by the area of the horizon. The problem of computing the entropy of a black hole reduces then to the problem of computing the number of states of a two-dimensional surface with area A.

This number diverges in the classical theory, but in loop gravity it is finite because of the discreteness of the geometry. The calculation can be done as follows. The Gibbs ansatz states that at equilibrium and at temperature $T = \frac{1}{\beta}$ the probability p_n for a system of being in a state with energy E_n is proportional to the Boltzmann factor $exp-\beta E_n$. In loop quantum gravity the contribution to the area of a single link of the graph with spin j arriving crossing

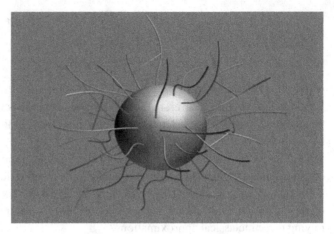

Figure 10.1 Black hole horizon punctured by the spin-network links.

the horizon (see Figure 10.1) is given by

$$A_j = 8\pi G\gamma \sqrt{j(j+1)}. \tag{10.17}$$

Since this link carries a spin j representation, there are in fact $d_j = 2j+1$ orthogonal states in which the system can be. The probability for having spin j at inverse temperature β, taking the degeneracy into account, is therefore

$$p_j(\beta) \sim (2j+1)e^{-\beta E}, \tag{10.18}$$

where E is the energy. Using the energy (10.11) and the explicit values of the area (7.30) for a link of spin j, this reads

$$p_j(\beta) \sim (2j+1)e^{-\beta \frac{aA}{8\pi G}} = (2j+1)e^{-\beta a\gamma \sqrt{j(j+1)}}. \tag{10.19}$$

The proportionality factor

$$p_j(\beta) = Z^{-1}(\beta)\,(2j+1)e^{-\beta a\gamma \sqrt{j(j+1)}} \tag{10.20}$$

is fixed by $\sum_j p_j = 1$, and is the partition function

$$Z(\beta) = \sum_j (2j+1)e^{-\beta a\gamma \sqrt{j(j+1)}}. \tag{10.21}$$

This is the partition function describing the Gibbs state of a single link at inverse temperature β, if the system is in equilibrium with respect to the flow of the accelerated observer. The statistical entropy of the Gibbs state is by definition

$$S = -\sum_j p_j \log p_j, \tag{10.22}$$

and it is immediate, taking derivatives, to see that it satisfies the standard thermodynamical relation

$$E = TS + F, \tag{10.23}$$

where

$$F = -T \log Z \tag{10.24}$$

is the Helmholtz free energy. Let as *assume* that the system is at the Unruh temperature $\beta = 2\pi/a$ (the Unruh temperature is an independent input); then we have immediately from the Frodden–Gosh–Perez (10.11) relation,

$$\begin{aligned} S &= \frac{2\pi}{a} \frac{aA}{8\pi G} + \log Z \\ &= \frac{A}{4G} + \log \sum_j (2j+1) e^{-2\pi\gamma\sqrt{j(j+1)}}. \end{aligned} \tag{10.25}$$

The solution for γ of the equation

$$\sum_j (2j+1) e^{-2\pi\gamma\sqrt{j(j+1)}} = 1 \tag{10.26}$$

is called γ_0 in the quantum gravity literature, and can be easily found numerically to be ~ 0.274. Thus, if

$$\gamma = \gamma_0 \sim 0.274 \tag{10.27}$$

then the free energy vanishes and we have that the statistically entropy is precisely the Bekenstein–Hawking entropy.[1]

More refined calculations of the number of quantum-geometry states with given area have been developed extensively using canonical methods, and we refer to the literature for the details (Rovelli 1996a) and (Ashtekar *et al.* 1998, 2005). The general result is that the entropy is finite and is proportional to the area, fully confirming the Bekenstein–Hawking result and deriving it from first principles, provided that the Barbero–Immirzi constant is chosen as in the last equation.

An alternative path that avoids this restriction by using a grand-canonical framework, where the number of punctures on the black hole is governed by a chemical potential, has been recently developed in Frodden *et al.* (2011); Ghosh *et al.* (2013). We refer the reader to the current literature for the present state of this approach, and in the next section we focus instead on an alternative approach that takes the dynamics more explicitly into account and is based on the covariant theory presented in this book.

---

[1] If γ is not equal to γ_0, we can guess a closed form of the partition function to be

$$Z = e^{-\frac{A}{8\pi G}(\beta a - 2\pi \frac{\gamma_0}{\gamma})}, \tag{10.28}$$

(see Bianchi (2012a)). This gives the entropy

$$S = \frac{E}{T} + \log Z = \frac{A}{4G} - \frac{A}{4G}(1 - \frac{\gamma_0}{\gamma}) = \frac{\gamma_0}{\gamma} \frac{A}{4G}, \tag{10.29}$$

which is the general result in the loop quantum cosmology (LQC) literature.

10.4 Dynamical derivation of the entropy

The core of the covariant derivation of the Bekenstein–Hawking formula is the observation that this relates extensive quantities, and therefore is local on the horizon. This fact is made explicit by the Frodden–Ghosh–Perez relations given above. The observer near the horizon interacts at most with a small patch of the horizon. Therefore the problem can be studied at the level of a single elementary horizon-surface triangle, which is to say, at the level of a single link of a spin-network graph crossing (or, in the jargon of trade, "puncturing") the horizon.

The second idea is to identify the entropy appearing in the Hawking–Bekenstein formula as entanglement entropy. The idea is old, and has been defended in the past by Jacobson, Parentani, and others. It had difficulties being accepted for two reasons. The first is that the entanglement entropy seems at first to be dependent on the number of existing fields, and this appears to be in conflict with the universality of the Bekenstein–Hawking formula. We will see that this objection is wrong: the entanglement entropy does not depend on the number of fields, because the fields are not independent at the Planck scale, where gravity dominates. The second difficulty has been the reluctance to accept the idea that entanglement entropy could give rise to thermodynamcal relations. But it does, and this is nowadays becoming ever more clear in various fields of physics.

On the basis of this, the idea is to recover the ingredients of the Frodden–Ghosh–Perez derivation presented above in Section 10.2, but in the context of the full quantum theory, using its dynamics restricted to a very small patch of spacetime, and the properties of entangled states.

The classical local geometry at a scale much smaller than the black-hole radius $2GM$ is flat, with the static Frodden–Ghosh–Perez observer moving along an accelerated trajectory with acceleration a. (See Exercise 10.2 in Section 10.5.1.) As we have seen, a trajectory at constant acceleration is generated by the boost generator K_z in the direction normal to the horizon surface. If we multiply K_z by a we obtain a transformation whose parameter along the trajectory is the proper time. Therefore $H = aK_z$ is the generator of proper time evolution, that is the hamiltonian, for the accelerated observer. This is the operator that moves the observer along a hyperbolic trajectory, or, equivalently, determines the change of the geometry as perceived from the static observer at fixed distance from the horizon (Figure 10.2).

The boost generator K_z is a well-defined operator in quantum gravity because the states sit in representations of the Lorentz group. Consider a link crossing the horizon. This is described by a state in the representation of SU(2) associated with the corresponding spin. Say it is an eigenstate of the area with spin j. Assume $j \gg 1$ since we are interested in the macroscopic limit. We are looking at regions small with respect to $2GM$ but large with respect to the Planck scale. The horizon is, by assumption, in the plane normal to the z direction. Therefore consider the coherent state with this orientation, which is $|j,j\rangle$. (Here we have gauge-fixed the internal frame to the external directions, for simplicity.) The area

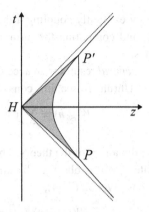

Figure 10.2 The wedge is bounded by a uniformly accelerated trajectory between P and P' and the two segments connecting P and P' to the origin H on the horizon. The wedge amplitude is a function of the boost parameter and evolves from P to P'.

operator is

$$A = 8\pi G\hbar\gamma|L^2|. \tag{10.30}$$

Therefore this state has area ($j \ll 1$)

$$A = 8\pi G\hbar\gamma j. \tag{10.31}$$

The evolution in spacetime of this state is governed by the map Y_γ, which maps it to

$$Y_\gamma|j,j\rangle = |\gamma j,j;j,j\rangle. \tag{10.32}$$

Its energy with respect to the accelerated observer is

$$E = \langle \gamma j,j;j,j|H|\gamma j,j;j,j\rangle = \langle \gamma j,j;j,j|a\hbar K_z|\gamma j,j;j,j\rangle. \tag{10.33}$$

But since we are in a γ-simple representation, this is

$$E = \langle \gamma j,j;j,j|a\hbar\gamma L_z|\gamma j,j;j,j\rangle = a\,\hbar\gamma\,j. \tag{10.34}$$

This can be rewritten in the form

$$E = a\hbar\gamma\,\frac{A}{8\pi G\hbar\gamma} = \frac{aA}{8\pi G}, \tag{10.35}$$

which is precisely the Frodden–Ghosh–Perez energy.

This result, obtained by Eugenio Bianchi (Bianchi 2012a), is very remarkable. To appreciate it, observe that one needs the Einstein equations to derive the Frodden–Ghosh–Perez relation. Here, the same relation emerges from the quantum theory using only the map Y_γ and the SU(2) and SL(2,$\mathbb{C}$) representation theory.

This confirms, indeed, that the information about the Einstein equations is encoded in the Y_γ map. Notice the manner in which the Barbero–Immirzi constant cancels.

The second main ingredient for deriving the Bekenstein–Hawking relation in the quantum theory is the observation that $|\gamma j,j;j,j\rangle$ is a thermal state at temperature $T = \hbar/2\pi$ for the flow generated by the operator K_z. This is shown by Bianchi in the original paper

(Bianchi 2012a) by explicitly coupling a thermometer to this state, following the accelerated trajectory, and computing the ratio between up and down transitions among the thermometer states.

In fact, this is a *general* result for an accelerated observer in a Lorentz invariant context, as first realized by Unruh. To see this, consider the *wedge* amplitude (7.62)

$$W(g,h) = \sum_j (2j+1)\, Tr_j[Y_\gamma^\dagger g Y_\gamma h], \qquad (10.36)$$

which defines the dynamics of the theory. The amplitude for a transition between the state $|j,m\rangle$ and the state $|j,m'\rangle$ as observed by an accelerated observer undergoing a boost of lorentzian angle η is given in (7.64),

$$W_{j,m,m'}(\eta) = \langle j,m'|Y^\dagger e^{iK\eta} Y|j,m\rangle, \qquad (10.37)$$

together with its explicit form. The correlation function between an observable A and an observable B respectively at the beginning and the end of this evolution is

$$G_{AB}(\eta) = \langle W(\eta)|AB|\Psi_\eta\rangle, \qquad (10.38)$$

where Ψ_η is the boundary state. What is the boundary state representing a state that is locally Lorentz invariant? To answer this, consider a key property of the Lorentz group: a boost rotates a spacelike vector into a spacelike vector and never moves a vector out of the wedge. But a boost with imaginary parameter is in fact a rotation, and can move a vector across the light cone. In particular, a boost with imaginary parameter $2\pi i$ moves a vector around the Minkowski plane, and a boost with imaginary parameter $2\pi i - \eta$ rotates the η half-line to the $\eta = 0$ one. If we take the boundary state to be given by the Lorentz dynamics outside the wedge, we can therefore postulate

$$\Psi_\eta = e^{iK(2\pi-\eta)}. \qquad (10.39)$$

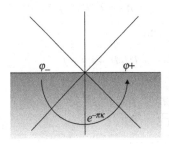

This is in fact nothing else than the quantum gravity version of the old formal argument that gives an intuitive version of the Bisognano–Wichmann theorem (Bisognano and Wichmann 1976), which gives the mathematics underlying the Unruh effect. The standard formal argument runs as follows. Consider a quantum field on Minkowski space. The vacuum state $|0\rangle$ can be written as the path integral on the fields of the lower half-plane.

$$\langle \varphi|0\rangle = \int D[\phi]\, e^{iS}. \qquad (10.40)$$

If we split the field into its components $\varphi_\pm$ at the left and the right of the origin, we have

$$\langle \varphi_-, \varphi_+|0\rangle = \int D[\phi]\, e^{iS}. \qquad (10.41)$$

But his can be read as the evolution generated by a π rotation from the negative to the positive x axis, thus

$$\langle \varphi_-, \varphi_+|0\rangle = \langle \varphi_-|e^{\pi K}|\varphi_+\rangle, \qquad (10.42)$$

where K is the boost generator. If we now trace over φ_- to get the density matrix representing the statistical state restricted to the positive axis, we have

$$\rho = Tr_{\varphi_-}[\langle\varphi_-,\varphi_+|0\rangle\langle 0|\varphi_-,\varphi_+\rangle] = Tr_{\varphi_-}[\langle\varphi_-|e^{\pi K}|\varphi_+\rangle\langle\varphi_+|e^{\pi K}|\varphi_-\rangle] = e^{2\pi K}. \quad (10.43)$$

Evolving this with $e^{-iK\eta}$ to the boundary of the wedge, we have the wedge boundary state (10.39).

Bringing together the amplitude and the state, we have

$$G_{AB}(\eta) = Tr[e^{iK(2\pi-\eta)}Be^{iK\eta}A], \quad (10.44)$$

where the trace is taken in the unitary Lorentz representation. This correlation function satisfies the Kubo–Martin–Schwinger (KMS) property, which is the mark of a thermal configuration (see the Complements to this chapter for a simple introduction to these notions), with dimensionless temperature

$$T = \frac{1}{2\pi} \quad (10.45)$$

with respect to the dimensionless evolution parameter η. Scaling the evolution parameter to the proper time $s = a\eta$, where a is the acceleration of the observer gives the Unruh temperature

$$T_H = \frac{\hbar a}{2\pi}. \quad (10.46)$$

Above we have seen that the covariant dynamics gives the Frodden–Ghosh–Perez energy. Here we have seen that it gives the Unruh temperature. The two together, as we have seen above, give the Bekenstein–Hawking entropy, with the correct factor. This shows that the Bekenstein–Hawking entropy S_{BH} can be derived entirely in the quantum theory. It also shows that it can be viewed as an effect of the quantum correlations across the horizon, because this is the source of the entropy captured by the thermal state of the Bisognano–Wichmann theorem.

The calculation can be summarized as follows. The quantum state of spacetime near the horizon has a form that is compatible with local Lorentz invariance. If we restrict it to the algebra of observables of an observer who keeps herself outside the hole, this is represented by the state

$$\rho = e^{-2\pi K}, \quad (10.47)$$

where K is the generator of boosts normal to the horizon. This state has an entropy

$$S = -Tr[\rho\log\rho]. \quad (10.48)$$

Now suppose this state is modified by an arbitrary process that changes it by an amount $\delta\rho$. Then the change in entropy is

$$\delta S = -\delta Tr[\rho\log\rho] = -Tr[(\delta\rho)\log\rho] - \delta Tr\left[\rho\frac{1}{\rho}\delta\rho\right]. \quad (10.49)$$

The second term vanishes because ρ must remain normalized in the change. The first term, using (10.47) and the relation between the boost generator and the area derived above, gives

$$\delta S = Tr[(\delta\rho)2\pi K] = 2\pi \, Tr[(\delta\rho)\frac{A}{8\pi \, G\hbar}] = 2\pi \frac{\delta A}{8\pi \, G\hbar} = \frac{\delta A}{4\pi \, G\hbar} = \delta S_{BH}. \qquad (10.50)$$

So we find the (differential form of the) Bekenstein–Hawking relation.

Say for simplicity we consider a local change of state given by a change δj in the spin of a link crossing the horizon. Then we have the change in the energy

$$\delta E = a \, \hbar\gamma \, \delta j \qquad (10.51)$$

and the change in entropy

$$\delta S = \delta E / T = \frac{a \, \hbar\gamma \, \delta j}{\hbar a / 2\pi} = 2\pi \, \gamma \delta j. \qquad (10.52)$$

Thus up to a possible additive constant, the entropy associated to a link is a simple function of the spin

$$S = 2\pi \gamma j \qquad (10.53)$$

and since the area of a link is $A = 8\pi \gamma \, G\hbar j$, this gives

$$S = \frac{2\pi \gamma A}{8\pi \gamma \, G\hbar} = \frac{A}{4\hbar G}; \qquad (10.54)$$

again, the Bekenstein–Hawking formula.

10.4.1 Entanglement entropy and area fluctuations

The two previous sections seem to point to different physical interpretations for the origin of the black-hole entropy. The first is the idea that this is the effect of the quantum fluctuations *of* the horizon. The second points to the idea that the black-hole entropy is given by the entanglement entropy *across* the horizon. Is there a contradiction?

No. The two interpretations are not in contradiction with one another; they are two faces of the same phenomenon, the local quantum fluctuations (of anything) near the horizon. On the one hand, these fluctuations generate geometry fluctuations, and therefore can be visualized as a quantum "shaking" of the horizon, or a Planck-scale "thickness" of the horizon, due to the fact that this cannot be resolved at a scale smaller than the Planck scale. On the other hand, we can hold a mental picture wherein the horizon is fixed in some (imaginary) background coordinates, and the quantum field theoretical fluctuations have finite wavelengths that correlate the value of any field across the horizon.

For some time the possibility that the black-hole entropy is related to entanglement entropy across the horizon has been considered implausible because of an obvious objection: the horizon fluctuations are a purely geometrical phenomenon and therefore can justify a universal formula like the Bekenstein–Hawking formula. But the horizon entanglement is a feature of all quantum fields that are present, and therefore the entropy should increase if the number of fields increases. Why then does the Bekenstein–Hawking formula not depend on the number of fields? The solution to this puzzle has recently been found by Eugenio Bianchi (Bianchi 2012b), who computed the variation of the entanglement entropy S_{ent} and of the horizon area A in a physical process, using standard quantum field theory and classical general relativity. He has found that these variation are finite and given by

$$\delta S_{\text{ent}} = \frac{\delta A}{4\hbar G}, \tag{10.55}$$

independently from the cut-off and from the number of fields. The key to the mystery is that δS_{ent} does not increase with the number of fields, because at small scale the fields are not independent. The cut-off on the field modes goes into effect when the *total* energy density of the fields reaches the Planck scale. The total energy density depends on the individual fields, fluctuations plus their (negative) potential gravitational energy. Therefore the cut-off scale on the wavelength of an individual field depends on the presence of other fields, which are fluctuating at small scale as well. Altogether, the result is that the entanglement entropy does not depend on the number of fields. (On this, see also the intriguing recent results in Ghosh *et al.* (2013), Frodden *et al.* (2012) and Han (2014).)

The issue of understanding black-hole entropy directly in quantum gravity has been largely clarified by these loop gravity results. The entropy of a black hole is a local phenomenon due to the fact that *stationary* observers are *accelerating*. It is related to quantum fluctuations of the horizon or, equivalently, to entanglement across the horizon. The entropy can be computed from first principles, including the famous 1/4 Hawking factor, it is finite, and the calculation can be done for realistic black holes, such as a Schwartzschild or Kerr hole. This is a beautiful achievement of the theory which for the moment is not matched by any of the other tentative quantum theories of gravity.

10.5 Complements

10.5.1 Accelerated observers in Minkowski and Schwarzschild metrics

Exercise 10.1 *Consider a uniformly accelerated particle in Minkowski space, following the trajectory*

$$t = l\sinh\eta, \tag{10.56}$$

$$x = l\cosh\eta. \tag{10.57}$$

Show that the acceleration is constant and is $a = 1/l$, the proper time among the trajectory is $s = \eta l$, the distance between the particle and the origin is constant and given by l, the null line $x = t$ is a horizon for this particle, and no signal emitted at $t = 0, x < 0$ can reach the particle. What is the Unruh temperature felt by this particle?

Exercise 10.2 *Consider a uniformly accelerated particle in Schwarzschild spacetime, following the trajectory*

$$r = 2GM + \epsilon, \tag{10.58}$$

where $\epsilon \ll 2GM$. Show that the acceleration is $a = 1/l$ where l is the physical distance of the particle from the horizon. Show that this distance is constant (do not confuse it with ϵ)! Compute the temperature felt by this particle using the Hawking temperature at infinity and the Tolman law. Compare with the previous exercise.

10.5.2 Tolman law and thermal time

If we ignore general relativity, two systems are in equilibrium when they have the same temperature. An isolated column of gas in a (newtonian) gravitational field, for instance, will thermalize to an equilibrium state where the pressure decreases with altitude but the temperature is constant. This, however, is no longer true when general relativity is taken into account. At equilibrium, the temperature T (measured by a standard thermometer) of the gas will decrease with the altitude as

$$\frac{\Delta T}{T} = \frac{g}{c^2},$$
(10.59)

where $g \sim 9.8$ m/s^2 is the Galileo acceleration. The covariant version of this relation is

$$T||\xi|| = \text{constant},$$
(10.60)

where $||\xi||$ is the norm of the Killing field ξ along which equilibrium is established. This fact, which at first is quite surprising, was first derived theoretically by Tolman and Ehrenfest in the early 1930s (Tolman 1930; Tolman and Ehrenfest 1930), and then re-derived in a number of ways by many authors.

The simplest way to understand it is simply to observe that gravity slows down time. In a stationary gravitational field, stationary clocks do not remain synchronized. Therefore there are two distinct notions of time flow in a stationary field: the common time τ given by the Killing field, along which spacetime is stationary; and the local time s, given by the proper time along each trajectory at rest. The ratio of the two is given by the norm of the Killing field. A clock is a device that measures the transition probability between energy levels, and these levels are determined by the proper time, not by τ. But an equilibrium configuration is in equilibrium with respect to the flow of the Killing field, namely in τ, not in s. Therefore the different regions in a global equilibrium configuration are red-shifted with respect to the other, by the ratio between τ and s.

Another way if viewing the same effect is to realize that in general relativity energy "weighs," and therefore accumulates at lower altitude. More precisely, when energy leaves a lower part of the system and reaches a higher part of it, it arrives red-shifted, therefore diminished. It follows that we can no longer maximize entropy by assuming that the energy lost by one subsystem is gained by the other: we must include the red-shift factor in the calculation, and this gives (10.59)

The equilibrium time parameter τ is an example of thermal time (Rovelli 1993a; Connes and Rovelli 1994). Temperature can be defined by the ratio between the thermal time and the proper time $T \sim d\tau/ds$. The proportionality factor fixes the scale of the Killing field and thermal time (Rovelli and Smerlak 2010). Fixing it to

$$T = \frac{\hbar}{k}\frac{d\tau}{ds}$$
(10.61)

is particularly convenient because it makes τ dimensionless and it identifies it with the number of individual quantum cells jumped over by a system at equilibrium: see Haggard and Rovelli (2013).

10.5.3 Algebraic quantum theory

We describe here the basic formalism of thermal states in the algebraic language, used in the text. This is more general and more powerful than the standard hamiltonian language for describing thermal states.

We also give a direct physical interpretation of the KMS condition, in terms of thermometer coupling.

Consider a quantum system with Hilbert space of states $\mathcal{H}$, hamiltonian H and an algebra $\mathcal{A}$ of self-adjoint observables $A, B, \ldots$. Every state $\psi \in \mathcal{H}$ defines a positive linear function on the algebra by

$$\omega_\psi(A) = \langle\psi|A|\psi\rangle \tag{10.62}$$

and more generally so does any density matrix ρ

$$\omega_\rho(A) = Tr[\rho A]. \tag{10.63}$$

The hamiltonian defines a flow on the algebra by

$$\alpha_t(A) = e^{iHt}Ae^{-iHt}. \tag{10.64}$$

The algebra $\mathcal{A}$ with the flow α_t and the set of its positive states ω provides a language for talking about quantum theory that is more general than the Hilbert-space language (see below). Let us use this language to describe thermal phenomena.

10.5.4 KMS and thermometers

Consider a quantum system described by an observable algebra $\mathcal{A}$, with observables $A, B, \ldots$, which is in a state $\omega : \mathcal{A} \to \mathbb{C}$. Let $\alpha_t : \mathcal{A} \to \mathcal{A}$ with $t \in \mathbb{R}$ be a flow on the algebra. We say that the state ω is an equilibrium state with respect to α_t if for all t,

$$\omega(\alpha_t A) = \omega(A), \tag{10.65}$$

and that its temperature is $T = 1/\beta$ if

$$f_{AB}(t) = f_{BA}(-t + i\beta), \tag{10.66}$$

where

$$f_{AB}(t) = \omega(\alpha_t A B). \tag{10.67}$$

An equivalent form of (10.66) is given in terms of the Fourier transform $\tilde{f}_{AB}$ of f_{AB} as

$$\tilde{f}_{AB}(\omega) = e^{-\beta\omega}\tilde{f}_{AB}(-\omega). \tag{10.68}$$

The canonical example is provided by the case where $\mathcal{A}$ is realized by operators on a Hilbert space $\mathcal{H}$ where a hamiltonian operator H with eigenstates $|n\rangle$ and eigenvalues E_n is defined, and the flow and the state are defined in terms the evolution operator $U(t) = e^{-iHt}$ and the density matrix $\rho = e^{-\beta H} = \sum_n e^{-\beta E_n}|n\rangle\langle n|$ by

$$\alpha_t(A) = U^{-1}(t)AU(t) \tag{10.69}$$

and

$$\omega(A) = tr[\rho A], \tag{10.70}$$

respectively. In this case, it is straightforward to verify that Eqs. (10.65) and (10.66) follow.

However, the scope of Eqs. (10.65) and (10.66) is wider than this canonical case. For instance, these equations permit the treatment of thermal quantum field theory, where the hamiltonian is ill-defined because of the infinite energy of a thermal state in an infinite space. It is indeed important to remark that these two equations capture immediately the physical notions of equilibrium and temperature. For Eq. (10.65), this is pretty obvious: the state is in equilibrium with respect to a flow of time if the expectation value of any observable is time-independent.

The direct physical interpretation of Eq. (10.66) is less evident: it describes the coupling of the system with a thermometer.

To see this, consider a simple thermometer formed by a two-state system with an energy gap ϵ, coupled to a quantum system S by the interaction term

$$V = g(|0\rangle\langle1| + |1\rangle\langle0|)A, \tag{10.71}$$

where g is a small coupling constant. The amplitude for the thermometer to jump up from the initial state $|0\rangle$ to the final state $|1\rangle$ while the system moves from an initial state $|i\rangle$ to a final state $|f\rangle$ can be computed using the Fermi golden rule to first order in g:

$$
\begin{aligned}
W_+(t) &= g \int_{-\infty}^{t} dt \left((\langle 1| \otimes \langle f|) \, \alpha_t(V) \, (|0\rangle + |i\rangle) \right) \\
&= g \int_{-\infty}^{t} dt \, e^{it\epsilon} \langle f| \alpha_t(A)|i\rangle.
\end{aligned}
\tag{10.72}
$$

The probability for the thermometer to jump up is the squared modulus of the amplitude, summed over the final state. This is

$$
P_+(t) = g^2 \int_{-\infty}^{t} dt_1 \int_{-\infty}^{t} dt_2 \, e^{i\epsilon(t_1 - t_2)} \omega(\alpha_{t_2}(A^\dagger)\alpha_{t_1}(A)),
$$

where we have used the algebraic notation $\omega(A) = \langle i|A|i\rangle$. If the initial state is an equilibrium state,

$$
P_+(t) = g^2 \int_{-\infty}^{t} dt_1 \int_{-\infty}^{t} dt_2 \, e^{i\epsilon(t_1 - t_2)} f_{A^\dagger A}(t_1 - t_2)
$$

and the integrand depends only on the difference of the times. The transition probability per unit time is then

$$
p_+ = \frac{dP_+}{dt} = g^2 \tilde{f}_{A^\dagger A}(\epsilon),
\tag{10.73}
$$

which shows that (10.67) is precisely the quantity giving the transition rate for a thermometer coupled to the system. It is important to repeat the calculation immediately for the probability to jump down, which gives

$$
p_- = g^2 f_{AA^\dagger}(-\epsilon).
\tag{10.74}
$$

And therefore (10.68) expresses precisely the fact that the thermometer thermalizes at temperature β, that is $p_+/p_- = e^{-\beta\epsilon}$

These observations show that Eqs. (10.65) and (10.66) define a generalization of the standard quantum statistical mechanics that fully captures thermal properties of a quantum system. The structure defined by these equations is called a modular flow in the mathematical literature: α_t is a modular flow, or a Tomita flow, for the state ω, and is the basic tool for the classification of the C^* algebras. In the physical literature, the state ω is called a KMS state (for the time flow). In order to show that a system is in equilibrium and behaves thermally in a certain state, it is sufficient to show that the state satisfies these two equations.

Notice that in the context of Hilbert-space quantum mechanics, a state can satisfy these two equations also if it is a pure state. The standard example is provided by the vacuum state of a Poincaré-invariant quantum field theory, which is KMS with respect to the modular flow defined by the boost in a given direction. Being Poincaré-invariant, the vacuum is invariant under this flow, and a celebrated calculation by Unruh shows that it is KMS at inverse temperature 2π.

In this case, the physical interpretation is simply given by the fact that an observer stationary with respect to this flow, namely an accelerated observer at unit acceleration, will measure a temperature $1/2\pi$. Unruh's original calculation, indeed, follows precisely the steps above in Eqs. (10.71–10.74).

10.5.5 General covariant statistical mechanics and quantum gravity

The relations between gravity, quantum, and statistics form a net of intertwined problems, where much of our confusion on the fundamental aspects of nature lies. At the core of the difficulty is the fact that we lack a coherent statistical theory for the gravitational field (as we have one for the electromagnetic field), even in the classical case. The reason is related to the peculiar manner in which time appears in general relativity, which makes the standard tools of statistical mechanics useless. The effort of extending the power of statistical reasoning to general covariant theories is, in our opinion, among the deepest and most beautiful problems open at the core of our present understanding of the physical world – as beautiful as, and most likely strictly related to, quantum gravity. Here we only sketch briefly a few results and ideas in a research line in this direction under development by the authors, pointing to the relevant literature.

Thermal time in the classical theory

If all the variables are on the same footing, it is not clear what equilibrium means. In such a situation (generically) any statistical state ρ can be seen as an equilibrium state, because it is invariant under a flow in phase space: the hamiltonian flow generated by $H = -\ln \rho$. Since ρ is a density on phase space, it has the dimension of inverse action, therefore we need a constant to fix its logarithm. Looking ahead, we do so by taking $\hbar = 1$ unit. The quantity H is called the thermal hamiltonian of the state ρ and the (dimensionless) flow parameter τ is called thermal time (Rovelli 1993a). The thermal hamiltonian of a Gibbs state is the standard hamiltonian scaled by the temperature, and the temperature is the ratio between the thermal time τ and the physical time t:

$$T = \frac{\hbar}{k}\frac{\tau}{t}. \tag{10.75}$$

An expression we have already seen in (10.61) generalizes to gravity, where t becomes the proper time s which varies, as does the temperature, from point to point.

Mean geometry

If we study the thermal fluctuations of the gravitational field, then it is reasonable to call "equilibrium" a state where the these fluctuations are centred around a "mean geometry" having a Killing field that can be identified with the thermal time flow. This construction is developed in Rovelli (2013a).

Thermal time in quantum field theory

The idea of thermal time gets bite in quantum field theory, where the thermal time flow is the modular flow of the state, which is independent of the state up to unitaries (Connes and Rovelli 1994). The modular flow of the vacuum state of a quantum field theory, restricted to the algebra of the observables on the Rindler wedge $x > |t|$, is the flow of the boost, the thermal time is the boost parameter, and the local temperature defined by (10.75) is the Unruh temperature. But the thermal flow makes sense also in non-stationary situations, such as a Friedmann cosmology, where the thermal time of the cosmic background radiation state turns out to be the cosmological time (Rovelli 1993c).

The zeroth principle

Remarkably, when two systems interact, equilibrium is given by the equality of their thermal times (Haggard and Rovelli 2013). This has a physical interpretation: thermal time is time in the units given by the time a quantum state moves to a state orthogonal to itself. Therefore two interacting systems in equilibrium interact with the same number of distinct states: the net flux of information they exchange vanishes. This notion of equilibrium, defined in terms of information rather than in terms of a preferred time variable, generalizes to the covariant systems where no preferred time is available (Haggard and Rovelli 2013). In turn, the combined state of the two systems itself can select a time variable among the combination of the individual times (Chirco et al. 2013).

Figure 10.3   Lens-shaped spacetime region with spacelike boundaries and corners (filled circles).

Statistical states in the boundary formalism

In Section 2.4.2 we argued that in gravity we need to associate states to the full boundary Σ of a spacetime region. If Σ can be split into a past and a future component, this Hilbert space has the structure $\mathcal{H} = \mathcal{H}_{past} \otimes \mathcal{H}_{future}$. This space contains states that are not of the tensor product of a state in $\mathcal{H}_{past}$ and a state in $\mathcal{H}_{future}$. What do these represent? It is not difficult to see that they represent statistical states (Bianchi *et al.* 2013).

Statistical versus quantum fluctuations

In particular, if there is a corner joining the future and the past components of the boundary, as in Figure 10.3, and if a quantum version of the equivalent principle holds, implying that the corner is locally like a Minkowski state in the UV, then it follows that *all* boundary states are mixes. This is because locally in the corner the state has the form

$$\rho = e^{-2\pi K}, \tag{10.76}$$

where K is the boost generator at the corner. This supports the idea that statistical fluctuations and thermal fluctuations become indistinguishable when gravity and horizons are in play.

These brief notes are of course far from exhausting this fascinating topic, which is still seeking clarity.

11 Cosmology

The major application of loop gravity to a realistic physical phenomenon is early cosmology. This is also the most important application, because it appears to be the most likely to lead to predictions that could confirm the theory. Observational cosmology is advancing rapidly and there is hope that loop gravity could lead to specific testable predictions in cosmological observations, in particular in cosmic background radiation.

As for the case of black-hole thermodynamics, there are two main avenues for applying loop gravity to cosmology. The first is based on the canonical formulation of the theory, the second on the covariant formulation. Both are based on the same idea as classical relativistic cosmology: focusing on the dynamics of the very large-scale degrees of the universe (Einstein 1917). Here we touch on the first only briefly, and we discuss in a bit more detail the second. We begin by reviewing the classical formulation of a simple cosmological model, concentrating on the aspects relevant for the quantum theory.

11.1 Classical cosmology

Cosmology is *not* the study of the totality of the things in the universe. It is the study of a few very large-scale degrees of freedom. It is a fact (intuited by Einstein) that on a very large scale the universe appears to be approximately homogeneous and isotropic. Therefore it admits an approximate description obtained by expanding in modes around homogeneous-isotropic geometries.

The spatial curvature of the universe appears to be very small at the scale we access. One must *not* conclude from this observation that the universe is spatially flat. Humanity has already made this mistake once: the Earth's average curvature is very small at our scales and we concluded that the Earth is flat. Let us not repeat this mistake (many do). But to a good approximation a spatially flat universe approximates the patch of the Universe that we see, and since a spatially flat universe is a bit easier to describe than a closed one, at least for some purposes, it makes sense to use a spatially flat metric here.

Thus, consider a Friedmann–Lemaître metric, where the function $a(t)$ is the scale factor, and write it in terms of tetrads:

$$ds^2 = -N(t)dt^2 + a(t)^2 d\vec{x}^2 = \eta_{IJ}\, e^I_\mu(t) e^J_\nu(t)\, dx^\mu dx^\nu, \tag{11.1}$$

where, introducing $e^I_\mu dx^\mu = e^I$,

$$e^0 = N(t)\, dt, \qquad\qquad e^i = a(t)\, dx^i. \tag{11.2}$$

Inserting this metric into the action gives

$$S[a] = \frac{3}{16\pi G} \int dt \, N \left(\dot{a}^2 a - \frac{\Lambda}{3} a^3 \right) + S_{matter}. \tag{11.3}$$

The momentum conjugate to the scale factor is

$$p_a = \frac{\partial L}{\partial \dot{a}} = \frac{3V_o}{8\pi G} N \dot{a} a, \tag{11.4}$$

where L is the lagrangian and V_o is the coordinate volume of the space considered. This volume cannot be taken as infinite without introducing a confusing infinity in the formalism. So we take it as finite. Normalizing the coordinates so that $V_o = 1$ has the consequence that a^3 is the actual volume of the universe. With this choice, the boundary term in the gravitational action is then

$$S_{boundary} = p_a a = \frac{3}{8\pi G} N \dot{a} a^2. \tag{11.5}$$

The dynamics of the scale factor defines a one-dimensional dynamical system, which is the ground for studying physical cosmology. Solutions to this system and perturbations around these solutions describe the large-scale dynamics of the universe quite well. The recent observations by the Planck satellite (Planck Collaboration 2013) have confirmed the credibility of this picture to a surprising degree of accuracy.

The equation of motion obtained by varying the Lapse function N is an equation that ties a and $\dot{a}$. In the presence of a matter energy density

$$\rho(t) = a^3 \frac{\partial S_{matter}}{\partial N(t)}, \tag{11.6}$$

the variation of the Lapse gives

$$\left(\frac{\dot{a}}{a} \right)^2 = \frac{\Lambda}{3} + \frac{8}{3} \pi G \rho, \tag{11.7}$$

which is the Friedmann equation, the main equation governing the large-scale dynamics of the universe. After taking this equation into account, we can fix the Lapse, say at $N = 1$. The Friedmann equation gives $\dot{a}$ as a function of a at every time. Therefore we can view the boundary term simply as a function of a. It follows that the Hamilton function is a pure boundary term of the form

$$S(a_i, a_f) = S(a_f) - S(a_i), \tag{11.8}$$

where

$$S(a) = \frac{3}{8\pi G} \dot{a} a^2 \tag{11.9}$$

and $\dot{a}$ depends on a via the Friedmann equation. Therefore we must expect the transition amplitude of the quantum theory in the small $\hbar$ limit to factorize

$$W(a_i, a_f) \sim e^{\frac{i}{\hbar} S(a_i, a_f)} = e^{\frac{i}{\hbar} S(a_f) - \frac{i}{\hbar} S(a_i)} = e^{\frac{i}{\hbar} S(a_f)} e^{-\frac{i}{\hbar} S(a_i)} \sim \overline{W(a_f)} W(a_i), \tag{11.10}$$

where

$$W(a) \sim e^{-\frac{i}{\hbar} \dot{a} a^2}. \tag{11.11}$$

This is the classical limit to which a cosmological quantum transition amplitude must converge, if it is to agree with standard cosmology.

Let us translate this to the Ashtekar variables defined in Section 3.4.3. The Ashtekar–Barbero connection $A_a^i = \Gamma_a^i + \gamma \; k_a^i$ conjugate to the densitized inverse triad $E_i^a = \frac{1}{2} \, \epsilon^{abc} \epsilon_{ijk} e_b^j e_c^k$ is the sum of the spin-connection Γ_a^i of the triad and the extrinsic curvature of the spacial slide k_a^i multiplied by the Barbero–Immirzi constant β, which we here take equal to γ, as usual in loop quantum cosmology. In the flat case $\Gamma_a^i = 0$ because space is flat and the extrinsic curvature is diagonal and a multiple of the time derivative of the scale factor

$$k_a^i = \dot{e}_a^i = \dot{a}\delta_a^i. \tag{11.12}$$

Therefore the Ashtekar–Barbero variables are reduced to

$$A_a^i = \gamma \, \dot{a} \, \delta_a^i = c\delta_a^i, \qquad\qquad E_i^a = a^2\delta_i^a = p\delta_i^a, \tag{11.13}$$

which satisfy

$$\{c,p\} = \frac{8\pi G}{3V_o}\gamma. \tag{11.14}$$

Inserting these variables into the hamiltonian constraint (3.86) gives (up to an irrelevant overall constant)

$$H = \left(\frac{cp}{\gamma}\right)^2 - \frac{\Lambda}{3}p^3. \tag{11.15}$$

$H = 0$ is the Friedmann equation for pure gravity. Adding the matter energy density ρ (assumed constant in space) we obtain

$$\frac{c^2}{\gamma^2 p} = \frac{\Lambda}{3} + \frac{8}{3}\pi G\rho, \tag{11.16}$$

which translated back into the $a(t)$ variable is (11.7), namely, the standard Friedmann equation for a flat universe with cosmological constant. The Hamilton function factorizes as above and reads

$$S(p) \sim \frac{c(p)\,p}{\gamma} \tag{11.17}$$

where $c(p)$ is the solution of the Friedmann equation. Notice that γ cancels. For instance, if there is no matter, we expect

$$S(p) \sim \sqrt{\frac{\Lambda}{3}}\, p^{\frac{3}{2}}. \tag{11.18}$$

Thus for the classical limit to hold, the transition amplitude must have the small $\hbar$ behavior

$$W(p) \sim e^{\frac{i}{\hbar}\sqrt{\frac{\Lambda}{3}}\, p^{\frac{3}{2}}}. \tag{11.19}$$

11.2 Canonical loop quantum cosmology

A large body of work has been developed to study the canonical quantization of the dynamical system defined by the dynamics of the scale factor, as well as the perturbations around it. We refer the interested reader to recent reviews, such as those by Ashtekar and Singh (2011) and Bojowald (2011), which follow slightly different paths, and references therein.

The main result of this work is that the dynamics of the *quantum* dynamical system describing the evolution of the scale factor is well defined also around what appears in the classical approximation as the initial singularity. That is, the Big Bang singularity does not appear to be a physical event, but only an artifact of the classical approximation. In this, it is analogous to the possibility for an electron to fall into the nucleus of the atom, which is predicted by the classical approximation but not by the quantum theory of the electron. The quantum evolution of the universe near the Big Bang is well defined, and no singularity develops.

Interestingly, the quantum corrections can be approximated in terms of a modified classical dynamical evolution for a wave packet. This gives a modified Friedmann equation. In the case $\Lambda = 0$ and $k = 0$, for instance, one obtains

$$\left(\frac{\dot{a}}{a}\right)^2 = \frac{8\pi G}{3}\rho\left(1 - \frac{\rho}{\rho_c}\right), \tag{11.20}$$

where the critical density ρ_c is a constant with a value that is approximatetively half of the Planck density ($\rho_{\text{Pl}} \approx 5.1 \times 10^{96}\,\text{kg/m}^3$). In the absence of the correction term, $\dot{a}$ cannot change sign at small scales. In the presence of the correction term, $\dot{a}$ can vanish, and the dynamics of the scale factor "bounces": semiclassical trajectories describe a collapsing universe that bounces into an expanding one. We do not give a derivation of (11.20) here: see Ashtekar and Singh (2011), but a few words on how this happens may be useful.

The idea is to study a quantum system with Friedmann cosmology as classical limit using two key inputs from the full theory.

1. The first is the idea that the proper operators are not the local ones but the integrated ones. In particular, the relevant operator is not the connection but rather its holonomy. This translates into cosmology in the requirement that the operator well defined in the quantum theory is not c but rather $e^{i\mu c}$, for suitable μ.
2. The second is that the dynamics is modified to take into account the quantum discreteness of space. This is obtained by replacing the curvature in the hamiltonian constraint with the holonomy around a small circuit encircling the smallest possible quantum of area.

The first input allows the definition of a suitable Hilbert space where the operators $e^{i\mu c}$ are defined. The second permits the definition of the dynamics. A detailed analysis of the resulting system and its semiclassical limit then yields the above result. The resolution of cosmological singularities can be seen to be quite generic (Singh 2009; Singh and Vidotto 2011).

The Poisson brackets (11.14) become commutators in the quantum theory, and yield Heisenberg uncertainty relations

$$\Delta c \Delta p > \frac{8\pi\hbar G}{3V_o}\gamma.$$ (11.21)

By taking V_o arbitrarily large, the uncertainties can be made arbitrarily small. This is analogous to the fact that by taking an object made by sufficiently many atoms we can have its position and velocity arbitrarily well defined. In other words, loop quantum cosmology looks only at the large modes of the field, which are averages in space of the gravitational field, and is blind to the local quantum fluctuations of the gravitational field, which are dominated by shorter-wavelength modes. It follows that the theory, so far, does not describe the actual fluctuating geometry at the bounce. The bounce is not determined by the universe being small, but by the matter energy density reaching Planck scale, and this can happen, depending on the matter content of the universe, at any size of the universe.

We also report here, without deriving it, the effect of quantum gravity on cosmological perturbations. Remarkably, this too can be summarized in a correction to the standard cosmological perturbation theory. The Mukanov equation that governs the dynamics of the perturbations (Mukhanov *et al.* 1992),

$$\ddot{v} - \nabla^2 v - \frac{\ddot{z}}{z}v = 0,$$ (11.22)

is corrected to Cailleteau *et al.* (2012a,b)

$$\ddot{v} - \left(1 - 2\frac{\rho}{\rho_c}\right)\nabla^2 v - \frac{\ddot{z}}{z}v = 0.$$ (11.23)

An electron does not fall into the Coulomb potential because of the uncertainty principle, which forbids concentrating it into too small a region of space without having a large momentum that allows it to escape. In a sense, quantum mechanics provides an effective repulsive force at short distance. The same happens for the universe: when too much energy density is concentrated into too small a region, the uncertainty principle prevents further collapse and acts as an effective strong repulsive force.

The precise interpretation of this semiclassical bounce is in our opinion not completely clear yet, because more understanding of the quantum fluctuations of spacetime at the bounce is needed. The full physics of the gravitational field near the bounce is not yet well understood. Is spacetime still approximately classical at the bounce or do the full quantum fluctuations make the very notion of spacetime ill defined around the bounce, as the notion of a classical trajectory is ill defined for an electron falling into a Coulomb potential? In the second case, the picture of a "previous" contracting phase of the universe bouncing into a "later" expanding one might be naive for some purposes: spacetime itself becomes ill defined in early cosmology.

11.3 Spinfoam cosmology

The approach to quantum cosmology based on the covariant formulation of loop gravity was introduced in Bianchi *et al.* (2010d) and is called "spinfoam cosmology." The expansion at the basis of the covariant theory is suitable for cosmology, because it is an expansion in the number of degrees of freedom. Since a cosmological model is the reduction of general relativity to a small number of degrees of freedom, it can be naturally described at low order in the spinfoam expansion.

Consider a closed homogeneous and isotropic universe, with the spatial geometry of a 3-sphere $\mathbb{S}^3$ of radius a, evolving in time. We want to study the quantum dynamics of the degree of freedom $a(t)$ as well as the quantum dynamics of the large-scale deformations of the 3-sphere.

In a finite lapse of time, a compact universe evolves from a 3-sphere to a 3-sphere. Therefore we want the transition amplitude from a state of quantum space with the topology of a 3-sphere to a different quantum state of space with the topology of a 3-sphere. To represent one such state, we have to choose a discretization of a 3-sphere. Spinfoam cosmology was first studied using the simplest cellular decomposition of $\mathbb{S}^3$. This can be obtained by gluing two tetrahedra by all their faces.[1] The graph dual to this triangulation of $\mathbb{S}^3$ is obtained by placing a node on each tetrahedron and connecting the two nodes across each of the four faces. (Figure 11.1). This graph is called Γ_2, or "dipole graph": it is formed by 2 nodes connected by 4 links. Quantum cosmology on this graph was first studied in the hamiltonian language (Rovelli and Vidotto 2008; Battisti *et al.* 2010; Borja *et al.* 2010, 2011). More general regular graphs were studied in Vidotto (2011). For instance, we can obtain a 3-sphere by pairwise gluing five tetrahedra, as in the boundary of a 4-simplex (Figure 11.2).[2] In the following we use a generic regular graph Γ. The first step to describe the large-scale geometry of the universe is to write the boundary states in the Hilbert space $\mathcal{H}_\Gamma$, representing semiclassical configurations. Then we can take a couple of these states, which can be thought of as *in* and *out* states, and compute the corresponding transition amplitude.

s   *t*

Figure 11.1   Two tetrahedra, glued by all their faces, triangulate a 3-sphere.

---

[1] The common sphere $\mathbb{S}^2$ can be obtained by taking two triangles and gluing them by their sides.

[2] The common sphere $\mathbb{S}^2$ can be obtained by taking four triangles and gluing them by their sides, as in the boundary of a tetrahedron.

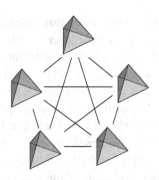

Figure 11.2 Five tetrahedra triangulate a 3-sphere.

11.3.1 Homogeneous and isotropic geometry

Fix a regular graph, namely, a graph where all the nodes have the same number of links.[3] As an example, we can think about Γ_5, the graph formed by five nodes all connected. Pick a cellular decomposition of a (metric) 3-sphere formed by equal cells and dual to the regular graph. Approximate each cell with a flat cell having the same volume, face area, and average area normal. This defines normals $\vec{n}_l$ associated with each oriented link of the graph. By regularity, all areas and extrinsic curvatures on the graph are equal. We can construct a coherent state on the graph, based on these geometrical data, using the coherent states developed in Section 8.4. The coherent state is determined by an $SL(2, \mathbb{C})$ element H_ℓ for each link, where H_ℓ is determined by the two (distinct) normals associated with two cells bounding the face, and a complex number z, which is the same for all links. The real and imaginary parts of z are determined by the area and the extrinsic curvature on the face. We assume on the basis of (8.121), that the relation between these and the coefficients of an isotropic geometry are given simply by

$$z = c + it\frac{p}{l_o^2}. \tag{11.24}$$

(For a finer analysis, see Magliaro *et al.* (2011a, b)). Notice that the Planck length in the denominator is imposed by (8.121) as well as by dimensions.

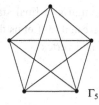

Γ_5

---

[3] There are an infinite number of such graphs. For instance, two nodes can be connected by an arbitrary number of links. Examples of regular graphs with $N > 2$ are given by the (dual of) the Platonic solids.

Let us now seek quantum states that describe the geometry of these triangulated spaces. For this, consider the extrinsic coherent states $\psi_{H_\ell}(h_\ell)$ in the Hilbert space $\mathcal{H}_\Gamma$ introduced in Section 8.4. Recall that these are determined by the data $(c_l, p_l, n_{s(l)}, n_{t(l)})$.

With these assumptions, a homogeneous isotropic coherent state on a regular graph is described by a single complex variable z, whose imaginary part is proportional to the area of each regular face of the cellular decomposition (and it can be put in correspondence with the total volume) and whose real part is related to the extrinsic curvature (Rovelli and Speziale 2010). We denote by $\psi_{H_\ell(z)}$ this state, and by $\psi_{H_\ell(z,z')} = \psi_{H_\ell(z)} \otimes \psi_{H_\ell(z')}$ the state on two copies of the regular graph, obtained by tensoring an "in" and an "out" homogeneous isotropic state. These states are peaked on a homogeneous and isotropic geometry, but since they are genuine coherent states, they include the quantum fluctuations around this geometry, for the degrees of freedom captured by the graph.

11.3.2 Vertex expansion

For a boundary state formed by two disconnected components, the transition amplitude determines the probability to go from one state to the other. For the coherent states that have just been introduced, this reads

$$W_C(z', z) = \langle z' \,|\, W \,|\, z \rangle = \int dh_\ell \int dh'_\ell \ \overline{\psi_{z'}(h'_\ell)} \ W_C(h'_\ell, h_\ell) \ \psi_z(h'_\ell). \tag{11.25}$$

The transition amplitude can be computed by approximating it to a truncated version of the theory, namely, by choosing a 2-complex C bounded by the boundary graph, that is by two Γ_5 graphs. At the lowest order, we have have the contribution of a 2-complex without vertices:

$$\langle z' \,|\, W_{C_o} \,|\, z \rangle \sim \tag{11.26}$$

This vanishes if the initial and final states are sufficiently far apart, which is the nontrivial case we are interested in. Let us therefore study the first-order term, determined by a 2-complex with a single vertex:

$$\langle z' \,|\, W_{C_1} \,|\, z \rangle \sim \tag{11.27}$$

We write this transition amplitude using: (7.67) and (7.68):

$$W_{C_1}(h'_\ell, h_\ell) = W(h'_\ell)W(h_\ell), \tag{11.28}$$

where

$$W(h_\ell) = \int_{SL(2,\mathbb{C})} dg'_e \prod_{\ell=1}^{L} P(h_\ell, g_e). \tag{11.29}$$

The integration over the $SL(2,\mathbb{C})$ elements h_n associated with the edges imposes Lorentz invariance. It is over all the g_n but one. Using these, and the definition (8.127) of the coherent states, a short calculation gives

$$W_{C_1}(z',z) = W(z)W(z'), \tag{11.30}$$

where

$$W(z) = \int dh_\ell \int dg'_n \prod_{\ell=1}^{L} K_t(h_\ell, H_\ell(z)) \, P(g_n, h_\ell)$$

$$= \int dg'_n \prod_{\ell=1}^{L} \sum_{j_\ell} (2j_\ell+1) \, e^{-2tj_\ell(j_\ell+1)} \, D_{mm'}^{(j_\ell)}(H_\ell(z)) \, D_{jm'jm}^{(\gamma j_\ell, j_\ell)}(G_\ell). \tag{11.31}$$

The transition amplitude factorizes, as we expected from the classical theory. Each individual term $W(z)$ can be interpreted as a Hartle–Hawking "wavefunction of the universe" determined by a *no-boundary* initial condition (Hartle and Hawking 1983), namely, the amplitude to go from nothing to a given state. The reason this is equivalent to the transition amplitude from a given geometry to another is a reflection of the fact that in classical theory the dynamics can be expressed as a relation between the scale factor and its momentum, which is a relation at a single time. The probability of measuring a certain "out" coherent state does not depend on the "in" coherent state.

Let us now evaluate $W(z)$ in the semiclassical regime

11.3.3 Large-spin expansion

We are interested in (11.31) in the regime where the volume of the 3-sphere (which is determined by p) is large compared with the Planck scale. By (11.24), this corresponds to the regime where the imaginary part of z is large. From (8.130), we have

$$D^{(j)}(H_\ell(z)) = D^{(j)}(R_{\vec{n}_s}) \, D^{(j)}(e^{-iz\frac{\sigma_3}{2}}) \, D^{(j)}(R_{\vec{n}_n}^{-1}). \tag{11.32}$$

When the imaginary part of z is large, $\mathrm{Im}\, z \gg 1$, the Wigner matrix

$$D^{(j)}(e^{-iz\frac{\sigma_3}{2}}) = \sum_m e^{-izm} \, |m\rangle\langle m| \tag{11.33}$$

is dominated by the term $m = j$, therefore

$$D^{(j)}(e^{-iz\frac{\sigma_3}{2}}) \approx e^{-izj} \, |j\rangle\langle j|, \tag{11.34}$$

where $|j\rangle$ is the eigenstate of L_3 with maximum eigenvalue $m = j$ in the representation j. Inserting this into (11.32) gives

$$D^{(j)}(H_\ell(z)) = D^{(j)}(R_{\vec{n}_s}) \; |j\rangle \; e^{-izj} \; \langle j| D^{(j)}(R_{\vec{n}_n}^{-1}), \qquad (11.35)$$

and recalling the definition (8.23) of the intrinsic coherent states this is

$$D^{(j)}(H_\ell(z)) = e^{-izj} |j, \vec{n}_\ell\rangle \langle j, \vec{n}_\ell|. \qquad (11.36)$$

Inserting this into (11.31) gives

$$W(z) = \int dg'_n \prod_{\ell=1}^{L} \sum_{j_\ell} (2j_\ell+1) \, e^{-2tj_\ell(j_\ell+1)} \, e^{-izj_\ell} \, \langle j_\ell, \vec{n}_\ell| D_{jm'jm}^{(\gamma j_\ell, j_\ell)}(G_\ell)|j_\ell, \vec{n}_\ell\rangle. \quad (11.37)$$

Each of the sums is a gaussian sum that peaks the spin on a large value, which, using (11.24), is

$$j_o = \frac{p}{4l_o^2}, \qquad (11.38)$$

the same for all ℓ's. Therefore we can write

$$W(z) = \left(\sum_j (2j+1) \, e^{-2tj(j+1)} \, e^{-izj} \right)^L \Omega(j_o) \qquad (11.39)$$

where

$$\Omega(j_o) = \int dg'_n \prod_{\ell=1}^{L} \langle j_o, \vec{n}_\ell| D_{jm'jm}^{(\gamma j_o, j_o)}(G_\ell)|j_o, \vec{n}_\ell\rangle. \qquad (11.40)$$

This integral can be computed in the large-spin regime using saddle-point methods and the formulas of the previous chapter. The matrix elements are simply peaked around the origin $g_n = \mathbb{1}$, and the result of the integration is determined by the hessian, which is a rational function of j_o. Here we are only interested in the phase of $W(s)$ and therefore we can write

$$W(z) \sim \left(\sum_j (2j+1) \, e^{-2tj(j+1)} \, e^{-izj} \right)^L \sim e^{-L\frac{z^2}{8t}}. \qquad (11.41)$$

The classical limit is obtained by the rapidly oscillating phase of the amplitude, as explained in Section 2.2. From the definition (11.24) of z, this is

$$W(z) \sim e^{-i\frac{Lcq}{4l_o^2}}. \qquad (11.42)$$

Translated into standard variables this is

$$W(z) \sim e^{-i\frac{L\dot{a}a^2}{8\pi G\hbar}}, \qquad (11.43)$$

which is the expected behavior (11.11). This confirms that the transition amplitudes computed in spinfoam cosmology yield the classical result in the appropriate limit.

For more details and more results in spinfoam cosmology, including the inclusion of the cosmological constant, see Bianchi *et al.* (2011a); Vidotto (2011); Borja *et al.* (2011).

11.4 Maximal acceleration

Spinfoam cosmology is not yet developed to the point of allowing us to describe the quantum physics of the near-black-hole region directly. However, it does provide an indirect element of evidence for the resolution of the classical singularity, because it directly predicts a limit on the acceleration. The idea that quantum gravity may limit the acceleration is old (Caianiello 1981; Caianiello and Landi 1985; Brandt 1989; Toller 2003). In covariant loop quantum gravity, it can be seen as a direct consequence of the quantization of the area and the simplicity relations that relate area and boost (Rovelli and Vidotto 2013). To see this, consider a uniformly accelerated observer in Minkowski space, with acceleration a. The wedge defined by a portion of its trajectory is depicted in Figure 10.2. Specifically, taking the boost angle of this wedge $\eta = 1$, it is an easy calculation to see that the lorentzian area of the wedge is

$$A_l = \frac{1}{2a^2} \tag{11.44}$$

Therefore the area of this wedge measures the acceleration of the observer. But in turn, the lorentzian area of this wedge is tied to the euclidean area A_e by the linear simplicity constraint

$$A_l = \frac{1}{\gamma} A_e, \tag{11.45}$$

and since the minimum non-vanishing eigenvalue of A_e is $8\pi G\hbar\gamma \sqrt{1/2(1/2+1)}$, it follows that the maximal value of the acceleration is

$$a_{max} = \sqrt{\frac{1}{8\pi G\hbar}}. \tag{11.46}$$

In turn, a maximal acceleration screens the classical singularity. The resolution of classical singularities under the assumption of a maximal acceleration has been studied using canonical methods for Rindler (Caianiello et al. 1990), Schwarzschild (Feoli et al. 1999), Reissner–Nordstrom (Bozza et al. 2000), Kerr–Newman (Bozza et al. 2001) and Friedmann–Lemaître (Caianiello et al. 1991) metrics. Here we sketch a simple argument using a homogeneous and isotropic cosmological model, with vanishing spatial curvature and pressure. The dynamics is governed by the Friedmann equation (we here use R for the scale factor in order not to confuse it with the acceleration a),

$$\frac{\dot{R}^2}{R^2} = \frac{8\pi G}{3}\rho. \tag{11.47}$$

$\rho \sim R^{-3}$ is the matter energy density. Any co-moving observer is accelerating with respect to his neighbours in this geometry. The growing acceleration approaching a classical singularity is bounded by the existence of a maximal acceleration $a \sim \sqrt{\ddot{R}/R}$. This gives a maximal value of the energy density

$$\rho_{max} \sim \left. \frac{3}{8\pi G} \frac{\dot{R}^2}{R^2} \right|_{max} = \frac{3}{8\pi G} \ell_{min}^{-2} = \frac{3}{\hbar(8\pi G)^2}. \tag{11.48}$$

So one recovers a Planck-scale maximal energy density as in the canonical loop theory.

11.5 Physical predictions?

The possibility of seeing quantum gravitational effects in the cosmic microwave background (CMB) is presently under intense study. We do not enter into any details here, since this would take us too far into cosmology and the subject is rapidly evolving. We only indicate a recent result, from which the interested reader can follow the field. To study quantum gravitational effects on the CMB one needs to study both the effects of the modified dynamics of the scale factor and those of the quantum fluctuations of the geometry themselves. The net effect of these is equivalent to having a "dressed" smooth geometry (Ashtekar *et al.* 2009). Using this result, one can repeat the standard cosmological derivation of the power spectrum of the CMB, but extend it to the pre-inflationary era. A recent result, for instance, is the computation of the power spectrum by Agullo *et al.* (2012). We give an example of the results in Figure 11.3, taken from that paper.

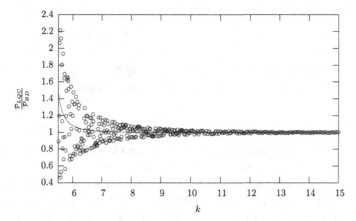

Figure 11.3 Ratio of predicted power spectrum with and without quantum gravity effects, as a function of the mode (Agullo *et al.* 2012).

For a detailed review, see for instance Agullo and Corichi (2013). This is a field that is rapidly growing and is of major importance, in view of the possibility of testing the theory.

12

Scattering

In the two previous chapters we have discussed applications of loop gravity to physically relevant, but specific situations – black holes and early cosmology. How to extract the entire information from the theory systematically, and compare it with the usual way of doing high-energy physics?

In conventional field theory, knowledge of the n-point functions

$$W(x_1,\ldots,x_n) = \langle 0 \,|\, \phi(x_n)\ldots\phi(x_1)\,|\,0\rangle, \tag{12.1}$$

amounts to the complete knowledge of the theory, as emphasized by Arthur Wightman in the 1950s (Wightman 1959). From these functions we can compute the scattering amplitudes and everything else. Can we recover the value of all *these* functions, from the theory we have defined in this book? This, for instance, would allow us to compare the theory with the effective perturbative quantum theory of general relativity, which, although non-renormalizable is nevertheless usable at low energy. More generally, it would connect the abstruse background-independent formalism needed for defining quantum gravity in general with the tools of quantum field theory that we are used to, from flat space physics.

The answer is yes, the value of the n-point functions can be computed from the theory we have defined in this book. This requires a careful understanding of how the information about the background around which the n-point functions are defined is dealt with in the background-independent theory. This is done in this chapter.

12.1 n-Point functions in general covariant theories

The n-point functions (12.1) should not be confused with the transition amplitudes. Transition amplitudes, such as

$$W[\varphi_{in},\varphi_{out}] = \langle\varphi_{out}|e^{-iHt}|\varphi_{in}\rangle \tag{12.2}$$

are defined between arbitrary (fixed-time) field configurations φ. While n-point functions, such as (12.1), are amplitudes for *quanta* (excitations) of the field over a preferred state (the vacuum, $|0\rangle$).

To illustrate the distinction between these two quantities and clarify the relation between the two, consider a simple system with one variable $q(t)$ that evolves in time. The variable $q(t)$ plays here the role of the fields $\phi(x)$. The transition amplitude is

$$W(q,t,q',t') = \langle q' \,|\, e^{iH(t'-t)} \,|\, q\rangle. \tag{12.3}$$

This determines the probability of going from $|q\rangle$ to $|q'\rangle$ in a time $t' - t$. The 2-point function, on the other hand, is

$$W(t,t') = \langle 0 | q(t') q(t) | 0 \rangle. \tag{12.4}$$

It is a function that depends only on t and t' and not on q and q'. It is the transition amplitude between the states $q(t)|0\rangle$ and $q(t')|0\rangle$, where $q(t)$ are (Heisenberg) operators at different times. The transition amplitude carries information about the fields, while the n-point function carries information about the quanta of the fields on a given state. The transition amplitude is the probability to see the system in a configuration $q'(t')$ if we have seen it in the configuration $q(t)$. If the system is in the vacuum state and we create a quantum at time t, then the 2-point function gives the probability to see a quantum at a later time t'.

The relation between these two quantities is as follows. Write the Heisenberg operators in terms of Schrödinger operators in the expression for the 2-point function,

$$W(t,t') = \langle 0 | q(t') q(t) | 0 \rangle = \langle 0 | e^{+iHt'} q e^{-iHt'} e^{+iHt} q e^{-iHt} | 0 \rangle. \tag{12.5}$$

The vacuum state is invariant under time evolution. So we can write

$$W(t,t') = \langle 0 | q \, e^{-iH(t'-t)} q | 0 \rangle$$

$$= \int dq \int dq' \, \langle 0 | q' \rangle q' \, W(q,t,q',t') \, q \, \langle q | 0 \rangle \tag{12.6}$$

$$= \int dq \int dq' \, W(q,t,q',t') \, qq' \, \psi_o(q) \, \overline{\psi_o(q')}, \tag{12.7}$$

where now q and q' are just numbers. The quantity $W(q,t,q',t')$ is the transition amplitude: it does not carry information about any particular state. It characterizes the full dynamics of the theory. On the other hand, $W(t,t')$ characterizes the dynamics of the theory in the particular state ψ_o. The n-point functions can be written in terms of the transition amplitude, the field operators, *and* a state ψ_o, to be specified.

There is another way to write a 2-point function, in terms of a path integral:

$$W(t,t') = \int Dq \, q(t) q(t') e^{iS[q]}. \tag{12.8}$$

In this expression the state is implicitly fixed by the boundary conditions at infinity in the path integral. The relation between this formula and the previous one is interesting. The path integral is (the limit of) a multiple integral in variables $q(t)$. Let us partition these into five groups: the single variables $q(t)$ and $q(t')$ for fixed values of t and t', and the three classes of times in which t and t' partition the real line.

Then clearly (12.8) reduces to (12.7) where the transition amplitude is the result of the integration in the region $[t,t']$,

$$W(q,t,q',t') = \int Dq \, e^{i \int_t^{t'} L}, \tag{12.9}$$

and the two states are the result of the integrations in the past and future regions

$$\psi_o(q) = \int Dq \, e^{i \int_{-\infty}^{t} L} \qquad (12.10)$$

with appropriate boundary conditions at infinity.

Let us now promote these same considerations to field theory. Let us fix an (increasing) time ordering for the points x_n, so we do not have to worry here about T-products. The n-point function can also be written as a path integral.

$$W(x_1,\dots,x_n) = \langle 0 \, | \, \phi(x_n),\dots,\phi(x_1) \, | \, 0 \rangle = \int D\phi \, \phi(x_n)\dots\phi(x_1) \, e^{iS[\phi]}. \qquad (12.11)$$

For simplicity, let us first restrict to a 2-point function. We can still break the integration into the five regions as above, obtaining

$$W(x,x') = \langle 0 \, | \, \phi(x')\phi(x) \, | \, 0 \rangle = \int D\phi D\phi' \, W[\phi,t,\phi',t'] \, \phi(\vec{x}) \, \phi'(\vec{x}') \, \Psi_o[\bar{\phi}']\Psi_o[\phi], \qquad (12.12)$$

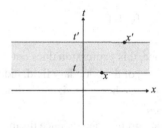

where now spacetime splits into five regions and the transition amplitude $W[\phi,t,\phi',t']$ is the field propagator in the intermediate band, obtained by integrating on the fields in that band. Observe that in order to perform this integration we have to fix the boundary values of the fields at the initial slice, at the final one, and also at spatial infinity.

Once this is understood, it is natural to consider also a different possibility: to split (12.11) differently. Instead of selecting a band bounded by two equal time surfaces, let us select an arbitrary *compact* region $\mathcal{R}$, as in the figure below.

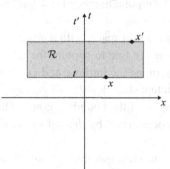

Then we can write

$$W_{\Psi_b}(x,x') = \int D\phi_b \, W[\varphi_b] \, \varphi(\vec{x})\varphi(\vec{x}') \, \Psi_b[\varphi_b], \qquad (12.13)$$

where φ_b are fields on the boundary of the region, $W[\phi_b]$ is a functional integral that depends only on the boundary fields, $\varphi(\vec{x})$ and $\varphi(\vec{x}')$ are field operators in two distinct points on the boundary, and $\Psi_b[\varphi_b]$ is the state of the field on the boundary that describes the integration outside the region.

In this manner we have found a connection with the boundary formalism of Section 2.4.2. The n-point functions can be written using the boundary formalism: the background state determines the boundary state, the field operators sit on the boundary.

Now let us return to quantum gravity. In covariant loop quantum gravity we have all the ingredients needed to write the last equation. We have the transition amplitude for a field on a boundary, the field operator, and a notion of coherent states that can describe the quantum gravitational field on the boundary of a region corresponding to a given classical configuration. By combining these ingredients, we can reconstruct n-point functions over a background from the background-independent quantum gravity theory. The background enters in determining the boundary state.

The region considered not need to be a square, in fact, we can take it to be a region bounded by spacelike hypersurfaces, as below:

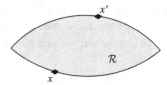

The crucial observation, now, was already made in Section 2.4.2: in a background-dependent theory, in addition to the boundary value of the field, we must *also* fix the shape of the boundary regions, its geometry, its size, the time lapsed from the beginning to the end of the process, and so on, in order to define the boundary amplitude:

$$W[\phi_b] = \int D\phi \, e^{i\int_{\mathcal{R}} L[\phi]}. \tag{12.14}$$

But in quantum gravity, due to general covariance, this expression does not depend on the shape or dimension of $\mathcal{R}$. Assigning the state $\Psi_b[\varphi]$ of the boundary field, indeed, *amounts to giving the geometry of the boundary*, including its shape, its size, the time lapsed from the beginning to the end of the process, and so on!

This is the true magic of quantum gravity: background dependence disappears entirely, and the transition amplitude is a function only of the field: there is nothing else in the universe to be considered besides the dynamical fields in the process.

The position of the n-point arguments of the n-point function is well defined *with respect to the background metric around which the boundary state is peaked*.

Thus, pick for instance a region $\mathcal{R}$ of Minkowski space, with a given metric geometry, and approximate it to a triangulation sufficiently fine to capture the relevant dynamical scale of the phenomenon to be studied. Determine the intrinsic and extrinsic geometry of the boundary of this region, and pick a quantum state $\Psi \in \mathcal{H}_\Gamma$ of gravity picked on these data. Then the two points $\vec{x}$ and $\vec{x}'$ sit on nodes of the boundary graph. The operators $\vec{E}_\ell$ associated with these nodes and directions determined by the links describe quanta over the boundary state.

This is the solution of a long-standing confusion in quantum gravity: formally, if $S[g]$ is a general covariant action and $D[g]$ a generally covariant measure, the quantity

$$W(x_1, \ldots, x_n) = \int D[g] \, g(x_1) \ldots g(x_n) \, e^{iS[g]} \tag{12.15}$$

is independent of the position of the points $(x_1, \ldots, x_n)$ (as long as they do not overlap), because general covariance can move the points around while leaving everything else invariant. Therefore (12.15) is an incorrect definition of n-point functions in a generally covariant theory. It has nothing to do with the quantity (12.1) computed by expanding on a background. The mistake is that the x_n in (12.1) are *physical distances*, computed in the background metric, while the x_n in (12.15) are *coordinate distances*. To have a quantity depending on physical distances, we must explicitly specify the background state on a boundary surface, and have the x_n defined by positions in this metric.

In more formal terms, (12.13) is non-trivial because both the positions of the points and Ψ_b transform together under a diffeomorphism.

12.2 Graviton propagator

Let us focus on the 2-point function, which in perturbation theory is defined by

$$W_{abcd}(x,y) = \langle 0 | g_{ab}(x) g_{cd}(y) | 0 \rangle, \tag{12.16}$$

where $g_{ab}(x)$ is the metric field. It is convenient to contract this with two couples of vectors at x and y, which we shall choose appropriately in a moment,

$$W(x,y) = \langle 0 | g_{ab}(x) g_{cd}(y) | 0 \rangle \, n^a(x) m^b(x) n^c(y) m_d(y). \tag{12.17}$$

We fix a compact metric region M in the background metric, having x and y on its boundary $\Sigma = \partial M$. Then Σ has an intrinsic and an extrinsic geometry associated with it. We choose vectors $\vec{n}$ and $\vec{m}$ tangent to Σ. Passing to triads on Σ we can write

$$W(x,y) = \langle 0 | \vec{E}_a(x) \cdot \vec{E}_b(x) \vec{E}_c(y) \cdot \vec{E}_d(y) | 0 \rangle \, n^a(x) m^b(x) n^c(y) m_d(y) \tag{12.18}$$

and translate this expression to one based on M,

$$W(x,y) = \langle W | \vec{E}_a(x) \cdot \vec{E}_b(x) \vec{E}_c(y) \cdot \vec{E}_d(y) | \Psi \rangle \, n^a(x) m^b(x) n^c(y) m_d(y), \tag{12.19}$$

where now Ψ is a state on a surface with the topology of a 3-sphere, peaked on the geometry of Σ. This state brings the background information with it.

To compute the last expression we resort to approximations. To this aim, we replace M with a triangulated space Δ. The finer the triangulation Δ, the better will be the approximation. The triangulation Δ induces a triangulation on the boundary and we search the state Ψ in the spin-network Hilbert space $\mathcal{H}_\Gamma$ where Γ is the graph dual to the boundary triangulation. We choose Ψ as the extrinsic coherent state (8.127) determined by the (discretized) intrinsic and extrinsic geometry of Σ. We identify the points x and y as nodes in the graph, or tetrahedra in the boundary of the triangulation, and fix the vectors n and m to be normal to the faces of the corresponding tetrahedra.

Then (12.19) becomes a well-defined expression: the boundary state Ψ is the state (8.127) chosen as mentioned, the operators $\vec{E}_b(x) n^b(x)$ are the triad operators (7.56) associated with the corresponding links of the spin network. The transition amplitude W is the amplitude (7.57) associated with the 2-complex dual to the bulk triangulation.

At the lowest non-trivial approximation we can take the triangulation formed by a single 4-simplex. Since we are expanding around flat space, this is reasonable, as there is no background curvature. That is, we are in the right regime for this expansion to have a chance to work. Then the boundary graph is the regular graph formed by five connected tetrahedra and W is directly given by the vertex amplitude (7.58). Thus, using Eqs. (8.127), (7.56), and (7.58), the 2-point function (12.19) becomes a completely explicit expression. Computing it is just a matter of technical ability.

The computation for arbitrary boundary size is hard, but it simplifies in the limit where the boundary geometry of Σ is large with respect to Planck scale. In this case it is possible to use the techniques developed in Barrett *et al.* (2010), replacing the vertex amplitude (7.58) with its saddle-point approximation (8.104)–(8.105).

Let us see this in a bit more detail, following Bianchi and Ding (2012). The expectation value of an operator is given in (7.60). The operator we are here interested in is the (density-two inverse-) metric operator $q^{ab}(x) = \delta^{ij} E_i^a(x) E_j^b(x)$. We focus on the *connected* 2-point correlation function $G^{abcd}(x,y)$ on a semiclassical boundary state $|\Psi_0\rangle$. It is defined as

$$G^{abcd}(x,y) = \langle q^{ab}(x)\, q^{cd}(y) \rangle - \langle q^{ab}(x) \rangle \langle q^{cd}(y) \rangle . \qquad (12.20)$$

Expressing it in terms of triads contracted with the normals to the tetrahedra faces, this reads

$$G_{nm}^{abcd} = \langle E_n^a \cdot E_n^b\, E_m^c \cdot E_m^d \rangle - \langle E_n^a \cdot E_n^b \rangle \langle E_m^c \cdot E_m^d \rangle . \qquad (12.21)$$

The first problem is to construct the correct boundary state. This is a coherent state on the graph

$$\Gamma_5 = \qquad$$ 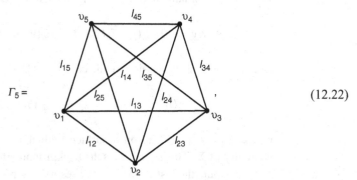 $$\qquad , \qquad (12.22)$$

with five nodes v_a dual to tetrahedra of the 4-simplex and ten links l_{ab}, $(a < b)$ dual to the corresponding meeting triangles. In constructing it, we should take care to keep track of the orientations of the time orientation of the tetrahedra and the Thin/Thick nature of the links discussed in Section 8.1.3. See Bianchi and Ding (2012). This allows us to define the intrinsic coherent states

$$|\Gamma_5, j_{ab}, \vec{n}_{ab}\rangle = \exp\left(-i \sum_{a<b} \Pi_{ab} j_{ab}\right) |\Gamma_5, j_{ab}, \vec{n}_{ab}\rangle \qquad (12.23)$$

that depend on a chosen temporal orientation of the boundary tetrahedra, on the spin of the links and the normals to the faces of the tetrahedra.

A lorentzian extrinsic coherent state peaked both on intrinsic and extrinsic geometry can be given by a superposition of lorentzian coherent spin-network states. Instead of using the full machinery of coherent states, it is simpler to just choose a wave packet

$$|\Psi_o\rangle = \sum_{j_{ab}} \psi_{j_o,\phi_o}(j)|j,\vec{n}\rangle \tag{12.24}$$

with coefficients $\psi_{j_o,\phi_o}(j)$ given by a gaussian times a phase,

$$\psi_{j_o,\phi_o}(j) = \exp\left(-i\sum_{ab}\gamma\phi_o^{ab}(j_{ab}-j_o)\right) \ \exp\left(-\sum_{ab,cd}\gamma\alpha^{(ab)(cd)}\frac{j_{ab}-j_o}{\sqrt{j_o}}\frac{j_{cd}-j_o}{\sqrt{j_o}}\right), \tag{12.25}$$

where ϕ_o labels the simplicial extrinsic curvature, which is an angle associated with the triangle shared by the tetrahedra; the 10×10 matrix $\alpha^{(ab)(cd)}$ is assumed to be complex with positive definite real part (Bianchi et al. 2009). The quantity j_o is an external input in the calculation and sets the scale of the background geometry of the boundary, on which the boundary state is fixed. It therefore establishes the distance between the two points of the 2-point function we are computing.

We can now insert the state (12.24), the form of the vertex amplitude given in (8.95) and (8.93), into the expectation values (7.60) for the 2-point function (12.21). This gives

$$G_{nm}^{abcd} = \frac{\sum_j \psi_j \langle W|E_n^a \cdot E_n^b\, E_m^c \cdot E_m^d|j,\vec{n}\rangle}{\sum_j \psi_j \langle W|j,\vec{n}\rangle} - \frac{\sum_j \psi_j \langle W|E_n^a \cdot E_n^b|j,\vec{n}\rangle}{\sum_j \psi_j \langle W|j,\vec{n}\rangle}\ \frac{\sum_j \psi_j \langle W|E_m^c \cdot E_m^d|j,\vec{n}\rangle}{\sum_j \psi_j \langle W|j,\vec{n}\rangle}. \tag{12.26}$$

The matrix elements $\langle W|E_n^a \cdot E_n^b|j,\Upsilon(\vec{n})\rangle$ and $\langle W|E_n^a \cdot E_n^b\, E_m^c \cdot E_m^d|j,\Upsilon(\vec{n})\rangle$ can be given an explicit integral expression following Bianchi et al. (2009), by introducing an "insertion"

$$Q_{ab}^i \equiv \langle j_{ab}, -\vec{n}_{ab}(\xi)|\, Y^\dagger g_a^{-1} g_b Y(E_b^a)^i|j_{ab},\vec{n}_{ba}(\xi)\rangle, \tag{12.27}$$

and expressing them in terms of this insertion. With some work (Bianchi and Ding 2012), we arrive at the expression

$$G_{nm}^{abcd} = \frac{\sum_j \psi_j \int d^4g\, d^{10}z\, q_n^{ab} q_m^{cd} e^S}{\sum_j \psi_j \int d^4g\, d^{10}z\, e^S} - \frac{\sum_j \psi_j \int d^4g\, d^{10}z\, q_n^{ab} e^S}{\sum_j \psi_j \int d^4g\, d^{10}z\, e^S}\ \frac{\sum_j \psi_j \int d^4g\, d^{10}z\, q_m^{cd} e^S}{\sum_j \psi_j \int d^4g\, d^{10}z\, e^S}, \tag{12.28}$$

where S is given in (8.93),

$$q_n^{ab} \equiv A_n^a \cdot A_n^b, \tag{12.29}$$

and

$$A_{ab}^i \equiv \gamma j_{ab}\frac{\langle\sigma^i Z_{ba},\xi_{ba}\rangle}{\langle Z_{ba},\xi_{ba}\rangle}. \tag{12.30}$$

The low-energy behavior of the 2-point function is determined by the large-j_o asympototics of the correlation function (12.28) because j_o is the parameter in the boundary state that determines the distance between the two points in the background metric. This can be done (Bianchi *et al.* 2009) by rescaling the spins j_{ab} and $(j_o)_{ab}$ by an integer λ so that $j_{ab} \rightarrow \lambda j_{ab}$ and $(j_o)_{ab} \rightarrow \lambda (j_o)_{ab}$; thus the 2-point function (12.28) can be re-expressed as $G(\lambda)$. Study of the large-spin limit thus turns into study of the large-λ limit, via stationary phase approximation. The stationary phase approximation of the integral has been studied in detail and we refer to the literature: Alesci and Rovelli (2008); Bianchi *et al.* (2009); Magliaro and Perini (2012); Rovelli and Zhang (2011), and in particular to Bianchi and Ding (2012). The result is that in the classical limit, introduced in Bianchi *et al.* (2009), where the Barbero–Immirzi parameter is taken to zero, $\gamma \rightarrow 0$, and the spin of the boundary state is taken to infinity, $j \rightarrow \infty$, keeping the size of the quantum geometry $A \sim \gamma j$ finite and fixed (the finite and fixed area A corresponds to the finite and fixed distance between the two points where the correlation functions are defined), the 2-point function exactly matches the one obtained from lorentzian Regge calculus (Regge 1961). In turn, in this limit the 2-point function matches that of general relativity (12.16).

Therefore the semiclassical limit of the theory gives back the usual linearized gravity, indicating that the theory *is* general relativity in an appropriate limit.

It is important also to go beyond first order in the vertex expansion. The asymptotic analysis of spinfoams with an arbitrary number of vertices is studied in $\gamma \rightarrow 0$ limit in Magliaro and Perini (2011a,b, 2013). Without taking $\gamma \rightarrow 0$, the large-j of spinfoams with an arbitrary number of vertices is studied in Conrady and Freidel (2008) with a closed manifold. An extensive discussion and analysis is given by Han (2013).

Some aspects of this result are not fully clear. Among these is the peculiar limit in which the classical results emerge. The same limit was considered in Bojowald (2001) in the context of loop quantum cosmology, and the reason is probably related to the relation (8.108) illustrated in Figure 8.8. But the situation is not yet completely clear. Also, the current calculation is involved. One has the impression that a simpler version of it should exist, especially since most of the complications lead to terms that eventually disappear.

The quantum corrections are contained in higher orders. At the time of writing, only preliminary calculations of radiative corrections with bubbles have been obtained (Riello 2013). This is an extremely important open direction of investigation in the theory. A number of questions regarding for instance the proper normalization of the n-point functions and similar, are still unclear, and much more work is needed.

Ideally, one would expect that the low-energy behavior of the n-point functions computed in the non-perturbative theory would agree with that computed using the non-renormalizable perturbative expansion, while at higher orders the calculations in the non-perturbative theory would amount to fixing all infinite free parameters of the non-renormalizable theory. Whether this is the case is still an open question.

The true importance of these results, on the other hand, is in the fact that they indicate how to compute background-dependent quantities from the non-perturbative theory, a problem that has long been a source of great confusion in quantum gravity.

Final remarks

We close with a short historical note on the development of the theory presented, and a few considerations on the main problems that remain open.

13.1 Brief historical note

The quantum theory of gravity presented in this book is the result of a long path to which many have contributed. We collect here a few notes about this path and a few references for the orientation of the reader. A comprehensive bibliography would be impossible.

The basic ideas are still those of John Wheeler (Wheeler 1968) and Bryce DeWitt (DeWitt 1967) concerning the state space, and those of Charles Misner (Misner 1957), Stephen Hawking (Hawking 1980), Jonathan Halliwell (Halliwell and Hawking 1985), and Jim Hartle (Hartle and Hawking 1983) concerning the sum over geometries. The theory would not have developed without Abhay Ashtekar's introduction (Ashtekar 1986) of his variables and grew out of the "loop space representation of quantum general relativity," which was conceived in the late 1980s (Rovelli and Smolin 1988, 1990; Ashtekar *et al.* 1991). Graphs were introduced by Jurek Lewandowski in (Lewandowski 1994); spin networks and discretization emerged from loop quantum gravity in the 1990s (Rovelli and Smolin 1995a,b; Ashtekar and Lewandowski 1997). The geometrical picture used here derives from Bianchi *et al.* (2011b). The idea of using spinfoams for quantum gravity derives from the work of Hirosi Ooguri (Ooguri 1992), who generalized to four-dimensional the Boulatov model (Boulatov 1992), which in turn is a generalization to three-dimensional of the two-dimensional matrix models. Ooguri theory in four-dimensional is not general relativity, but it has a structure that matches remarkably well that of loop quantum gravity and therefore it appeared natural to try to import his technique into gravity (Rovelli 1993b), yielding the spinfoam ideas (Reisenberger and Rovelli 1997). The problem of adapting the Ooguri model to gravity by suitably constraining it was much discussed in the 1990s and finally brilliantly solved by John Barrett and Louis Crane (Barrett and Crane 1998, 2000), who devised the first spinfoam model for four-dimensional quantum gravity. The model prompted the start of the group field theory technology (De Pietri *et al.* 2000), but it turned out to have problems, which became clear when it was used to compute the graviton 2-point function, which did not turn out to be correct (Alesci and Rovelli 2007). The effort to correct it led to the model presented here, which was developed in a joint effort involving numerous people: Engle *et al.* (2007, 2008a); Freidel and Krasnov (2008); Livine and Speziale (2007). Its euclidean version was found independently and simultaneously by two groups. The lorentzian version appeared in Engle *et al.* (2008b). The extension of the theory to arbitrary complexes was rapidly developed by the Warsaw group (Kaminski *et al.* 2010a,b).

The theory was puzzling and seen with suspicion until John Barrett and his collaborators succeeded in the tour de force of computing the semiclassical approximation of the amplitude (Barrett *et al.* 2010) and this turned out to match the Regge theory. Shortly after, the version with cosmological constant was introduced (Han 2011a; Fairbairn and Meusburger 2012), building on the mathematical work in Noui and Roche (2003), and the theory proved finite. The coherent states techniques have been developed independently by many people: Thiemann (2006); Livine and Speziale (2007); Bianchi *et al.* (2010a); Freidel and Speziale (2010b). The importance of bubbles for the study of divergences was understood early (Perez and Rovelli 2001), but radiative calculations started only recently (Perini *et al.* 2009; Riello 2013). Fermion coupling is found in Bianchi *et al.* (2010b). Spinfoam cosmology was introduced in Bianchi *et al.* (2010d) and extended to the case with many nodes and many links in Vidotto (2011). The application of spinfoams to black holes was developed in Bianchi (2012a). The technique for computing n-point functions in a background-independent context was introduced in Rovelli (2006), using the boundary formalism developed by Robert Oeckl (Oeckl 2003). The lorentzian 2-point function computed in Bianchi and Ding (2012). n-Point functions for higher n have been computed in Rovelli and Zhang (2011). The asymptotic of the theory has been studied recently in Han and Zhang (2013) and in Han (2013).

This is far from being a complete bibliography of loop quantum gravity, and we sincerely apologize to the many scientists who have contributed importantly to the theory but are not mentioned here or elsewhere in the book; there are many of them.

13.2 What is missing

The main open questions in the theory are two. The first regards its consistency, the second the possibility of making predictions. In addition, the theory still needs to be further developed, or perhaps modified, to become complete.

Consistency

1. With the cosmological constant, the transition amplitudes are finite at all orders and the classical limit of each converges to the truncation of the classical limit of GR on a finite discretization of spacetime; in turn, these converge to classical GR when the discretization is refined. This gives a coherent approximation scheme. However the approximation scheme may go wrong if the quantum part of the corrections that one obtains in refining the discretization is large. These can be called "radiative corrections," since they are somewhat analogous to standard QFT radiative corrections: possibly large quantum effects that appear on taking the next order in an approximation. It is not sufficient for these radiative corrections to be finite, for the approximation

to be viable: they must also be small. Since the theory includes a large number – the ratio of the cosmological constant scale over the Planck scale (or over the observation scale) – these radiative corrections a priori could be large. Understanding whether these are under control (in suitable physical regimes) is the key open problem on which the consistency of theory hinges today.

Since large contributions to amplitudes come from large spins and these are cut off by the cosmological constant, these potentially dangerous large numbers can be studied as divergences in the $\Lambda = 0$ theory. This is the path taken by Aldo Riello (Riello 2013), where some bubble divergences have been computed. The results so far are reassuring, as the cut-off comes in only logarithmically in the divergent terms, suggesting at worse a logarithm of the cosmological constant to enter the amplitudes. But a general analysis is needed, and the problem is fully open.

It is then essential to study whether *other* bubble divergences appear in the $\Lambda = 0$ theory, besides those considered in Riello (2013).

2. Can a general result about these divergences be obtained?
3. A separate question is posed by the contribution of very fine discretizations. Does something non-trivial happen to theory in approaching the continuum limit? Are there phase transitions? Although this seems unlikely at present, this is a possibility. A large amount of work is being developed for study of the continuum limit in model theories that share similarities with quantum gravity; see for instance Bonzom and Livine (2011), Dittrich *et al.* (2013) and references therein. The results of these works may be of great relevance for information on the continuum limit of quantum gravity.

Completing the theory

1. The matter sector of the theory has not been sufficiently developed. Can we obtain the Dirac propagator from the vertex?
2. The q-deformed version of the theory has been sketched (Han 2011a,b; Fairbairn and Meusburger 2012) but it is very little developed (Ding and Han 2011). It deserves to be studied and better understood. In particular, can we reobtain in this theory the results obtained in the $\Lambda = 0$ theory?

Modifying the theory?

1. The two versions of the Y map with $p = \gamma j$ and $p = \gamma(j + 1)$ are equally viable and interesting for the moment. Is there something that favors one of the two theories?
2. The possibility of expressing the theory as a group field theory has been studied in Ben Geloun *et al.* (2010). While transition amplitudes on individual discretizations are unaffected by this, this formulation leads to a different definition of the continuum limit, where contributions of increasingly refined discretizations are summed. This is a possible alternative definition of the theory, certainly of interest. In this context, a possible alternative to the formulation given here is also the theory defined by Daniele Oriti and Aristide Baratin in Baratin *et al.* (2010).

Physical predictions

1. The most likely window to observational support to the theory seems at present to come from cosmology. While canonical loop quantum cosmology is well developed, the covariant theory is only in its infancy (Bianchi *et al.* 2010d; Vidotto 2011; Rennert and Sloan 2013) and much needs to be done to bring it to the same level. Computing cosmological transition amplitudes to the next relevant orders in the lorentzian theory and for less trivial 2-complexes is a necessary step. The recent results in writing cosmological states obtained by Alesci and Cianfrani (2013) could provide useful tools for this. Comparison with canonical loop quantum cosmology (Ashtekar 2009; Bojowald 2005) also remains to be done.

2. The idea that propagation can be affected by quantum gravity – considered and then partially discarded some years ago because of the proof that the theory is Lorentz invariant – is not necessarily wrong, as Lorentz–invariant corrections to the propagator are possible. This is also a direction to explore. For this, one needs to study the higher-order corrections to the graviton propagator computed in Bianchi *et al.* (2009); Bianchi and Ding (2012). Is there a Planck correction to the form of the classical propagator?

3. As mentioned, Riello's results in Riello (2013) indicate that radiative corrections are proportional to the logarithm of the cosmological constant. Can this be used to derive some physics? The scaling of G or Λ? The vacuum energy renormalization?

4. A recent suggestion (Rovelli and Vidotto 2014) is that a star that collapses gravitationally can reach a further stage of its life, where quantum-gravitational pressure counteracts weight, as happens in loop cosmology. This is called a "Planck star." The duration of this stage is very short in the star's proper time, yielding a bounce, but is extremely long seen from the outside, because of the huge gravitational time dilation. Loop quantum gravity indicates that the onset of quantum-gravitational effects is governed by energy density, or by acceleration – not by size – therefore the star can be much larger than planckian in this phase. The object emerging at the end of the Hawking evaporation of a black hole can then be larger than planckian by a factor $(m/m_P)^n$, where m is the mass that has fallen into the hole, m_P is the Planck mass, and n is positive. These objects could have astrophysical and cosmological interest: primordial black holes ending their evaporation now could produce a detectable signal, of quantum gravitational origin. The large factor amplifying quantum gravitational effects is here the ratio of the Hubble time t_H (the life of the primordial black hole) to the Planck time t_{Pl}, giving a wavelength (Rovelli and Vidotto 2014):

$$\lambda \sim \sqrt[3]{\frac{t_H}{t_{Pl}}}\, L_{Planck} \sim 10^{-14}\,\text{cm} \tag{13.1}$$

which is well within the window of currently detectable signals.[1] Is this suggestion viable? Can the dynamics of a Planck star be studied with the loop theory?

---

[1] In (Haggard and Rovelli 2014) the quantum transition from a black to white hall was studied, and the possibly of a signal at much larger wavelength was considered. The phenomenology of the two cases was studied in (Barrau and Rovelli 2014) and (Barrau, Rovelli and Vidotto 2015).

There is much yet to do in quantum gravity. While we wait for some empirical support, we cannot be sure we are steering the right course. But the sky is much clearer than just a few years ago in quantum gravity. So, here is the final task:

Exercise 13.1

- *Show that the theory defined in this book is fully consistent; if it is not, correct it.*
- *Find observable and testable consequences.*
- *Using these, test the theory.*
- *If the theory is confirmed, we will feel good, and feeling good is good enough ...*

References

Adler, S. L. 1982. Einstein gravity as a symmetry breaking effect in quantum field theory. *Rev. Mod. Phys.*, **54**, 729.

Agullo, I., and Corichi, A. 2013. Loop quantum cosmology. arXiv:1302.3833.v1 [gr-gc]. [In Ashtekar, A. and Petkov, V. (eds.). 2014. *The Springer Handbook of Spacetime.* Springer.]

Agullo, I., Ashtekar, A., and Nelson, W. 2012. A quantum gravity extension of the inflationary scenario. *Phys. Rev. Lett.*, **109**, 251301, arXiv:1209.1609.

Alesci, E., and Cianfrani, F. 2013. Quantum-reduced loop gravity: cosmology. *Phys. Rev.*, **D87**, 83521, arXiv:1301.2245.

Alesci, E., and Rovelli, C. 2007. Complete LQG propagator: difficulties with the Barrett–Crane vertex. *Phys. Rev.*, **D76**(10), 104012, arXiv:0708.0883.

Alesci, E., and Rovelli, C. 2008. The complete LQG propagator: II. Asymptotic behavior of the vertex. *Phys. Rev.*, **D77**, 44024, arXiv:0711.1284.

Alexandrov, S. 2010. New vertices and canonical quantization. *Phys. Rev.*, **D82**(July), 24024, arXiv:1004.2260.

Aquilanti, V., Haggard, H. M., Hedeman, A., Jeevanjee, N., Littlejohn, R. G., and Yu, L. 2010. Semiclassical mechanics of the Wigner 6j-symbol. arXiv:1009.2811.

Arkani-Hamed, N., Dubovsky, S., Nicolis, A., Trincherini, E., and Villadoro, G. 2007. A measure of de Sitter entropy and eternal inflation. *J. High. Energy. Phys.*, **0705**, 55, arXiv:0704.1814.

Arnold, V. I. 1989. *Mathematical Methods of Classical Mechanics.* Springer Verlag.

Arnowitt, R. L., Deser, S., and Misner, C. W. 1962. The dynamics of general relativity. arXiv:0405109 [gr-qc].

Ashtekar, A. 1986. New variables for classical and quantum gravity. *Phys. Rev. Lett.*, **57**, 2244–2247.

Ashtekar, A. 1991. *Lectures on Non-Perturbative Canonical Gravity.* World Scientific.

Ashtekar, A. 2009. Loop quantum cosmology: an overview. *Gen. Relat. Gravit.*, **41**, 707–741, arXiv:0812.0177.

Ashtekar, A. 2011. Introduction to loop quantum gravity. In: *Proceedings of the 3rd Quantum Gravity and Quantum Geometry School.* PoS (QGQGS2011)001.

Ashtekar, A., and Krasnov, K. 1998 Quantum geometry and black holes. In: B. R. Iyer and B. Bhawal (eds.), *'Black Holes, Gravitational Radiation and the Universe', Essays in Honor of C. V. Vishveshwara* (pp. 149–170). Kluwer.

Ashtekar, A., and Lewandowski, J. 1995. Projective techniques and functional integration for gauge theories. *J. Math. Phys.*, **36**, 2170–2191, arXiv:9411046 [gr-qc].

Ashtekar, A., and Lewandowski, J. 1997. Quantum theory of geometry. I: Area operators. *Classical Quant. Grav.*, **14**, A55–A82, arXiv:9602046 [gr-qc].

Ashtekar, A., and Lewandowski, J. 1998. Quantum theory of geometry. II: Volume operators. *Adv. Theor. Math. Phys.*, **1**, 388–429, arXiv:9711031 [gr-qc].

Ashtekar, A., and Lewandowski, J. 2004. Background independent quantum gravity: a status report. *Classical Quant. Grav.*, **21**, R53, arXiv:0404018 [gr-qc].

Ashtekar, A., and Petkov, V. 2014. *Springer Handbook of Spacetime.* Springer.

Ashtekar, A., and Singh, P. 2011. Loop quantum cosmology: a status report. *Classical Quant. Grav.*, **28**, 213001, arXiv:1108.0893.

Ashtekar, A., Rovelli, C., and Smolin, L. 1991. Gravitons and loops. *Phys. Rev.*, **D44**, 1740–1755, arXiv:9202054 [hep-th].

Ashtekar, A., Rovelli, C., and Smolin, L. 1992. Weaving a classical geometry with quantum threads. *Phys. Rev. Lett.*, **69**(Mar.), 237–240, arXiv:9203079 [hep-th].

Ashtekar, A., Baez, J., Corichi, A., and Krasnov, K. 1998. Quantum geometry and black hole entropy. *Phys. Rev. Lett.*, **80**, 904–907, arXiv:9710007 [gr-qc].

Ashtekar, A., Engle, J., and Van Den Broeck, C. 2005. Quantum horizons and black hole entropy: inclusion of distortion and rotation. *Classical Quant. Grav.*, **22**, L27, arXiv:0412003 [gr-qc].

Ashtekar, A., Kaminski, W., and Lewandowski, J. 2009. Quantum field theory on a cosmological, quantum space-time. *Phys. Rev.*, **D79**, 64030, arXiv:0901.0933 [gr-qc].

ATLAS Collaboration. 2012. Observation of a new particle in the search for the Standard Model Higgs boson with the ATLAS detector at the LHC. *Phys. Lett. B*, **716**(1), 1–29.

Baez, J. C. 1994a. Generalized measures in gauge theory. *Lett. Math. Phys.*, **31**, 213–223.

Baez, J. C. (ed.). 1994b. *Knots and Quantum Gravity.* Oxford Lecture Series in Mathematics and its Applications, vol. 1. Clarendon Press.

Baez, J. C., and Munian, J. P. 1994. *Gauge Fields Knots, and Gravity.* World Scientific, Singapore.

Bahr, B., and Thiemann, T. 2009. Gauge-invariant coherent states for loop quantum gravity II: Non-abelian gauge groups. *Classical Quant. Grav.*, **26**, 45012, arXiv:0709.4636.

Bahr, B., Hellmann, F., Kaminski, W., Kisielowski, M., and Lewandowski, J. 2011. Operator spin foam models. *Classical Quant. Grav.*, **28**, 105003, arXiv:1010.4787.

Baratin, A., Dittrich, B., Oriti, D., and Tambornino, J. 2010. Non-commutative flux representation for loop quantum gravity. arXiv:1004.3450 [hep-th].

Barbero, J. F. 1995. Real Ashtekar variables for Lorentzian signature space times. *Phys. Rev.*, **D51**, 5507–5510, arXiv:9410014 [gr-qc].

Barbieri, A. 1998. Quantum tetrahedra and simplicial spin networks. *Nucl. Phys.*, **B518**, 714–728, arXiv:9707010 [gr-qc].

Bardeen, J. M., Carter, B., and Hawking, S. W. 1973. The four laws of black hole mechanics. *Commun. Math. Phys.*, **31**, 161–170.

Barrau, A., and Rovelli, C. 2014. Fast Radio Bursts and White Hole Signals, arXiv:1409.4031.

Barrau, A., and Rovelli, C. 2014. Planck star phenomenology. arXive:1404.5821.

Barrett, J. W., and Crane, L. 1998. Relativistic spin networks and quantum gravity. *J. Math. Phys.*, **39**, 3296–3302, arXiv:9709028 [gr-qc].

Barrett, J. W., and Crane, L. 2000. A Lorentzian signature model for quantum general relativity. *Classical Quant. Grav.*, **17**, 3101–3118, arXiv:9904025 [gr-qc].

Barrett, J. W., and Naish-Guzman, I. 2009. The Ponzano–Regge model. *Classical Quant. Grav.*, **26**, 155014, arXiv:0803.3319.

Barrett, J. W., Dowdall, R. J., Fairbairn, W. J., Hellmann, F., and Pereira, R. 2010. Lorentzian spin foam amplitudes: graphical calculus and asymptotics. *Classical Quant. Grav.*, **27**, 165009, arXiv:0907.2440.

Barrett, J., Giesel, K., Hellmann, F., Jonke, L., Krajewski, T., *et al.* 2011a. Proceedings of 3rd Quantum Geometry and Quantum Gravity School. *Proceedings of Science, PoS*, (**QGQGS2011**).

Barrett, J. W., Dowdall, R. J., Fairbairn, W. J., Hellmann, F., and Pereira, R. 2011b. Asymptotic analysis of Lorentzian spin foam models. In: *Proceedings of the 3rd Quantum Gravity and Quantum Geometry School*. PoS **QGQGS2011**009.

Battisti, M. V., Marciano, A., and Rovelli, C. 2010. Triangulated loop quantum cosmology: Bianchi IX and inhomogenous perturbations. *Phys. Rev.*, **D81**, 64019.

Bekenstein, J. D. 1973. Black holes and entropy. *Phys. Rev.*, **D7**, 2333–2346.

Ben Geloun, J., Gurau, R., and Rivasseau, V. 2010. EPRL/FK group field theory. *Europhys. Lett.*, **92**, 60008, arXiv:1008.0354 [hep-th].

Bertone, G. 2010. *Particle Dark Matter*. Cambridge University Press.

Bianchi, E. 2012a. Entropy of non-extremal black holes from loop gravity. arXiv:1204.5122.

Bianchi, E. 2012b. Horizon entanglement entropy and universality of the graviton coupling. arXiv:1211.0522.

Bianchi, E., and Ding, Y. 2012. Lorentzian spinfoam propagator. *Phys. Rev.*, **D86**, 104040, arXiv:1109.6538.

Bianchi, E., and Rovelli, C. 2010a. Is dark energy really a mystery? *Nature*, **466**, 321, arXiv:1003.3483.

Bianchi, E., and Rovelli, C. 2010b. Why all these prejudices against a constant?, arXiv:1002.3966.v3.

Bianchi, E., and Wieland, W. 2012. Horizon energy as the boost boundary term in general relativity and loop gravity. arXiv:1205.5325.

Bianchi, E., Magliaro, E., and Perini, C. 2009. LQG propagator from the new spin foams. *Nucl. Phys.*, **B822**, 245–269, arXiv:0905.4082.

Bianchi, E., Magliaro, E., and Perini, C. 2010a. Coherent spin-networks. *Phys. Rev.*, **D82**, 24012, arXiv:0912.4054.

Bianchi, E., Han, M., Magliaro, E., Perini, C., and Wieland, W. 2010b. Spinfoam fermions. arXiv:1012.4719.

Bianchi, E., Magliaro, E., and Perini, C. 2010c. Spinfoams in the holomorphic representation. *Phys. Rev.*, **D82**, 124031, arXiv:1004.4550 [gr-qc].

Bianchi, E., Rovelli, C., and Vidotto, F. 2010d. Towards spinfoam cosmology. *Phys. Rev.*, **D82**, 84035, arXiv:1003.3483.

Bianchi, E., Krajewski, T., Rovelli, C., and Vidotto, F. 2011a. Cosmological constant in spinfoam cosmology. *Phys. Rev.*, **D83**, 104015, arXiv:1101.4049 [gr-qc].

Bianchi, E., Donà, P., and Speziale, S. 2011b. Polyhedra in loop quantum gravity. *Phys. Rev.*, **D83**, 44035, arXiv:1009.3402.

Bianchi, E., Haggard, H. M., and Rovelli, C. 2013. The boundary is mixed. arXiv:1306.5206.

Bisognano, J. J., and Wichmann, E. H. 1976. On the duality condition for quantum fields. *J. Math. Phys.*, **17**, 303–321.

Bohr, N., and Rosenfeld, L. 1933. Det Kongelige Danske Videnskabernes Selskabs. *Mathematiks-fysike Meddeleser*, **12**, 65.

Bojowald, M. 2001. The semiclassical limit of loop quantum cosmology. *Classical Quant. Grav.*, **18** (Sept.), L109–L116, arXiv:0105113 [gr-qc].

Bojowald, M. 2005. Loop quantum cosmology. *Living Rev. Relativ.*, **8**, 11, arXiv:0601085 [gr-qc].

Bojowald, M. 2011. Quantum cosmology. *Lect. Notes Phys.*, **835**, 1–308.

Bonzom, V. 2009. Spin foam models for quantum gravity from lattice path integrals. *Phys. Rev.*, **D80**, 64028, arXiv:0905.1501.

Bonzom, V., and Livine, E. R. 2011. Yet another recursion relation for the 6j-symbol. arXiv:1103.3415 [gr-qc].

Borja, E. F., Diaz-Polo, J., Garay, I., and Livine, E. R. 2010. Dynamics for a 2-vertex quantum gravity model. *Classical Quant. Grav.*, **27**, 235010.

Borja, E. F., Garay, I., and Vidotto, F. 2011. Learning about quantum gravity with a couple of nodes. *SIGMA*, **8**, 15.

Born, M, and Jordan, P. 1925. Zur Quantenmechanik. *Zeitschrift für Physik*, **34**, 858–888.

Born, M, Jordan, P, and Heisenberg, W. 1926. Zur Quantenmechanik II. *Zeitschrift für Physik*, **35**, 557–615.

Boulatov, D. V. 1992. A model of three-dimensional lattice gravity. *Mod. Phys. Lett.*, **A7**, 1629–1646, arXiv:9202074 [hep-th].

Bozza, V., Feoli, A., Papini, G., and Scarpetta, G. 2000. Maximal acceleration effects in Reissner–Nordstrom space. *Phys. Lett.*, **A271**, 35–43, arXiv:0005124 [gr-qc].

Bozza, V. Feoli, A. Lambiase, G. Papini, G. and Scarpetta, G. 2001. Maximal acceleration effects in Kerr space. *Phys. Lett.*, **A283**, 53–61, arXiv:0104058 [gr-qc].

Brandt, H. E. 1989. Maximal proper acceleration and the structure of space-time. *Found. Phys. Lett.*, **2**, 39.

Bronstein, M. P. 1936a. Kvantovanie gravitatsionnykh voln (Quantization of gravitational waves). *Zh. Eksp. Teor. Fiz.*, **6**, 195.

Bronstein, M. P. 1936b. Quantentheorie schwacher Gravitationsfelder. *Phys. Z. Sowjetunion*, **9**, 140–157.

Buffenoir, E., and Roche, P. 1999. Harmonic analysis on the quantum Lorentz group. *Commun. Math. Phys.*, **207**(3), 499–555, arXiv:9710022 [q-alg].

Caianiello, E. R. 1981. Is there a maximal acceleration? *Lett. Nuovo Cim.*, **32**, 65.

Caianiello, E. R., and Landi, G. 1985. Maximal acceleration and Sakhrov's limiting temperature. *Lett.Nuovo Cimento.*, **42**, 70.

Caianiello, E. R., Feoli, A., Gasperini, M., and Scarpetta, G. 1990. Quantum corrections to the space-time metric from geometric phase space quantization. *Int. J. Theor. Phys.*, **29**, 131.

Caianiello, E. R., Gasperini, M., and Scarpetta, G. 1991. Inflation and singularity prevention in a model for extended-object-dominated cosmology. *Classical and Quant. Grav.*, **8**(4), 659.

Cailleteau, T., Mielczarek, J., Barrau, A., and Grain, J. 2012a. Anomaly-free scalar perturbations with holonomy corrections in loop quantum cosmology. *Classical Quant. Grav.*, **29**, 95010, arXiv:1111.3535 [gr-qc].

Cailleteau, T., Barrau, A., Grain, J., and Vidotto, F. 2012b. Consistency of holonomy-corrected scalar, vector and tensor perturbations in loop quantum cosmology. *Phys. Rev.*, **D86**, 87301, arXiv:1206.6736.

Carlip, S., and Teitelboim, C., 1995. The off-shell black hole. *Classical Quant. Grav.*, **12**, 1699–1704, arXiv:9312002 [gr-qc].

Carter, J. S., Flath, D. E., and Saito, M. 1995. *The Classical and Quantum 6j-Symbol*. Princeton University Press.

Chandia, O., and Zanelli, J. 1997. Torsional topological invariants (and their relevance for real life). arXiv:hep-th/9708138 [hep-th].

Chirco, G., Haggard, H. M., and Rovelli, C. 2013. Coupling and equilibrium for general-covariant systems. *Phys. Rev.* **D88**, 084027, arXiv:1309.0777.

Choquet-Bruhat, Y., and Dewitt-Morette, C. 2000. *Analysis, Manifolds and Physics, Part II*. North-Holland.

Choquet-Bruhat, Y., and Dewitt-Morette, C. 2004. *Analysis, Manifolds and Physics, Part I*. 2 edn. North-Holland.

Christodoulou, M., Riello, A., and Rovelli, C. 2012. How to detect an anti-spacetime. *Int. J. Mod. Phys.*, **D21**, 1242014, arXiv:1206.3903 [gr-qc].

Christodoulou, M., Langvik, M., Riello, A., Röken, C., and Rovelli, C. 2013. Divergences and orientation in spinfoams. *Classical Quantum. Grav.*, **30**, 055009, arXiv:1207.5156.

CMS Collaboration. 2012. Observation of a new boson at a mass of 125 GeV with the CMS experiment at the LHC. *Phys. Lett. B*, **716**(1), 30–61.

Colosi, D., and Rovelli, C. 2003. A simple background-independent Hamiltonian quantum model. *Phys. Rev.*, **D68**, 104008, arXiv:0306059 [gr-qc].

Colosi, D., and Rovelli, C. 2009. What is a particle? *Classical Quant. Grav.*, **26**, 25002, arXiv:0409054 [gr-qc].

Connes, A., and Rovelli, C. 1994. Von Neumann algebra automorphisms and time thermo-dynamics relation in general covariant quantum theories. *Classical Quant. Grav.*, **11**, 2899–2918, arXiv:9406019 [gr-qc].

Conrady, F., and Freidel, L. 2008. Path integral representation of spin foam models of 4d gravity. *Classical Quant. Grav.*, **25**, 245010, arXiv:0806.4640.

De Pietri, R., Freidel, L., Krasnov, K., and Rovelli, C. 2000. Barrett–Crane model from a Boulatov–Ooguri field theory over a homogeneous space. *Nucl. Phys.*, **B574**, 785–806, arXiv:9907154 [hep-th].

DeWitt, B. S. 1967. Quantum theory of gravity. 1. The canonical theory. *Phys. Rev.*, **160**, 1113–1148.

Ding, Y., and Han, M. 2011. On the asymptotics of quantum group spinfoam model. arXiv:1103.1597.

Ding, Y., and Rovelli, C. 2010a. Physical boundary Hilbert space and volume operator in the Lorentzian new spin-foam theory. *Classical Quant. Grav.*, **27**, 205003, arXiv:1006.1294.

Ding, Y., and Rovelli, C. 2010b. The volume operator in covariant quantum gravity. *Classical Quant. Grav.*, **27**, 165003, arXiv:0911.0543.

Dirac, P. A. M. 1950. Generalized Hamiltonian dynamics. *Can. J. Math.*, **2**, 129–148.

Dirac, P. A. M. 1956. Electrons and the vacuum. In: *Dirac Papers*. Florida State University Libraries, Tallahassee, Florida, USA.

Dirac, P. A. M. 1958. Generalized Hamiltonian dynamics. *Proc. R. Soc. Lond.*, **246**, 326–332.

Dirac, P. 2001. *Lectures on Quantum Mechanics*. Dover Books on Physics. New York: Belfer Graduate School of Science, Yeshiva University, New York.

Dittrich, B., Martin-Benito, M., and Steinhaus, S. 2013. Quantum group spin nets: refinement limit and relation to spin foams. arXiv:1312.0905.

Doná, P., and Speziale, S. 2013. Introductory lectures to loop quantum gravity. In: Bounames, A., and Makhlouf, A. (eds.), *TVC 79. Gravitation: théorie et expérience.* Hermann.

Durr, S., Fodor, Z., Frison, J., Hoelbling, C., Hoffmann, R., *et al.* 2008. Ab-initio determination of light hadron masses. *Science*, **322**, 1224–1227, arXiv:0906.3599 [hep-lat].

Einstein, A. 1905a. On a heuristic viewpoint of the creation and modification of light. *Annalen der Physik*, **1905** (17), 133–148.

Einstein, A. 1905b. Zur Elektrodynamik bewegter Körper. *Annalen der Physik*, **1905**(10), 891–921.

Einstein, A. 1915. Zur Allgemeinen Relativitätstheorie. *Sitzungsher. Preuss. Akad: Wiss. Berlin (Math. Phys.)*, **1915**, 778–786.

Einstein, A. 1916. Grundlage der allgemeinen Relativitätstheorie. *Annalen der Physik*, **49**, 769–822.

Einstein, A. 1917. Cosmological considerations in the general theory of relativity. *Sitzungsber. Preuss. Akad. Wiss. Berlin (Math. Phys.)*, **1917**, 142–152.

Engle, J., and Pereira, R. 2009. Regularization and finiteness of the Lorentzian LQG vertices. *Phys. Rev.*, **D79**, 84034, arXiv:0805.4696.

Engle, J., Pereira, R., and Rovelli, Carlo. 2007. The loop-quantum-gravity vertex-amplitude. *Phys. Rev. Lett.*, **99**, 161301, arXiv:0705.2388.

Engle, J., Pereira, R., and Rovelli, C. 2008a. Flipped spinfoam vertex and loop gravity. *Nucl. Phys.*, **B798**, 251–290, arXiv:0708.1236.

Engle, J., Livine, E., Pereira, R., and Rovelli, C. 2008b. LQG vertex with finite Immirzi parameter. *Nucl. Phys.*, **B799**, 136–149, arXiv:0711.0146.

Fairbairn, W. J., and Meusburger, C. 2012. Quantum deformation of two four-dimensional spin foam models. *J. Math. Phys.*, **53**, 22501, arXiv:1012.4784 [gr-qc].

Farmelo, G. 2009. *The Strangest Man. The Hidden Life of Paul Dirac, Quantum Genius.* Faber and Faber.

Feoli, A., Lambiase, G., Papini, G., and Scarpetta, G. 1999. Schwarzschild field with maximal acceleration corrections. *Phys. Lett.*, **A263**, 147–153, arXiv:9912089 [gr-qc].

Feynman, R. P. 1948. Space-time approach to nonrelativistic quantum mechanics. *Rev. Mod. Phys.*, **20**, 367–387.

Flanders, H. 1963. *Differential Forms: With Applications to the Physical Sciences*. Dover.

Folland, G. B. 1995. *A Course in Abstract Harmonic Analysis*. CRC Press.

Frankel, T. 2003. *The Geometry of Physics: An Introduction*. 2 edn. Cambridge University Press.

Freidel, L., and Krasnov, K. 2008. A new spin foam model for 4d gravity. *Classical Quant Grav.*, **25**, 125018, arXiv:0708.1595.

Freidel, L., and Speziale, S. 2010a. From twistors to twisted geometries. *Phys. Rev.*, **D82**, 84041, arXiv:1006.0199 [gr-qc].

Freidel, L., and Speziale, S. 2010b. Twisted geometries: A geometric parametrisation of SU(2) phase space. *Phys. Rev.*, **D82**, 84040, arXiv:1001. 2748.

Freidel, L., Krasnov, K., and Livine, E. R. 2009. Holomorphic factorization for a quantum tetrahedron. arXiv:0905.3627.

Frodden, E., Ghosh, A., and Perez, A. 2011. Black hole entropy in LQG: Recent developments. *AIP Conf. Proc.*, **1458**, 100–115.

Frodden, E., Geiller, M., Noui, K., and Perez, A. 2012. Black hole entropy from complex Ashtekar variables. arXiv:1212.4060.

Gambini, R., and Pullin, J. 2010. *Introduction to Loop Quantum Gravity*. Oxford University Press.

Gel'fand, I. M., and Shilov, G. E. 1968. *Generalized functions 1–5*. Academic Press.

Gel'fand, I. M., Minlos, R. A., and Shapiro, Z.Ya. 1963. *Representations of the Rotation and Lorentz Groups and Their Applications*. Pergamon Press.

Ghosh, A., and Perez, A. 2011. Black hole entropy and isolated horizons thermodynamics. *Phys. Rev. Lett.*, **107**, 241301, arXiv:1107.1320.

Ghosh, A., Noui, K., and Perez, A. 2013. Statistics, holography, and black hole entropy in loop quantum gravity. arXiv:1309.4563.

Gibbons, G. W., and Hawking, S. W. 1977. Action integrals and partition functions in quantum gravity. *Phys. Rev.*, **D15**(10), 2752–2756.

Gorelik, G., and Frenkel, V. 1994. *Matvei Petrovich Bronstein and Soviet Theoretical Physics in the Thirties*. Birkhäuser Verlag.

Haag, R. 1996. *Local Quantum Physics: Fields, Particles, Algebras*. Springer.

Haggard, H. M. 2011. Asymptotic analysis of spin networks with applications to quantum gravity. PhD dissertation, University of California, Berkeley. http://bohr.physics.berkeley.edu/hal/pubs/Thesis/.

Haggard, H. M., and Rovelli, C. 2013. Death and resurrection of the zeroth principle of thermodynamics. *J. Mod. Phys.*, **D22**, 1342007, arXiv:1302.0724.

Haggard, H. M., and Rovelli, C. 2014. Black hole fireworks: quantum-gravity effects outside the horizon spark black to white hole tunnelling. arXives 1407.0989.

Haggard, H. M., Rovelli, C., Wieland, W., and Vidotto, F. 2013. The spin connection of twisted geometry. *Phys. Rev.*, **D87**, 24038, arXiv:1211.2166.

Halliwell, J. J., and Hawking, S. W. 1985. The origin of structure in the universe. *Phys. Rev.*, **D31**, 1777.

Han, M. 2011a. 4-Dimensional spin-foam model with quantum lorentz group. *J. Math. Phys.*, **52**, 72501, arXiv:1012.4216 [gr-qc].

Han, M. 2011b. Cosmological constant in LQG vertex amplitude. *Phys. Rev.*, **D84**, 64010, arXiv:1105.2212.

Han, M. 2013. Covariant loop quantum gravity, low energy perturbation theory, and Einstein gravity. arXiv:1308.4063.

Han, M. 2014. Black hole entropy in loop quantum gravity, analytic continuation, and dual holography. arXiv:1402.2084.

Han, M., and Rovelli, C. 2013. Spin-foam fermions: PCT symmetry, Dirac determinant, and correlation functions. *Classical Quant. Grav.*, **30**, 75007, arXiv:1101.3264.

Han, M., and Zhang, M. 2013. Asymptotics of spinfoam amplitude on simplicial manifold: Lorentzian theory. *Classical Quant. Grav.*, **30**, 165012, arXiv:1109.0499 [gr-qc].

Hartle, J. B., and Hawking, S. W. 1983. Wave function of the universe. *Phys. Rev.*, **D28**, 2960–2975.

Hawking, S. W. 1974. Black hole explosions? *Nature*, **248**, 30–31.

Hawking, S.W. 1975. Particle creation by black holes. *Commun. Math. Phys.*, **43**, 199–220.

Hawking, S. W. 1980. The path integral approach to quantum gravity. In: Hawking, S.W., Israel, W. (eds.) *General Relativity*. Cambridge University Press; 746–789.

Hawking, S. W., and Ellis, G. F. R. 1973. *The Large Scale Structure of Space-time*. Cambridge Monographs on Mathematical Physics. Cambridge University Press.

Heisenberg, W. 1925. Uber quantentheoretische Umdeutung kinematischer und mechanischer Beziehungen. *Zeitschrift für Physik*, **33**(1), 879–893.

Hellmann, F., and Kaminski, W. 2013. Holonomy spin foam models: asymptotic geometry of the partition function. arXiv:1307.1679.

Hojman, R., Mukku, C., and Sayed, W. A. 1980. Parity violation in metric torsion theories of gravity. *Phys. Rev.*, **D22**, 1915–1921.

Hollands, S., and Wald, R. M. 2008. Quantum field theory in curved spacetime, the operator product expansion, and dark energy. *Gen. Relat. Gravit.*, **40**, 2051–2059, arXiv:0805.3419.

Holst, S. 1996. Barbero's Hamiltonian derived from a generalized Hilbert–Palatini action. *Phys. Rev.*, **D53**, 5966–5969, arXiv:9511026 [gr-qc].

Kaminski, W. 2010. All 3-edge-connected relativistic BC and EPRL spin networks are integrable. arXiv:1010.5384 [gr-qc].

Kaminski, W., Kisielowski, M., and Lewandowski, J. 2010a. Spin-foams for all loop quantum gravity. *Classical Quant. Grav.*, **27**, 95006, arXiv:0909.0939.

Kaminski, W., Kisielowski, M., and Lewandowski, J. 2010b. The EPRL intertwiners and corrected partition function. *Classical Quant. Grav.*, **27**, 165020, arXiv:0912.0540.

Kauffman, L. H., and Lins, S. L. 1994. *Temperley–Lieb Recoupling Theory and Invariants of 3-Manifolds*. Annals of Mathematics Studies, vol. 134. Princeton University Press.

Kiefer, C. 2007. *Quantum Gravity (International Series of Monographs on Physics)*. 2 edn. Oxford University Press.

Kobayashi, S., and Nomizu, K. *Foundations of Differential Geometry*, Interscience (John Wiley), vol. 1 (1963) and vol. 2 (1968).

Kogut, J. B., and Susskind, L. 1975. Hamiltonian formulation of Wilson's lattice gauge theories. *Phys. Rev.*, **D11**, 395.

Krajewski, T. 2011. Group field theories. In: *Proceedings of the 3rd Quantum Gravity and Quantum Geometry School*. PoS(**QGQGS2011**)005.

Krajewski, T., Magnen, J., Rivasseau, V., Tanasa, A., and Vitale, P. 2010. Quantum corrections in the group field theory formulation of the EPRL/FK models. *Phys. Rev.*, **D82**, 124069, arXiv:1007.3150 [gr-qc].

Landau, L. D., and Lifshitz, E. M. 1951. *The Classical Theory of Fields*. Pergamon Press.

Landau, L. D., and Lifshitz, E. M. 1959. *Quantum Mechanics*. Pergamon Press.

Landau, L., and Peierls, R. 1931. Erweiterung des Unbestimmheitsprinzips fur die relativistische Quantentheorie. *Zeitschrift für Physik*, **69**, 56–69.

Lewandowski, J. 1994. Topological measure and graph-differential geometry on the quotient space of connections. *Int. J. Mod. Phys. D*, **3**, 207–210.

Liberati, S., and Maccione, L. 2009. Lorentz violation: motivation and new constraints. *Annu. Rev. Nucl. Part. Sci.*, **59**, 245–267, arXiv:0906.0681 [astro-ph.HE].

Littlejohn, R. 2013. *Lectures in Marseille*. [To appear.]

Littlejohn, R. G. 1992. The Van Vleck formula, Maslov theory, and phase space geometry. *J. Stat. Phys.*, **68**, 7–50.

Livine, E. R., and Speziale, S. 2007. A new spinfoam vertex for quantum gravity. *Phys. Rev.*, **D76**, 84028, arXiv:0705.0674.

Magliaro, E., and Perini, C. 2011a. Emergence of gravity from spinfoams. *EPL (Europhysics Letters)*, **95**(3), 30007, arXiv:1108.2258.

Magliaro, E., and Perini, C. 2011b. Regge gravity from spinfoams. arXiv:1105.0216.

Magliaro, E., and Perini, C. 2012. Local spin foams. *Int. J. Mod. Phys.*, **D21**, 1250090, arXiv:1010.5227 [gr-qc].

Magliaro, E., and Perini, C. 2013. Regge gravity from spinfoams. *Int. J. Mod. Phys. D*, **22**(Feb.), 1350001, arXiv:1105.0216.

Magliaro, E., Marciano, A., and Perini, C. 2011. Coherent states for FLRW space-times in loop quantum gravity. *Phys. Rev. D*, **83**, 44029.

Misner, C. W. 1957. Feynman quantization of general relativity. *Rev. Mod. Phys.*, **29**, 497.

Misner, C. W., Thorne, K., and Wheeler, J. A. 1973. *Gravitation (Physics Series)*. W. H. Freeman.

Mizoguchi, S., and Tada, T. 1992. Three-dimensional gravity from the Turaev–Viro invariant. *Phys. Rev. Lett.*, **68**, 1795–1798, arXiv:9110057 [hep-th].

Morales-Tecotl, H. A., and Rovelli, C. 1994. Fermions in quantum gravity. *Phys. Rev. Lett.*, **72**, 3642–3645, arXiv:9401011 [gr-qc].

Morales-Tecotl, H. A., and Rovelli, C. 1995. Loop space representation of quantum fermions and gravity. *Nucl. Phys.*, **B451**, 325–361.

Mukhanov, V. F, Feldman, H. A., and Brandenberger, R. H. 1992. Theory of cosmological perturbations. Part 1. Classical perturbations. Part 2. Quantum theory of perturbations. Part 3. Extensions. *Phys. Rep.*, **215**, 203–333.

Noui, K., and Roche, P. 2003. Cosmological deformation of Lorentzian spin foam models. *Classical Quant. Grav.*, **20**, 3175–3214, arXiv:0211109 [gr-qc].

Oeckl, R. 2003. A 'general boundary' formulation for quantum mechanics and quantum gravity. *Phys. Lett.*, **B575**, 318–324, arXiv:0306025 [hep-th].

Oeckl, R. 2008. General boundary quantum field theory: foundations and probability interpretation. *Adv. Theor. Math. Phys.*, **12**, 319–352, arXiv:0509122 [hep-th].

Ooguri, H. 1992. Topological lattice models in four-dimensions. *Mod. Phys. Lett.*, **A7**, 2799–2810, arXiv:9205090 [hep-th].

Perez, A. 2012. The spin foam approach to quantum gravity. *Living Rev. Relativ.*, **16**, 3, arXiv:1205.2019.

Perez, A., and Rovelli, C. 2001. A spin foam model without bubble divergences. *Nucl. Phys.*, **B599**, 255–282, arXiv:0006107 [gr-qc].

Perini, C., Rovelli, C., and Speziale, S. 2009. Self-energy and vertex radiative corrections in LQG. *Phys. Lett.*, **B682**, 78–84, arXiv:0810.1714.

Planck Collaboration. 2013. Planck 2013 results. I. Overview of products and scientific results. arXiv:1303.5062.

Ponzano, G., and Regge, Tullio. 1968. Semiclassical limit of Racah coefficients. In: Bloch, F. (ed.), *Spectroscopy and Group Theoretical Methods in Physics*. North-Holland.

Regge, T. 1961. General relativity without coordinates. *Nuovo Cimento.*, **19**, 558–571.

Reisenberger, M. P., and Rovelli, C. 1997. "Sum over surfaces" form of loop quantum gravity. *Phys. Rev.*, **D56**, 3490–3508, arXiv: 9612035.

Rennert, J., and Sloan, D. 2013. Towards anisotropic spinfoam cosmology. arXiv:1304.6688 [gr-qc].

Riello, A. 2013. Self-energy of the Lorentzian EPRL-FK spin foam model of quantum gravity. *Phys. Rev.*, **D88**, 24011, arXiv:1302.1781.

Roberts, J. 1999. Classical 6j-symbols and the tetrahedron. *Geom. Topol.*, **3**, 21–66, arXiv:9812013 [math-ph].

Rovelli, C. 1993a. Statistical mechanics of gravity and the thermodynamical origin of time. *Classical Quant. Grav.*, **10**, 1549–1566.

Rovelli, C. 1993b. The basis of the Ponzano–Regge–Turaev–Viro–Ooguri model is the loop representation basis. *Phys. Rev.*, **D48**, 2702–2707, arXiU:math_ph:9304164v1.

Rovelli, C. 1993c. The statistical state of the universe. *Classical Quant. Grav.*, **10**, 1567.

Rovelli, C. 1996a. Black hole entropy from loop quantum gravity. *Phys. Rev. Lett.*, **77**(16), 3288–3291, arXiv:9603063 [gr-qc].

Rovelli, C. 1996b. Relational quantum mechanics. *Int. J. Theor. Phys.*, **35**(9), 1637, arXiv:9609002 [quant-ph].

Rovelli, C. 2004. *Quantum Gravity*. Cambridge University Press.

Rovelli, C. 2006. Graviton propagator from background-independent quantum gravity. *Phys. Rev. Lett.*, **97**, 151301, arXiv:0508124 [gr-qc].

Rovelli, C. 2011. Zakopane lectures on loop gravity. In: *Proceedings of the 3rd Quantum Gravity and Quantum Geometry School*. PoS **(QGQGS2011)**003.

Rovelli, C. 2013a. General relativistic statistical mechanics. *Phys, Rev, D.*, **D87**, 84055, arXiV:1209.0065 [gr-gc].

Rovelli, C. 2013b. Why Gauge? arXiv:1308.5599.

Rovelli, C., and Smerlak, M. 2010. Thermal time and the Tolman–Ehrenfest effect: temperature as the "speed of time". *Classical Quant. Grav.* **28**, 075007, arXiv:1005.2985.

Rovelli, C., and Smerlak, M. 2012. In quantum gravity, summing is refining. *Classical Quant. Grav.*, **29**, 55004, arXiv:1010.5437 [gr-qc].

Rovelli, C., and Smolin, L. 1988. Knot theory and quantum gravity. *Phys. Rev. Lett.*, **61**, 1155.

Rovelli, C., and Smolin, L. 1990. Loop space representation of quantum general relativity. *Nucl. Phys.*, **B331**, 80.

Rovelli, C., and Smolin, L. 1995a. Discreteness of area and volume in quantum gravity. *Nucl. Phys.*, **B442**, 593–622, arXiv:9411005 [gr-qc].

Rovelli, C., and Smolin, L. 1995b. Spin networks and quantum gravity, arXiv:9505006 [gr-qc].

Rovelli, C., and Speziale, S. 2010. On the geometry of loop quantum gravity on a graph. *Phys. Rev.*, **D82**, 44018, arXiv:1005.2927.

Rovelli, C., and Speziale, S. 2011. Lorentz covariance of loop quantum gravity. *Phys. Rev.*, **D83**, 104029, arXiv:1012.1739.

Rovelli, C., and Vidotto, F. 2008. Stepping out of homogeneity in loop quantum cosmology, arXiv:0805.4585.

Rovelli, C., and Vidotto, F. 2010. Single particle in quantum gravity and BGS entropy of a spin network. *Phys. Rev.*, **D81**, 44038, arXiv:0905.2983.

Rovelli, C., and Vidotto, F. 2013. Evidence for maximal acceleration and singularity resolution in covariant loop quantum gravity. *Phys. Rev. Lett.*, **111**(9), 091303, arXiv:1307.3228.

Rovelli, C., and Vidotto, F. 2014. Planck stars. arXiv:1401.6562.

Rovelli, C., and Zhang, M. 2011. Euclidean three-point function in loop and perturbative gravity. *Classical Quant. Grav.*, **28**, 175010, arXiv:1105.0566 [gr-qc].

Ruhl, W. 1970. *The Lorentz Group and Harmonic Analysis*. W.A. Benjamin.

Schrödinger, E. 1926. Quantisierung als Eigenwertproblem. *Annalen der Physik*, **13**, 29.

Sen, A. 1982. Gravity as a spin system. *Phys. Lett. B*, **119**, 89–91.

Shannon, C. E. 1948. A mathematical theory of communication. *The Bell System Technical Journal*, **XXVII**(3), 379.

Singh, P. 2009. Are loop quantum cosmos never singular? *Classical Quant. Grav.*, **26**, 125005, arXiv:0901.2750.

Singh, P., and Vidotto, F. 2011. Exotic singularities and spatially curved loop quantum cosmology. *Phys. Rev.*, **D83**, 64027, arXiv:1012.1307.

Smolin, L. 1994. Fermions and topology. arXiv:9404010 [gr-qc].

Smolin, L. 2012. General relativity as the equation of state of spin foam. arXiv:1205.5529.

Sorkin, R. D. 1977. On the relation between charge and topology. *J. Phys. A: Math. Gen.*, **10**, 717.

Stone, D. 2013. *Einstein and the Quantum: The Quest of the Valiant Swabian*. Princeton University Press.

Sundermeyer, K. 1982. *Constrained Dynamics : With Applications to Yang–Mills Theory, General Relativity, Classical Spin, Dual String Model*. Springer-Verlag.

Thiemann, T. 2006. Complexifier coherent states for quantum general relativity. *Classical Quant. Grav.*, **23**, 2063–2118, arXiv:0206037 [gr-qc].

Thiemann, T. 2007. *Modern Canonical Quantum General Relativity*. Cambridge University Press.

Toller, M. 2003. Geometries of maximal acceleration. arXiv:0312016 [hep-th].

Tolman, R. C. 1930. On the weight of heat and thermal equilibrium in general relativity. *Phys. Rev.*, **35**, 904–924.

Tolman, R. C., and Ehrenfest, P. 1930. Temperature equilibrium in a static gravitational field. *Phys. Rev.*, **36**(12), 1791–1798.

Turaev, V. G., and Viro, O. Y. 1992. State sum invariants of 3 manifolds and quantum 6j symbols. *Topology*, **31**, 865–902.

Unruh, W. G. 1976. Notes on black hole evaporation. *Phys. Rev.*, **D14**, 870.

Vidotto, F. 2011. Many-nodes/many-links spinfoam: the homogeneous and isotropic case. *Classical Quant Grav.*, **28**(245005), arXiv:1107.2633.

Wald, R. M. 1984. *General relativity*. Chicago, University Press.

Wheeler, J. A. 1962. *Geometrodynamics*. New York: Academic Press.

Wheeler, J. A. 1968. Superspace and the nature of quantum geometrodynamics. In: DeWitt, B. S, and Wheeler, J. A. (eds.), *Batelles Rencontres*. Benjamin.

Wieland, W. 2012. Complex Ashtekar variables and reality conditions for Holst's action. *Annales Henri Poincare*, **13**, 425, arXiv:1012.1738 [gr-qc].

Wightman, A. 1959. Quantum field theory in terms of vacuum expectation values. *Phys. Rev.*, **101**(860).

Wilson, K. G. 1974. Confinement of quarks. *Phys. Rev.*, **D10**, 2445–2459.

York, J. W. 1972. Role of conformal three-geometry in the dynamics of gravitation. *Phys. Rev. Lett.*, **28**, 1082.

Index

Printed in the United States
by Baker & Taylor Publisher Services